SCIENTIFIC SOFTWARE DESIGN

THE OBJECT-ORIENTED WAY

This book is about software design. Although many current discussions of scientific programming focus on scalable performance, this book focuses on scalable design. The authors analyze how the structure of a package determines its developmental complexity according to such measures as bug search times and documentation information content. The work presents arguments for why these issues impact solution cost and time more than does scalable performance. The final chapter explores the question of scalable execution and shows how scalable design leads to scalable execution. The book's focus is on program organization, which has received considerable attention in the broader software engineering community, where graphical description standards for modeling software structure and behavior have been developed by computer scientists. These discussions might be enriched by engineers and scientists who write scientific codes. This book aims to bring such scientific programmers into discussion with computer scientists. The authors do so by introducing object-oriented software design patterns in the context of scientific simulation.

Dr. Damian Rouson is the manager of the Reacting Flow Research Department at Sandia National Laboratories. He was formerly Section Head of Combustion Modeling and Scaling at the U.S. Naval Research Laboratory. He was assistant professor of mechanical engineering at the City University of New York, visiting assistant professor at the University of Maryland, and Special Scientist at the University of Cyprus. Damian Rouson received his Bachelors in Mechanical Engineering at Howard University and his Masters and PhD in Mechanical Engineering from Stanford University.

Dr. Jim Xia is currently a software designer and tester at the IBM Canada Lab in Markham, Ontario, Canada. He received his PhD in Physics from the University of Western Ontario in 1997.

Dr. Xiaofeng Xu is currently an Analyst at General Motors Corp. in Pontiac, Michigan. At this job, he performs airflow and combustion CFD analysis to support base engine designs. He received his PhD in Mechanical Engineering (2003) from Iowa State University in Ames and is the author or co-author of 39 referred publications.

Scientific Software Design

THE OBJECT-ORIENTED WAY

Damian Rouson
Sandia National Laboratories

Jim Xia
IBM Canada Lab in Markham

Xiaofeng Xu
General Motors Corp

CAMBRIDGE
UNIVERSITY PRESS

32 Avenue of the Americas, New York NY 10013-2473, USA

Cambridge University Press is part of the University of Cambridge.

It furthers the University's mission by disseminating knowledge in the pursuit of education, learning and research at the highest international levels of excellence.

www.cambridge.org
Information on this title: www.cambridge.org/9781107415331

First published 2011
First paperback edition 2014

A catalogue record for this publication is available from the British Library

Library of Congress Cataloguing in Publication data
Rouson, Damian, 1967–
Scientific software design : the object-oriented way / Damian Rouson, Jim Xia, Xiaofeng Xu.
p. cm.
Includes bibliographical references and index.
ISBN 978-0-521-88813-4 (hardback)
1. Science–Data processing. 2. Engineering–Data processing. 3. Software engineering.
I. Xia, Jim, 1970– II. Xu, Xiaofeng, 1972– III. Title.
Q183.9.R68 2011
005.1–dc22 2011007351

ISBN 978-0-521-88813-4 Hardback
ISBN 978-1-107-41533-1 Paperback

For Zendo in exchange for the many books you have written for me. For Leilee as a tangible product of every moment of support.

DAMIAN ROUSON

To my wife, Lucy, and two lovely children, Ginny and Alan, for their encouragement and support.

JIM XIA

To my daughter Caroline, my wife Li and parents, for their love and support.

XIAOFENG XU

Contents

Figures

Tables

Preface

This book is about software design. We use "design" here in the sense that it applies to machines, electrical circuits, chemical plants, and buildings. At least two differences between designing scientific software and designing these other systems seem apparent:

1. The extent to which one communicates a system's structure by representing schematically its component parts and their interrelationships,
2. The extent to which such schematics can be analyzed to evaluate suitability and prevent failures.

Schematic representations greatly simplify communications between developers. Analyses of these schematics can potentially settle long-standing debates about which systems will wear well over time as user requirements evolve and usage varies.

This book does not chiefly concern high-performance computing. While most current discussions of scientific programming focus on scalable performance, we unabashedly set out along the nearly orthogonal axis of scalable design. We analyze how the structure of a package determines its developmental complexity according to such measures as bug search times and documentation information content. We also present arguments for why these issues impact solution cost and time more than does scalable performance.

We firmly believe that science is not a horse race. The greatest scientific impact does not necessarily go to the swiftest code. Groundbreaking results often occur at the intersection of multiple disciplines, where the basic science is so poorly understood that important insights can be gleaned from creative simulations of modest size. In multidisciplinary regimes, scaling up to hundreds of program units (e.g., procedures, modules, components, or classes) matters at least as much as scaling up to hundreds of execution units (e.g., processes, processors, threads, or cores). Put another way, distributed development matters as much as distributed computing.

Nonetheless, recent trends suggest that exploiting even the most modest hardware – including the laptops on which much of this book was written – requires parallelizing code across multiple execution units. After our journey down the path toward scalable design, the final chapter turns to the question of scalable execution, or how to go from running on a single execution unit to running on many. In the process, we demonstrate how scalable design illuminates a clear path to scalable execution.

Program organization – our main concern – has received considerable attention in the broader software engineering community, where computer scientists have developed graphical description standards for modeling software structure and behavior. They have used these standards to articulate diagrammatic templates, or *design patterns*, that solve recurring problems. They have also proposed metrics to describe software properties. However, these discussions have taken place largely outside the purview of the domain experts who write scientific code. This book aims to bring scientific programmers into the discussion. We do so by introducing object-oriented software design patterns in the context of scientific simulation.

We have used each of the presented patterns in production code across a range of applications. Since our primary expertise lies in fluid turbulence, that topic provides a common thread, but the range of phenomena we have coupled to the fluid motion should interest those working in other fields. These phenomena include electromagnetic field behavior, quantum effects, particle transport, and biomedical device interaction. Each is the subject of a brief case study.

Two languages dominate scientific computing: Fortran and C++. A 2006 survey of the United States Department of Defense High Performance Computing users indicated that approximately 85% write in C/C++/C#, while nearly 60% write in Fortran.[1] Clearly there is overlap between the two communities, so we present examples in both languages.

Historically, patterns grew out of the adaptation of design principles from building architecture to object-oriented software architecture. Since the Fortran 2003 standard provides explicit support for all of the concepts central to object-oriented programming (OOP), a unique opportunity has emerged to put Fortran and C++ on equal footing. With this in mind, we intend the book to be accessible to at least three audiences:

1. Fortran programmers who are unfamiliar with OOP,
2. Fortran programmers who emulate OOP using Fortran 90/95, and
3. C++ programmers who are familiar with OOP.

While we use Fortran 2003 to introduce object orientation, we incorporate notes that compare Fortran constructs to similar C++ constructs. This serves the dual purpose of introducing OOP in a context familiar to the Fortran programmers and introducing Fortran in a context familiar to the C++ programmers.

Part I introduces the object-oriented way of programming in Fortran 2003 along with various cost and complexity analyses of scientific programming. Chapter 1 couples computational complexity theory, Amdahl's law, and the Pareto principle to motivate the need to reduce data dependencies between lines of code and to balance scalability requirements against programmability. Chapter 2 details how the object-oriented and functional programming paradigms reduce complexity by reducing data dependencies. Chapter 3 motivates the need for a general calculus for abstract data types, describes the implementation of such a calculus, and further extends our complexity analysis.

[1] Simon Management Group, *HPC Survey Summary*, July 13, 2006.

Part II presents design patterns in Fortran 2003 and C++. Included in Chapters 5–9 are two domain-specific patterns for multiphysics modeling and five general patterns, including one language-specific Fortran pattern.

Part III surveys advanced topics. Chapter 10 augments our design pattern schematics with formal constraints that ensure desirable runtime behavior. The growing interest in integrating multiple software applications into multiphysics packages inevitably leads to questions of multilanguage development. Chapter 11 addresses multilanguage interoperability via standard language constructs as well as automated tools for compile-time and runtime integration. Chapter 12 charts a path toward runtime scalability and takes a preliminary jaunt down that path. Chapter 12 also surveys the multiphysics software architectures from which we culled the design patterns in Part II. We conclude Chapter 12 with a discussion of how these patterns are being incorporated into the Multiphysics Object-oriented Reconfigurable Fluid Environment for Unified Simulations (Morfeus) framework, an open-source software development project funded by the Office of Naval Research.

Much of Part I was written for a course on scientific software design taught by the first author at the University of Cyprus during a visiting faculty appointment in the autumn of 2006. Students in the course were beginning and advanced graduate students. The material is also suitable for a senior-level undergraduate course. The text assumes some introductory-level familiarity with partial differential equations, numerical analysis, and linear algebra. The examples are chosen to keep the expected experience with these subjects comparable to that of most senior-level engineering and physical science students. For those lacking this exposure, Appendix A summarizes the requisite mathematical background. The text should also prove useful as a resource for practicing computational scientists and engineers interested in learning design principles for object-oriented software construction.

Damian Rouson
Berkeley, California

Jim Xia
Toronto, Ontario

Xiaofeng Xu
Detroit, Michigan

Acknowledgments

This text is an outgrowth of research the authors have conducted with graduate students, postdoctoral researchers, and various collaborators past and present. In particular, Dr. Helgi Adalsteinsson contributed greatly to many of the C++ examples. Drs. Karla Moris, Yi Xiong, Irene Moulitsas, Hari Radhakrishnan, Ioannis Sarris, and Evangelos Akylas contributed to the research codes that inspired many of this book's Fortran examples.

Many examples draw from simulations we have performed in collaboration with several domain experts in fields ranging from quantum physics to magnetohydrodynamics, fire protection engineering, and atmospheric boundary layers. These collaborators include Profs. Joel Koplik of the City University of New York, Andre Marshall of the University of Maryland, Stavros Kassinos of the University of Cyprus, and Robert Handler of the Texas A&M University.

Much of the work presented in this book was funded by the Office of Naval Research (ONR) Automation in Ship Systems program under the direction of program manager Anthony Seman. The authors are also deeply indebted to Dr. Sameer Shende of the University of Oregon for providing the development platform on which the bulk of the examples in this text were written, and to Dr. Andrew McIlroy of Sandia National Laboratories for approving the purchase of the platform on which the remainder of the examples were developed. We thank Daniel Strong, also of Sandia, for his artistic expression of this book's theme in the cover art. Sandia National Laboratories is a multi-program laboratory managed and operated by Sandia Corporation, a wholly owned subsidiary of Lockheed Martin Corporation, for the U.S. Department of Energy's National Nuclear Security Administration under contract DE-AC04-94AL85000.

Disclaimer

The opinions expressed in this book are those of the authors and do not necessarily represent those of International Business Machines Corporation or any of its affiliates.

PART I

THE TAO OF SCIENTIFIC OOP

1 Development Costs and Complexity

"Premature optimization is the root of all evil."
Donald Knuth

1.1 Introduction

The past several decades have witnessed impressive successes in the ability of scientists and engineers to accurately simulate physical phenomena on computers. In engineering, it would now be unimaginable to design complex devices such as aircraft engines or skyscrapers without detailed numerical modeling playing an integral role. In science, computation is now recognized as a third mode of inquiry, complementing theory and experiment. As the steady march of progress in individual spheres of interest continues, the focus naturally turns toward leveraging efforts in previously separate domains to advance one's own domain or in combining old disciplines into new ones. Such work falls under the umbrella of *multiphysics modeling*.

Overcoming the physical, mathematical, and computational challenges of multiphysics modeling comprises one of the central challenges of 21st-century science and engineering. In one of its three major findings, the National Science Foundation Blue Ribbon Panel on Simulation-Based Engineering Science (SBES) cited "open problems associated with multiscale and multi-physics modeling" among a group of "formidable challenges [that] stand in the way of progress in SBES research." As the juxtaposition of "multiphysics" and "multiscale" in the panel's report implies, multi-physics problems often involve dynamics across a broad range of lengths and times.

At the level of the physics and mathematics, integrating the disparate dynamics of multiple fields poses significant challenges in simulation accuracy, consistency, and stability. Modeling thermal radiation, combustion, and turbulence in compartment fires, for example, requires capturing phenomena with length scales separated by several orders of magnitude. The room-scale dynamics determine the issue of paramount importance: safe paths for human egress. Determining such gross parameters while resolving neither the small length scales that control the chemical kinetics nor the short time scales on which light propagates forms an active area of research.

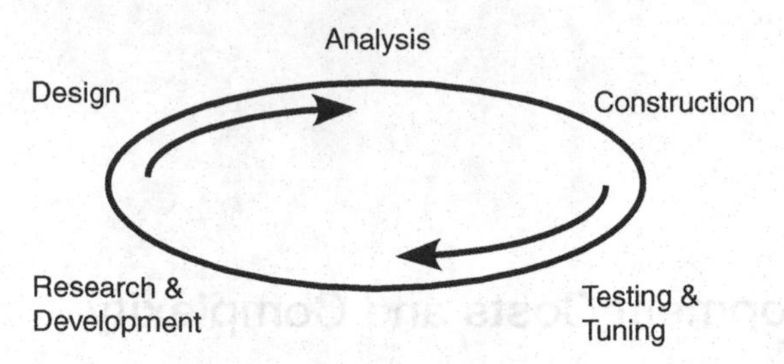

Figure 1.1. Scientific software life cycle.

At the level of the computation, the sheer size of multiscale, multiphysics simulations poses significant challenges in resource utilization. Whereas the largest length and time scales determine the spatial and temporal extent of the problem domain, the shortest determine the required resolution. The ratio of the domain size to the smallest resolvable features determines the memory requirements, whereas the ratio of the time window to the shortest resolvable interval determines the computing time. In modeling-controlled nuclear fusion, for example, the time scales that must be considered span more orders of magnitude than can be tracked simultaneously on any platform for the foreseeable future. Hence, much of the effort in writing related programs goes into squeezing every bit of performance out of existing hardware.

In between the physical/mathematical model development and the computation lies the software development process. One can imagine the software progressing through a life cycle much like any other product of human effort: starting with research and development and ending with the testing and fine-tuning of the final product (see Figure 1.1). However, as soon as one searches for evidence of such a life cycle in scientific computing, gaping holes appear. Research on the scientific software development process is rare. Numerous journals are devoted to novel numerical models and the scientific insights obtained from such models, but only a few journals focus on the software itself. Notable examples include the refereed journals *Scientific Programming* and *ACM Transactions on Mathematical Software.*

Further along in the life cycle, the terrain becomes even more barren. Discussions of scientific software design rarely extend beyond two categories: (1) explanations of a particular programming paradigm, which fall into the implementation stage of the life cycle, and (2) strategies for improving run-time efficiency, which arguably impacts all stages of the life cycle but comprises an isolated activity only once a prototype has been constructed in the testing and tuning phase. Articles and texts on the first topic include presentations of structured and object-oriented programming, whereas those on the second include discussions of parallel programming and high-performance computing.

Implementation-independent design issues such as the chosen modular decomposition and its developmental complexity have received scant attention in the scientific computing community. Not surprisingly then, quantitative analyses of software designs have been limited to run-time metrics such as speed and parallel efficiency. In presenting attempts to analyze static source code organization, the author often encounters a perception that program structure is purely stylistic – of the "You say 'toe-may-toe'; I say 'toe-mah-toe' " variety. Even to the extent it is recognized that program structure matters, no consensus is emerging on what structural precepts prove best.

That scientific program design warrants greater attention has been noted at high levels. The 1999 Presidential Information Technology Advisory Committee (PITAC) summarized the situation for developers of multidisciplinary scientific software:

> Today it is altogether too difficult to develop computational science software and applications. Environments and toolkits are inadequate to meet the needs of software developers in addressing increasingly complex interdisciplinary problems... In addition, since there is no consistency in software engineering best practices, many of the new applications are not robust and cannot easily be ported to new hardware.

Fortunately, the situation with regards to environments and toolkits has improved since the PITAC assessments. Using tools such as Common Component Architecture (CCA), for example, one can now link applications written by different developers in different languages using different programming paradigms into a seamless, scalable package without incurring the language interoperability and communication latency issues associated with non-scientific toolkits that facilitate similar linkages. Several groups have constructed development frameworks on top of CCA to facilitate distributed computing (Zhang et al. 2004), metacomputing (Malawski et al. 2005), and other computing models.

However, the situation with regard to software engineering best practices has seen far less progress. Most authors addressing scientific programming offer brief opinions to aid robustness before retreating to the comfortable territory of run-time efficiency or numerical accuracy. Without a healthy debate in the literature, it is unsurprising that no consensus has emerged. By contrast, in the broader software engineering community, consensus has been building for over a decade on the best practices at least within one development paradigm: object-oriented design (OOD). In this context, the best practices have been codified as *design patterns*, which are cataloged solutions to problems that recur across many projects.

Experience indicates many scientific programmers find the term "design pattern" vague or awkward on first hearing. However, its application to software development conforms with the first four definitions of "pattern" in the Merriam Webster dictionary:

1. a form or model proposed for imitation,
2. something designed or used as a model for making things,
3. an artistic, musical, literary, or mechanical design or form,
4. a natural or chance configuration.

Having proved useful in the past, design patterns are offered as models to be followed in future work. The models have artistic value in the sense of being elegant and in the sense that software engineering, like other engineering fields, is part science and part art. And much like mechanical design patterns, some software design patterns mimic patterns that have evolved by chance in nature.

Gamma et al. (1995) first collected and articulated object-oriented software design patterns in 1995 in their seminal text *Design Patterns: Elements of Reusable*

Object-Oriented Software. They drew inspiration from the 1977 text, *A Pattern Language*, by architects Alexander et al., who in turn found inspiration in the beauty medieval architects achieved by conforming to local regulations that required specific building features without mandating specific implementations of those features. This freedom to compose variations on a theme facilitated individual expression within the confines of proven forms.

Gamma et al. did not present any domain-specific patterns, although their book's introduction suggests it would be worthwhile for someone to catalog such patterns. Into this gap we thrust the current text. Whereas Part I of this text lays the object-oriented foundation for discussing design patterns and Part III discusses several related advanced topics, Part II aims to:

1. Catalog general and domain-specific patterns for multiphysics modeling,
2. Quantify how these patterns reduce complexity,
3. Present Fortran 2003 and C++ implementations of each pattern discussed.

Each of these objectives makes a unique contribution to the scientific computing field. The authors know of only a handful of publications on object-oriented software design patterns for scientific software (Blilie 2002; Decyk and Gardner 2006; Decyk and Gardner 2007; Gardner and Manduchi 2007; Markus 2006; Rouson et al. 2010) and even fewer quantitative analyses of how patterns reduce scientific programming complexity (Allan et al. 2008; Rouson 2008). Finally, published design pattern examples that take full advantage of the new object-oriented constructs of Fortran 2003 have only just begun to appear (Markus 2008; Rouson et al. 2010).

That Fortran implementations of object-oriented design patterns have lagged those in other languages so greatly is both ironic and understandable. It is ironic because object-oriented programming started its life nearly four decades ago with Simula 67, a language designed for physical system simulation, which is the primary use of Fortran. It is understandable because the Fortran standards committee, under heavy influence by compiler vendors, moved at glacial speeds to provide explicit support for object-orientation. (See Metcalf et al. 2004 for some history on this decision.) In the absence of object-oriented language constructs, a small cadre of researchers began developing techniques for emulating object-orientation in Fortran 90/95 in the 1990s (Decyk et al. 1997a, 1997b, 1998; Machiels and Deville 1997), culminating in the publication of the first text on OOP in Fortran 90/95 by Akin (2003).

The contemporaneous publication of the Fortran 2003 standard, which provides object-oriented language constructs, makes the time ripe for moving beyond the basic mechanics to higher-level discussions of objected-oriented design patterns for scientific computing. Because the availability of Fortran 2003 motivates this book, Fortran serves as the primary language for Part I of this text. Because most C++ programmers already know OOP, there is less need to provide C++ in this part of the text. Some of the complexity arguments in Part I, however, are language-independent, so we provide C++ translations to make this section accessible to C++ programmers. In Part II, greater effort is made to write C++ examples that stand on their own as opposed to being translations of Fortran.

A Pressing Need for 21st-Century Science

The 20th-century successes in simulating individual physical phenomena can be leveraged to simulate multiscale, multiphysics phenomena much more simply if we begin distilling, analyzing, and codifying the best practices in scientific programming.

1.2 Conventional Scientific Programming Costs

The word "costs" plays two roles in this section's title. As a noun, it indicates that the section discusses the costs associated with writing and running conventional scientific programs. As a verb, it indicates that the conventional approach to writing these programs costs projects in ways that can be alleviated by other programming paradigms. Whereas this section estimates those costs, section (1.4) presents alternative paradigms.

The software of interest is primarily written for research and development in the sciences and engineering sciences. Although many such codes later form the core of commercial packages used in routine design and analysis, all are grouped under the heading "scientific" in this text. During the early phases of the development, many scientific programming projects tackle problems that are sufficiently unique or demanding to preclude the use of commercial off-the-shelf software for the entire project. Although some commercial library routines might be called, these serve supporting roles such as solving linear and nonlinear systems or performing data analysis. Viewed from a client/server perspective, the libraries provide general mathematical services for more application-specific client code that molds the science of interest into a form amenable to analysis with the chosen library.

Some mathematical libraries are portable and general-purpose, such as those by the Numerical Algorithms Group (NAG) or the Numerical Recipes series. Others are tuned to exploit the features of specific hardware. The Intel Math Kernel Library (MKL) targets Intel processors and scales up to large numbers of processors on distributed-memory clusters, whereas the Silicon Graphics, Inc. (SGI) math library scales on distributed shared-memory supercomputers with SGI's proprietary high-bandwidth memory architecture.

Mathematical libraries attempt to embody the best available numerical algorithms without imposing any particular program organization. Responsibility for program organization, or *architecture*, thus rests with the developers. Although the definition of software architecture remains a subject of active debate, there is long-standing agreement on the attractiveness of modular design, so for present purposes, "architecture" denotes a specific modular decomposition. This decomposition would include both the roles of individual *modules* and the relationships between them. To avoid language- and paradigm-specificity, "module" here denotes any isolated piece of code whether said isolation is achieved by separation into its own file or by some other construct.

Most scientific programmers are trained in engineering or science. Typical curricula in these fields include only an introductory programming course that teaches the use of a particular programming language and paradigm. Teaching only the

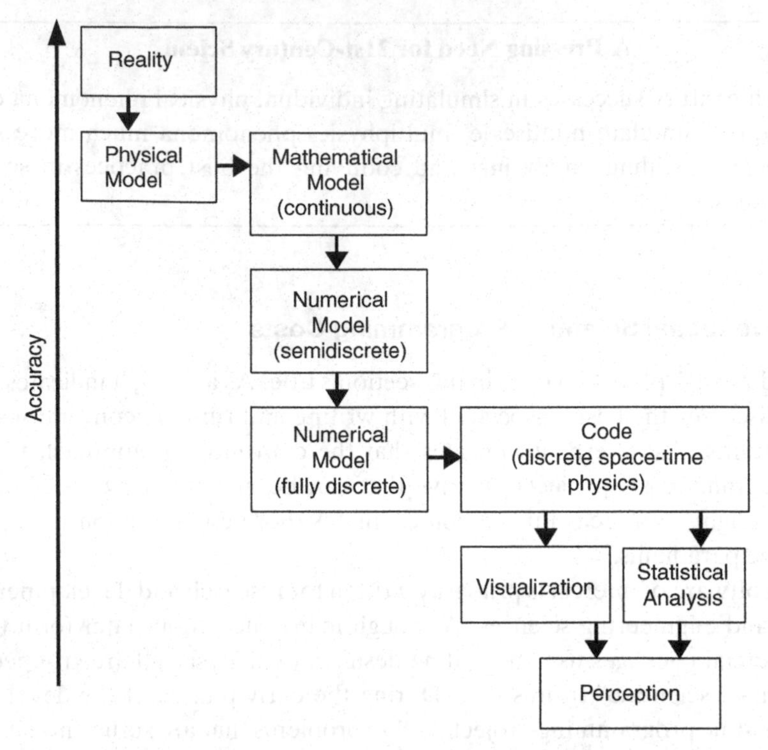

Figure 1.2. Sequence of models in the conventional scientific code development process.

implementation phase differs considerably from the rest of the curriculum, wherein students are encouraged to think abstractly about systems and to quantitatively evaluate competing models for describing those systems. In the automatic controls courses taken by many engineering students, for example, model elements are delineated precisely and their roles and relationships are specified mathematically, so the system behavior can be analyzed rigorously.

Figure 1.2 situates programming within the conventional modeling process. One starts off wishing to model physical reality. Based on observations of that reality, one constructs a physical model, which for present purposes comprises a conceptual abstraction of reality. Both "model" and "abstraction" here connote useful simplifications that retain only those elements necessary for a given context. For example, one might abstract a cooling fin on a microprocessor chip as a thin solid with constant properties transporting thermal energy in the direction orthogonal to the chip. One might further assume the energy transfer occurs across a distance that is large relative to the material's intermolecular spacing, so it can be considered a continuum.

Although physical models are typically stated informally, Collins (2004) recently proposed the development of a physics markup language (PML) based on the extensible markup language (XML). Formal specifications in such a markup language would complement the programming techniques presented in this book. Specifically, automatic code generation from a markup document might produce software counterparts of algebraic and differential operators specified in PML. Chapter 3 discusses the implementation of such operators.

The first formal step in conventional scientific work involves expressing the physical model in a mathematical model with continuous variation of the dependent variables in space and time. For the fin problem, this leads to the one-dimensional (1D), unsteady heat equation:

$$\begin{aligned} \frac{\partial T}{\partial t} &= \alpha \frac{\partial^2 T}{\partial x^2}, \quad \Omega \times T = (0, L_{fin}) \times (0, t_{final}) \\ T(0,t) &= T_{chip} \\ T(L_{fin},t) &= T_{air} \\ T(x,0) &= T_{air} \end{aligned} \tag{1.1}$$

where $T(x,t)$ is the temperature at time t a distance x from the chip, α is the fin's thermal diffusivity, $\Omega \times T$ is the space-time domain, L_{fin} is the fin length, t_{final} is the final time of the simulation, and where T_{chip} and T_{air} are boundary conditions.

Solving the mathematical model on a computer requires rendering equations (1.1) discrete. Most numerical schemes employ the *semidiscrete method*, which involves discretizing space first and time second. Most spatial discretizations require laying a grid over the spatial domain. Given a uniformly spaced grid overlaid on Ω, applying the central difference formula to the right-hand side of (1.1) leads to a set of coupled ordinary differential equations for the nodal temperatures:

$$\frac{d\vec{T}}{dt} \equiv \frac{\alpha}{\Delta x^2} \begin{bmatrix} -2 & 1 & & \\ 1 & -2 & 1 & \\ & \ddots & \ddots & \\ & & 1 & -2 \end{bmatrix} \vec{T} + \frac{\alpha}{\Delta x^2} \begin{Bmatrix} T_{chip} \\ 0 \\ \vdots \\ T_{air} \end{Bmatrix} \tag{1.2}$$

where $\Delta x \equiv L_{fin}/N$ is the uniform spacing in between the $N+1$ grid points covering Ω. One can derive the same discrete system of equations via a finite volume scheme or a Galerkin finite element method with linear-basis functions. We refer to the intervals between nodes generically as *elements* and assume the reader is familiar with one or another path that starts from equations (1.1) and leads to equations (1.2).

The semidiscrete model is rendered fully discrete by applying a time integration algorithm to equations (1.2). The simplest case is explicit Euler, which corresponds to a two-term Taylor series expansion:

$$\vec{T}^{n+1} = \vec{T}^n + \Delta t \left.\frac{d\vec{T}}{dt}\right|_{t_n} = \vec{T}^n + \Delta t \left(A\vec{T}^n + \vec{b} \right) \tag{1.3}$$

where $\vec{T}^n$ and $\vec{T}^{n+1}$ are vectors of $N-1$ internal grid point temperatures at time steps$t_n \equiv n\Delta t$ and $t_{n+1} \equiv t_n + \Delta t$, respectively.

Most scientific computing publications focus on the previously described modeling steps, treating their implementation in code as an afterthought, if at all. This has numerous adverse consequences. First, several very different concepts are hard-wired into equation (1.3) in ways that can be hard to separate should one stage in the modeling process require reconsideration. Equation (1.3) represents a specific numerical integration of a specific discrete algorithm for solving a specific partial differential equation (PDE). Without considerable forethought, the straightforward expression of this algorithm in software could force major code revisions if one later decides to change the discretization scheme, for example, the basis functions or the PDE itself. The next section explains why the conventional approach to separating these issues scales poorly as the code size grows.

A side effect of conflating logically separate modeling steps is that it becomes difficult to assign responsibility for erroneous results. For example, instabilities that exist in the fully discrete numerical model could conceivably have existed in continuous mathematical model. Rouson *et al.* (2008b) provided a more subtle example in which information that appeared to have been lost during the discretization process was actually missing from the original PDE.

As previously mentioned, the code writing is the first step with no notion of abstraction. Except for flow charts – which, experience indicates, very few programmers construct in practice – most discussions of scientific programs offer no formal description of program architecture other than the code itself. Not surprisingly then, scientific program architecture is largely ad hoc and individualistic. Were the problems being solved equally individualistic, no difficulty would arise. However, significant commonalities exist in the physical, continuous mathematical, and discrete numerical models employed across a broad range of applications. Without a language for expressing common architectural design patterns, there can be little dialogue on their merits. Chapter 2 employs just such a language: the Unified Modeling Language (UML). Part II presents design patterns that exploit commonalities among a broad range of problems to generate flexible, reusable program modules. Lacking such capabilities, conventional development involves the reinvention of architectural forms from scratch.

How does redevelopment influence costs? Let's think first in the time domain:

$$\text{Total solution time} = \text{development time} + \text{computing time} \tag{1.4}$$

Until computers develop the clairvoyance to run code before it is written, programming will always precede execution, irrespective of the fact that earlier versions of the program might be running while newer versions are under development. Since the development time often exceeds the computing time, the fraction of the total solution time that can be reduced by tuning the code is limited. This is a special case of Amdahl's law, which we revisit in Section 1.5 and Chapter 12.

Time is money, so equation (1.4) can be readily transformed into monetary form by assuming the development and computing costs are proportional to the development and computing times as follows:

$$\begin{aligned}\$_{solution} &= \$_{development} + \$_{computing} \\ &= N_{dev}p_{avg}t_{dev} + \frac{\$_{computer}}{N_{users}}\frac{t_{run}}{t_{useful}}\end{aligned} \tag{1.5}$$

where the \$ values are costs; N_{dev}, p_{avg}, and t_{dev} are the number of developers, their average pay rate, and the development time, respectively; and N_{users}, t_{run}, and t_{useful} are the number of computer users, the computing time, and the computer's useful life, respectively. In the last term of equation (1.5), the first factor estimates the fraction of the computer's resources available for a given user's project. The second factor estimates the fraction of the computer's useful life dedicated to that user's runs.

There are of course many ways equation (1.5) could be refined, but it suffices for current purposes. For a conservative estimate, consider the case of a graduate student researcher receiving a total compensation including tuition, fees, stipend, and benefits of \$50,000/yr for a one-year project. Even if this student is the sole developer of her code and has access to a \$150,000 computer cluster with a useful life of three years, she typically shares the computer with, say, four other users. Therefore, even if her simulations run for half a year on the wall clock, the solution cost breakdown is

$$\$_{solution} = (1 \cdot \text{developer})(\$50{,}000/\text{year})(1 \cdot \text{year}) \tag{1.6}$$

$$+ \frac{\$150{,}000}{5 \cdot \text{users}} \frac{0.5 \cdot \text{user} \cdot \text{years}}{3 \cdot \text{years}} \tag{1.7}$$

$$= \$50{,}000 + \$5{,}000 \tag{1.8}$$

so the development costs greatly exceed the computing costs. The fraction of costs attributable to computing decreases even further if the developer receives higher pay or works for a longer period – likewise if the computer costs less or has a longer useful life. By contrast, if the number of users decreases, the run-time on the wall clock decreases proportionately, so the cost breakdown is relatively insensitive to the resource allocation. Similar, the breakdown is insensitive to adding developers if the development time is inversely proportional to the team size (up to some point of diminishing returns).

For commercial codes, the situation is a bit different but the bottom line is arguably the same. The use and development of commercial codes occur in parallel, so the process speedup argument breaks down. Nonetheless, development time savings lead to the faster release of new features, providing customer benefit and a competitive advantage for the company.

This forms the central thesis on which this book rests:

> **Your Time Is Worth More Than Your Computer's Time**
>
> Total solution time and cost can be reduced in greater proportion by reducing development time and costs than by reducing computing time and costs.

1.3 Conventional Programming Complexity

Part of the reason scientific computing has traditionally focused on the earlier stages of Figure 1.2 is that the accuracy losses at these stages are well known and rigorously quantifiable. As indicated by the figure's vertical axis, each transition in the discretization process has an associated error. For example, the truncation error, e, of the spatial discretization inherent in equation (1.2) is bounded by a term proportional to the grid spacing:

$$e \leq C \Delta x^3 \tag{1.9}$$

where C typically depends on properties of the exact solution and its derivatives.

Program organization may seem orthogonal to accuracy – hence the horizontal line leading to the code in Figure 1.2. Complexity arguments to be presented in this section, however, lay the foundation for an OOP strategy in which memory management plays a critical role. Recalling that $\Delta x = L_{fin}/N$ leads directly to the fact that the number of points, N, determines the error:

$$e \leq C\left(\frac{L_{fin}}{N}\right)^3 \tag{1.10}$$

Furthermore, the maximum number of points is set by the available memory, M, and the chosen precision, p, according to:

$$N_{max} = \frac{M}{p} \tag{1.11}$$

where typical units for M and p would be bytes and bytes per nodal value, respectively. Substituting (1.11) into (1.10) leads to the lowest possible upper bound on the error:

$$\sup e \leq C\left(\frac{pL_{fin}}{M}\right)^3 \tag{1.12}$$

Whether a given program actually achieves this level of accuracy is determined by its dynamic memory utilization. Consider two ways of computing the right-hand side (RHS) of equation (1.3):

$$\vec{T}^{n+1} = \vec{T}^n + \Delta t\left(A\vec{T}^n + \vec{b}\right) \tag{1.13}$$

$$= (I + \Delta t A)\,\vec{T}^n + \vec{b} \tag{1.14}$$

where I is the identity matrix. Working outward from inside the parenthesis, evaluating the RHS in equation (1.13) requires keeping $\vec{T}^n$ intact while computing $A\vec{T}^n$, so a time exists when three vectors are stored in memory ($\vec{T}^n$, $A\vec{T}^n$, and $\vec{b}$). Adding this storage requirement to the additional fixed storage of A, Δt, and other values reduces the achievable N below the maximum allowed by equation (1.11) and thereby increases the error in (1.12).

By contrast, evaluating equation (1.14) allows for overwriting $\vec{T}^n$ with the result of the product $(I + \Delta t A)\,\vec{T}^n$, so at most two vectors need be stored at any given time. This facilitates increasing N closer to N_{max}, thereby decreasing the error. Thus, to the extent complexity arguments lead to a strategy where memory management plays a critical role, programming complexity plays a critical and quantifiable role in the accuracy of the results. Two programs that are algorithmically equivalent in terms of the numerical methods employed might have different levels of error if one program requires significantly more dynamic memory. Chapter 12 presents much more severe examples of this principle in which the solutions are three-dimensional (3D) and therefore place much greater demands on the available memory.

The other way that complexity affects accuracy is by increasing the likelihood of bugs. A statistical uncertainty argument could be constructed to quantify the influence of bugs on accuracy, but it seems more useful to return to the theme of the previous section by asking, "How can we quantify complexity in ways that reduce development time?" Several approaches of varying sophistication are employed throughout this book. For present purposes, let's break the development time into the four stages:

1. Development of the physical, mathematical, and numerical models,
2. Designing the program,
3. Writing the code,
4. Debugging the code.

Because this book focuses on the design stage, one wonders how program design influences the debugging process. Such considerations could easily lead us down a path toward analyzing the psychology and sociology of programmers and groups of programmers. These paths have been trod by other authors, but we instead turn to simple truisms about which we can reason quantitatively.

In their text *Design Patterns Explained*, Shalloway and Trott (2002) write:

> The overwhelming amount of time spent in maintenance and debugging is on finding bugs and taking the time to avoid unwanted side effects. The actual fix is relatively short!

Likewise, Oliveira and Stewart (2006), in their text *Writing Scientific Software: A Guide to Style*, write: "The hardest part of debugging is finding where the bug is." Accepting these arguments, one expects debugging time to increase monotonically with the number of program lines, λ, one must read to find a given bug:

$$t_{\text{debugging}} = f(\lambda), \; f(\lambda_1) > f(\lambda_2) \text{ if } \lambda_1 > \lambda_2 \tag{1.15}$$

in which case reducing the number of lines one must read reduces debugging time. This relationship argues not so much for reducing the absolute number of program lines but for reducing the dependencies between lines. Bugs typically present themselves as erroneous output, whether from the program itself or the operating system. Exterminating bugs requires tracing backward from the line where the symptom occurs through the lines on which the offending line depends.

To understand how dependencies arise, it helps to examine real code. Since this book focuses on design, we first construct the only design document one typically encounters for conventional scientific code. Figure 1.3 depicts a flow chart that uses the cooling fin numerical model from section 1.2 to estimate the time required for the fin to reach steady state. At long times, one expects the solution to reach a steady state characterized by $\partial T/\partial t = 0 = \alpha\partial^2 T/\partial x^2$, so the temperature varies linearly in space but ceases changing in time. The algorithm in Figure 1.3 senses this condition by checking whether the second derivative of the fin temperature adjacent to the chip drops below some tolerance.

Additional tertiary dependencies might be generated if any procedures listed previously call other procedures and so forth. Dissecting the temperature calculation according to the values it uses makes clear that shared data mediates the interactions between program lines.

Figure 1.4 provides a sample Fortran implementation of the algorithm described in Figure 1.3. Section 1.7 discusses several stylistic points regarding the sample code. In the current section, the emphasis remains on dependencies between lines. For discussion purposes, imagine that during the course of solution, program line 29 in `main` (Figure 1.4(c)) prints temperatures outside the interval (T_{air}, T_{chip}), which is

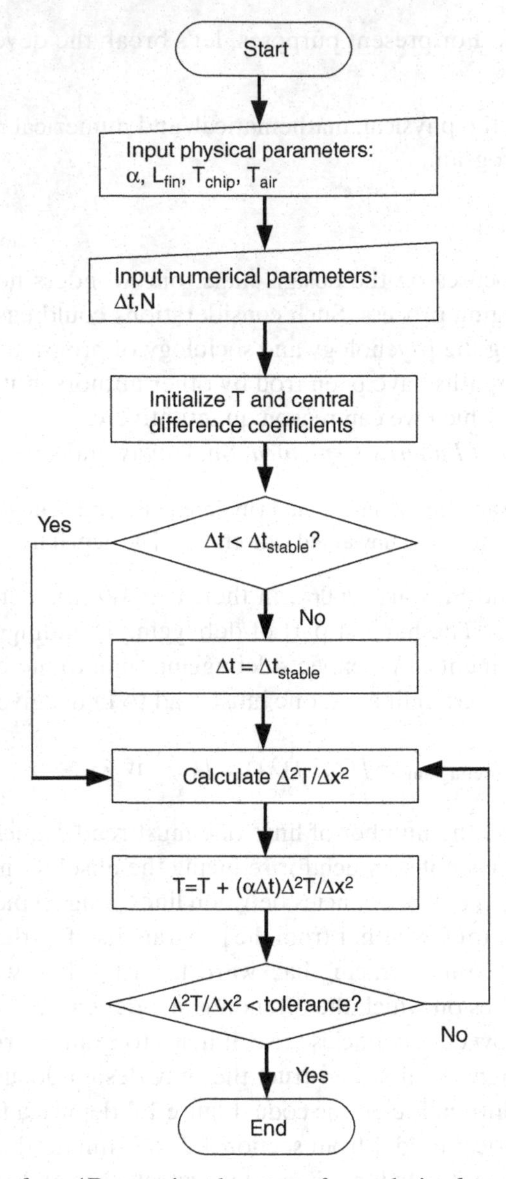

Figure 1.3. Flow chart for a 1D, transient heat transfer analysis of a cooling fin.

known to be physically impossible. Dissecting line 28 in `main` from left to right, one finds the temperature depends directly on the following:

1. The time step read in `read_numerics()` at line 18 in `main`,
2. The second derivative `T_xx` calculated by `differentiate()` at line 27 of `main`, and
3. The diffusivity initialized on line 16 of `main` and possibly overwritten on line 24 of `main`.

Furthermore, dissecting line 27 of `main` from left to right generates the following secondary dependencies of `T` on:

4. The `laplacian` differencing matrix created at line 22 of `main`,

5. Any modifications made to T by `differentiate()` (which can be determined to be none after noticing the procedure's `pure` attribute, which in turn necessitates the `intent(in)` attribute on line 54 in Figure 1.4(b)),
6. The boundary conditions `air_temp` and `chip_temp`, which are initialized at line 9 in `main` and possibly overwritten by `read_physics()` on line 16 in `main`.

Figure 1.5 diagrams a sample interaction for a code similar to that of Figure 1.4 but without dummy argument `intent` specifications. In programs large enough to be broken into multiple procedures, the mechanisms for sharing data include argument passing and file access. A certain number of data dependencies across procedures are inevitable in modular code. In the worst-case scenario, however, the global sharing facilitated by the Fortran `common` blocks in many scientific programs implies erroneous output from any one program line could send us on a wild-goose chase through all $n-1$ other lines.

With a bit of empirical data and straightforward reasoning, the previously described scenario leads to an estimate of conventional programming debugging time. First consider that an exhaustive study of a large commercial software project by Fenton and Ohlsson (2000) found that module defect density, r, was roughly constant across virtually the entire range of module sizes comprising the software. Contrary to conventional wisdom, this implies the number of defects in a modular program does *not* depend on the module granularity. In a collection of modules comprised of λ lines total, the number of defects has an expected value of $r\lambda$.

Figure 1.4(a)

```
1  module kind_parameters          ! Type kind parameter
2    implicit none
3    private
4    public :: rkind,ikind,ckind
5    integer ,parameter :: digits=8   ! num. digits of kind
6    integer ,parameter :: decades=9 ! num. representable decades
7    integer ,parameter :: rkind = selected_real_kind(digits)
8    integer ,parameter :: ikind = selected_int_kind(decades)
9    integer ,parameter :: ckind = selected_char_kind('default')
10 end module
```

Figure 1.4(b)

```
1  module conduction_module
2    use kind_parameters ,only : rkind,ikind
3    implicit none               ! Tri-diagonal array positions:
4    integer(ikind) ,parameter::diagonals=3,low_diag=1,diag=2,up_diag=3
5  contains
6    pure logical function found(namelist_io)
7      integer(ikind) ,intent(in) ::namelist_io
8      if (namelist_io==0) then
9        found=.true.
10     else
11       found=.false.
12     end if
13   end function
14
```

```
15   logical function read_physics(unit,alpha,L_fin,T_chip,T_air)
16     real(rkind)      ,intent(out) :: alpha,L_fin,T_chip,T_air
17     integer(ikind) ,intent(in)   :: unit
18     integer(ikind)                :: physics_io
19     namelist/physics/ alpha,L_fin,T_chip,T_air
20     read(unit,nml=physics,iostat=physics_io)
21     read_physics = found(physics_io)
22   end function
23
24   logical function read_numerics(unit,dt,nodes,dx,L_fin)
25     real(rkind)      ,intent(out) :: dt,dx
26     integer(ikind) ,intent(out) :: nodes
27     real(rkind)      ,intent(in)  :: L_fin
28     integer(ikind) ,intent(in)   :: unit
29     integer(ikind)                :: numerics_io,elements
30     namelist/numerics/ dt,nodes
31     read(unit,nml=numerics,iostat=numerics_io)
32     read_numerics = found(numerics_io)
33     elements = nodes+1
34     dx   = L_fin/elements
35   end function
36
37   pure real(rkind) function stable_dt(dx,alpha)
38     real(rkind) ,intent(in) :: dx,alpha
39     real(rkind) ,parameter  :: safety_factor=0.9
40     stable_dt = safety_factor*(dx**2/alpha)
41   end function
42
43   pure function differencing_stencil(dx,nodes) result(centralDiff)
44     real(rkind)      ,intent(in) :: dx
45     integer(ikind) ,intent(in) :: nodes
46     real(rkind),dimension(:,:),allocatable :: centralDiff
47     allocate(centralDiff(nodes,diagonals))
48     centralDiff(:,low_diag)=  1./dx**2
49     centralDiff(:,     diag)= -2./dx**2
50     centralDiff(:, up_diag)=  1./dx**2
51   end function
52
53   pure function differentiate(finiteDiff,T,T1st,Tlast)result(T_xx)
54     real(rkind),dimension(:)   ,intent(in) ::T
55     real(rkind),dimension(:,:),intent(in) ::finiteDiff!differentiation op
56     real(rkind)                  ,intent(in) :: T1st,Tlast
57     real(rkind),dimension(:)   ,allocatable:: T_xx
58     integer(ikind) :: nodes,i
59     nodes = size(T)
60     allocate(T_xx(nodes))
61     T_xx(1) =  finiteDiff(1,low_diag)*T1st &
62               +finiteDiff(1,     diag)*T(1) &
63               +finiteDiff(1, up_diag)*T(2)
64     forall(i=2:nodes-1)
65       T_xx(i) =  finiteDiff(i,low_diag)*T(i-1) &
66                 +finiteDiff(i,     diag)*T(i  ) &
67                 +finiteDiff(i, up_diag)*T(i+1)
68     end forall
69     T_xx(nodes) =  finiteDiff(nodes,low_diag)*T(nodes-1) &
```

```
70              +finiteDiff(nodes,    diag)*T(nodes  ) &
71              +finiteDiff(nodes, up_diag)*Tlast
72   end function
73 end module
```

Figure 1.4(c)

```
1  program main
2    use iso_fortran_env ,only : input_unit
3    use kind_parameters ,only : rkind,ikind
4    use conduction_module
5    implicit none
6    real(rkind)    ,parameter :: tolerance=1.0E-06
7    integer(ikind) ,parameter :: chip_end=1
8    integer(ikind)            :: elements=4,intervals
9    real(rkind) :: air_temp=1.,chip_temp=1. & !boundary temperatures
10     ,diffusivity=1.,fin_length=1.,time=0.  & !physical parameters
11     ,dt=0.01,dx                             !time step, grid spacing
12   real(rkind),dimension(:)  ,allocatable:: T,T_xx !temperature,2nd deriv
13   real(rkind),dimension(:,:),allocatable:: laplacian !differential op
14   dx=fin_length/elements                     !default element size
15   if (.not. &
16     read_physics(input_unit,diffusivity,fin_length,chip_temp,air_temp))&
17     print *,'In main: Using default physics specification.'
18   if (.not. read_numerics(input_unit,dt,elements,dx,fin_length)) &
19     print *,'In main: Using default numerical specification.'
20   allocate(T(elements),laplacian(elements,diagonals),T_xx(elements))
21   T         = air_temp
22   laplacian = differencing_stencil(dx,elements)
23   print *,'time,temperature=',time,T
24   dt = min(dt,stable_dt(dx,diffusivity))
25   do
26     time = time + dt
27     T_xx = differentiate(laplacian,T,chip_temp,air_temp)
28     T = T + dt*diffusivity*T_xx
29     print *,'time,temperature=',time,T
30     if (T_xx(chip_end)<tolerance) exit
31   end do
32   print *,'steady state reached at time ',time
33 end program
```

Figure 1.4(d)

```
1 &physics
2   alpha=1.,L_fin=1.,T_chip=1.,T_air=1.
3 /
4 &numerics
5   dt=0.01,nodes=9
6 /
```

Figure 1.4. Fin heat conduction program: (a) global constants, (b) solver, (c) driver, and (d) input file.

Assuming programmers are conscientious, a single symptom of a problem is sufficient to start the search for the bug that caused it, so there exists a one-to-one relationship between bugs and symptoms. Now imagine a chronological listing of

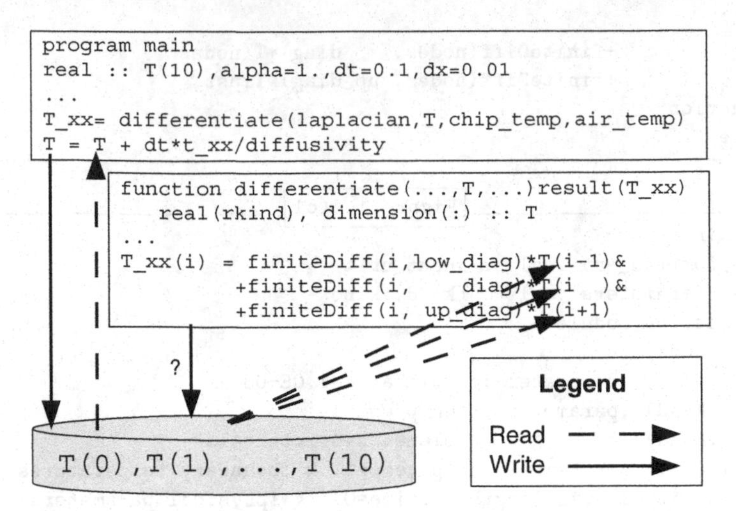

program main
real :: T(10),alpha=1.,dt=0.1,dx=0.01
...
T_xx= differentiate(laplacian,T,chip_temp,air_temp)
T = T + dt*t_xx/diffusivity
function differentiate(...,T,...)result(T_xx)
real(rkind), dimension(:) :: T
...
T_xx(i) = finiteDiff(i,low_diag)*T(i-1)&
+finiteDiff(i, diag)*T(i)&
+finiteDiff(i, up_diag)*T(i+1)
?
T(0),T(1),...,T(10)
Legend
Read
Write

Figure 1.5. A subset of the data dependencies for a temperature calculation without argument intent. Without intent, the entire differentiate() procedure must be reviewed to determine whether the procedure writes to T.

the λ program lines executed before the symptom occurs. Assume redundant occurrences are removed while ensuring that all lines that execute before the symptom occurs are listed prior to listing the symptomatic line. Some of the lines preceding the symptomatic line in such a listing might appear after the symptomatic line in the source code. For example, if the symptom occurs early in the second pass through a loop, all lines executed during the first pass through the loop would be listed prior to the symptomatic line.

The aforementioned assumptions are intended to avoid inflating λ above the total number of lines in the code and to ensure that causality confines the bug to the lines preceding the symptom on the list. Now if a given symptom is equally likely to occur anywhere on the list, then its expected location is $\lambda/2$ lines from the top, so the bug must be in the preceding $\lambda/2-1$ lines.

If the bug is equally likely to be found on any of these lines, the expected value of the number of lines that must be searched backward to find it is $(\lambda/2-1)/2$. For code with $r\lambda$ bugs, the search algorithm just described has an expected completion time of:

$$t_{find} = \frac{r\lambda}{2}\left(\frac{\lambda}{2}-1\right)t_{line} \tag{1.16}$$

where t_{line} is the average time to review one line. We define "review" to include reading it, testing it, and thinking about it over coffee. Since we do not include the time to revise it, which might also include revising other lines, equation (1.16) provides a lower bound on the debugging time, which provides a lower bound on the development time. Development time thus grows *at least* quadratically with code size.

To estimate r, consider that Hatton (1997) found an average of 8 statically detectable faults per 1,000 lines among millions of lines of commercially released scientific C code and 12 per 1,000 in Fortran 77 code. Separately, he apparently estimated that approximately two to three times as many errors occur in C++[1]. He

[1] Van Snyder reported Hatton's C++ results in a smartly and humorously written response to a petition to retire Fortran. See http://www.fortranstatement.com/Site/responses.html

defined a fault as "a misuse of the language which will very likely fail in some context." Examples include interface inconsistencies and using un-initialized variables. Hatton's average fault rate yields the estimate $r \approx 0.01$.

Equation (1.16) provides a polynomial time estimate for completing the debugging process. Even though polynomial time estimates fall under the purview of *computational complexity theory*, one must be careful to distinguish the complexity of the debugging process from the complexity of the code itself. Attempts to quantify code complexity in useful ways – particularly to predict fault rates – have a checkered past.

Exacerbating the problem is the reliance of most programs on externally defined libraries with which data must be shared. Truly pernicious bugs (which the vendors might call "features") result when code not written by the programmer modifies its arguments unexpectedly. Although such dependencies may seem inevitable, Chapter 6 explains how to construct a library that never gets its grubby hands on the data on which it operates.

As individual programs merge into complete software packages, Figure 1.5 becomes a plate of spaghetti, providing one interpretation of the infamous term "spaghetti code." The next section explores several philosophies for disentangling the knots in spaghetti code. The most popular modern alternative, OOP, reduces data dependencies by constructing modules with local data and allowing access to that data only to procedures inside the associated module. Chapter 2 refines the estimate in equation (1.16) by analyzing a more representative fault localization algorithm and demonstrates how OOP reduces the search time. Chapter 3 demonstrates how an OOP strategy that is broadly useful in scientific programming further limits search times.

The benefits of the above modularization strategy are manifold, but the strategy does have limits. Although strong evidence suggests defect rates are roughly independent of module size as assumed implicitly in equation (1.16) (Fenton and Ohlsson 2000), the same study provided evidence that defect rates sometimes spike upward for the smallest modules. As module granularity increases, one must explore additional avenues for reducing dependencies. The next section surveys a range of possibilities.

Ravioli Tastes Better Than Spaghetti Code

Spaghetti code results partly from spaghetti dependencies. One way to reduce the dependencies between program lines is to reorganize the program by partitioning the data into private, local pockets. To the extent that managing the memory in these pockets determines the maximum resolution, a program's organization determines its accuracy.

1.4 Alternative Programming Paradigms

Many programming paradigms represent communities as much as philosophies. In some cases, the community members have written hundreds of publications building veritable edifices of human achievement. It would be impossible to wander through one of these edifices in the allotted time and fully grasp how the community goes about its daily work. Cameras slung unabashedly around our necks, we hope to capture a few choice snapshots, to take in the sights of most interest, to discern the basic building blocks, to overhear a bit of the mindset expressed in conversation, and to return home changed forever. Although the following paragraphs read at first like a timeline, most of these communities remain vibrant in certain corners, with good reason and bad.

In the beginning, there was *unstructured programming*. Early machine language programming employed individual lines as the basic unit of decomposition. Although programmers were free to think in terms of higher levels of organization, no language support for such basic forms as loops existed. With the advent of Fortran as the first high-level programming came looping constructs (`do` loops) and conditional execution blocks (`if` constructs).

At this level of program organization, the only commonly employed program abstraction is the flow chart. With the resultant emphasis on program flow rather than data, the first few evolutionary steps beyond unstructured code involved new ways to express and decompose algorithms. The 1958 version of Fortran, for example, added the capability to define a `subroutine` or `function`. The procedures constructed therewith became the basic units of decomposition in *procedural programming*.

Whereas some paradigm shifts expand the developers' toolset, others prune away its rougher edges. Those aspects of *structured programming* and *functional programming* considered here involve pruning. Given the natural emphasis on a program's logical flow that stems from staying within the procedural paradigm, most structured programming methods emphasize the avoidance of arbitrarily employing discontinuous jumps from one region of code to another. Because the unconditional `goto` statement offers the most severe and abrupt such jump, the `goto` has been frequently maligned by structured programming adherents. One of the founders of the structured programming movement, Dijkstra (1968), summarized some of the basic reasons in an article entitled "GOTO statement considered harmful."

Knuth (1974) pointed out that the most harmful uses of `goto` were already beginning to be subsumed by newer, more specific constructs, a prime example being error handling, where it is desired to exit some branch in a program's call tree to dispatch a distant portion of code intended to deal gracefully with an aberrant situation such as reaching the end of an input file. Knuth attempted to redirect the structured programming dialogue toward more fundamental questions of composing programs from collections of entities about which formal correctness proofs can be made. He also provided a simple rule for avoiding complicated flow: It should be possible to draw all forward jumps in a flow chart on one side and all backward jumps on the other side with no jumps crossing each other. Most references to "spaghetti code" are to programs that violate these principles. Nonetheless, as late as 1987, the debate on the `goto` statement raged on, with Rubin (1987) writing an article entitled " 'GOTO considered harmful' considered harmful" that garnered multiple responding letters to the editor, including one from a disappointed Disktra.

In addition to eschewing `goto` statements in this book's source code, a recent incarnation of structured programming is adopted to facilitate splitting tasks across a number of hardware units. The Open Multi-Processing (OpenMP[2]) toolsets for parallelizing code operate on so-called structured blocks of code. The OpenMP standard

[2] http://www.openmp.org

defines a structured block as a contiguous set of statements with one point of entry and one point of exit.

Lest we pat ourselves on the back for advancing so far from the Stone Age, a resurgence of unstructured programming came with the advent of interactive mathematical packages. This trend has many benefits. For relatively simple tasks and even for some more complicated tasks with regularly spaced data and modest operation counts, the benefits of packages such as Matlab and Mathcad far outweigh their disadvantages. The ability to leverage hundreds of programmer-years with the press of a button, to integrate visualization into calculations, and especially to work with units that are automatically converted when necessary in Mathcad outweighs the difficulty of life without subroutines and functions for small code segments of a 10–20 lines. Of the approaches discussed in this section, functional programming represents the first that deals with the data dependencies discussed in the previous section. Functional programming emphasizes the specification of relationships between entities, rather than specifying steps that must be done. Mathematica is the most commonly employed functional programming language/environment with which most scientific programmers would be familiar.

The subset of functional programming that most influences this book is *purely functional programming*, in which side effects, including modifying procedure arguments, are completely ruled out. While we find it impractical to rule out side effects completely, partly for reasons explained in Chapter 3, minimizing side effects amounts to limiting data access. Limiting data access reduces the linkages in diagrams like Figure 1.5, which has an important impact on the debugging time estimate of equation (1.16). We discuss this impact in Chapter 3.

Object-based programming further limits data access by encouraging encapsulation and information hiding. Although Chapter 3 provides more complete explanations, here we define encapsulation as the act of partitioning data and procedures for operating on that data into modules. The data referenced here is inside a derived type of the form

```
type fluid
  real :: viscosity
  real, dimension(:,:,:), allocatable :: velocity
  character, allocatable :: chemical_formula
end type
```

With the object-based approach, one puts procedures into a given module only if they operate on instances (objects) of the type defined in that module or provide services to a procedure that does. Furthermore, one generally lets only those module procedures, typically termed *methods* in OOP discussions, operate on instances of the module's data type. (Chapter 3 details the construction of methods via Fortran 2003 type-bound procedures.) The latter relationship represents data privacy, which is one form of information hiding. Another form involves keeping some of the procedures in the module inaccessible to code outside the module. Each of these strategies limits data access and thereby helps reduce the debugging search time. Encapsulation and information hiding are so closely linked that they are often considered one concept. That concept is the first of the three main pillars of OOP.

The second pillar, *polymorphism*, has Greek roots meaning "many faces." One category of polymorphism takes the form of a one-to-many mapping from procedure names to procedure implementations. Here we define an interface as the information summarized in a Fortran interface body: a procedure's name, its argument types, and its return type, if any. Although polymorphic variables did not enter the language until Fortran 2003, Fortran programmers had already encountered one form of polymorphism when using intrinsic functions. For example, in writing the code:

```
real :: x=1.,sin_x
real ,dimension(10,10) :: a=0.,sin_a
sin_x = sin(x)
sin_a = sin(a)
```

where the function `sin` can accept a scalar argument or an array arguments and correspondingly return either a scalar or an array. The procedure that responds to each of the above calls must be somewhat different even though the call retains the same form. Chapter 3 explains other forms of polymorphism and how the programmer can construct her own polymorphic procedures. Chapter 3 also analyzes the influence polymorphism has on programming complexity.

One form of polymorphism is so important that it is typically given its own name: operator overloading. Again it is closely related to something Fortran 90 programmers have already encountered in writing expressions like:

```
real                   :: x=1.,y=2.,z
real ,dimension(10,10) :: a=0.,b=1.,c
z = x + y
c = a + b
```

where the "+" operator apparently has different meanings in different contexts: the first being to add two scalars and the second being to perform element-wise addition on two matrices. Operator overloading is when the programmer specifies how such operators can be applied to derived types created by the programmer. Chapter 3 explains how operator overloading plays such a critical role in reducing programming complexity that it could be considered the fourth pillar of *scientific OOP* (SOOP).

Object-oriented programming is object-based programming plus *inheritance*, the third pillar of OOP.[3] Inheritance increases code reuse by facilitating subtypes, to be defined based on existing types. In Fortran 2003, subtypes are termed *extended types* and take the form:

```
type ,extends(fluid) :: magnetofluid
  real ,dimension(:,:,:) ,allocatable :: magnetic_field
end type magnetofluid
```

which specifies that a `magnetofluid` is a `fluid` with an associated magnetic field. Every instance of a `magnetofluid` thus inherits a viscosity, velocity, and a chemical

[3] Perhaps the best definition is that object-based programming is OOP minus whatever one considers to be the most critical feature of OOP. For example, another interpretation is that object-based programming is OOP minus polymorphism (Satir and Brown 1995).

formula from the `fluid` type that `magnetofluid` extends. Equally important, languages that support OOP also provide facilities for the module implementing the magnetofluid to inherit various methods from the Fluid module. By analogy with biological inheritance, the extended type is often referred to as the "child," whereas the type it extends is termed the "parent."

Component-based software engineering (CBSE) can be viewed as the next evolutionary step beyond OOP. Where OOP informs how we construct and connect individual objects, CBSE works with components that can be comprised of collections of objects along with code not necessarily written in an object-oriented fashion. Most of the CBSE toolsets developed for mainstream software engineering have severe inefficiencies when applied to scientific problems. Where distributing applications across the Internet is increasingly the norm in nonscientific work, parallelizing applications across a set of closely linked computational nodes has become the norm in scientific work. The communication patterns and needs differ considerably in these two domains.

Fortunately, for high-performance scientific programming, the CCA facilitates the management of interprocedural communication at run-time among programs that might have been written in disparate languages employing equally disparate programming paradigms (Bernholdt et al. 2006). A concept closely linked with CBSE is *design by contract*, which expresses the notion that software clients write specifications with which software designers must comply. In fact, one definition of a component is an object with a specification. Chapter 11 discusses a toolset for writing specifications and how the specification process influences programming in ways that increase code robustness.

The *generic programming* paradigm enhances programmers' ability to write reusable algorithms. It is often referred to as programming with concepts, wherein procedures or algorithms are expressed with a high level of abstraction in the applicable data types. When sorting a set of objects, for example, whether they are of numeric type or the previously defined `fluid` type, the well-established quick-sort algorithm can be applied, providing a well-defined comparison method between any two objects. Appropriate use of this paradigm leads to highly reusable components, thus rendering simpler programs.

A number of modern programming languages explicitly support *generic programming*. Templates in C++, such as the container classes in the Standard Template Library (STL), are among the most commonly used generic constructs. Templates allow one to write code in which the type being manipulated can remain unspecified in the expression of the algorithm. Even though Fortran lacks templates, one can emulate the simpler template capabilities in Fortran. Part II discusses some template-emulation techniques. Furthermore, Fortran supports its own version of generic programming via parameterized derived types that allow type parameters to be generic in source. Type parameters specify attributes used in defining the derived type as in:

```
type vector(precision,length)
  real(precision) ,dimension(length) :: v
end type
```

where `precision` and `length` are type parameters that can be used in a subsequent declaration to specify the kind and dimension of `v`.

Although not precisely programming paradigms, *design patterns* and *formal methods* represent two software engineering subdisciplines that greatly influence designs. In light of the earlier presented definition of OOP, we can now state that design patterns specify the roles that the programmer-defined types play and how those roles mesh to form a solution to some recurring problem. Object-oriented scientific design patterns comprise the central focus of Part II of this text.

Closely related to the *design by contract* notion, formal methods represent mathematically precise specifications of algorithms and rigorous verifications that a particular implementation is correct with respect to the specification. Scientific programming is already formal in its numerical algorithms. The mathematical model to be approximated, for example, Maxwell's equations for electromagnetics, provides a formal specification. Convergence proofs for the discrete approximations employed provide a formal verification that the numerical algorithms conform to the specification. The field of formal methods brings this same level of rigor to nonscientific software. Chapter 10 discusses how scientific programmers can benefit from specifying formal constraints to the nonnumerical aspects of scientific software, for example, memory management.

You Say Tomato. We Say Baloney!

Although it is common in the scientific community to think of the differences in coding practices as purely stylistic ("You say toe-may-toe. I say toe-mah-toe."), quantitative arguments presented in the previous section suggest that structured programming scales poorly with increasing program size. Functional, object-based, object-oriented, generic, and component-based approaches each build on the structured paradigm while increasing its ability to scale with package size.

1.5 How Performance Informs Design

Since this book's development-time focus runs counter to the prevailing trends in the scientific computing literature, it is incumbent on us to explain how design relates to the more common concern: performance. Any discussion of performance ultimately turns on Amdahl's law, which is discussed in detail in Chapter 12. For present purposes, what matters is that Amdahl's law determines the maximum factor by which any process can be sped up. Specifically, it shows that the maximum fractional speedup is limited by the fraction of the process that can be sped up. If 80% of the process can be sped up, the maximum speedup is:

$$\frac{t_{run}}{(1-0.8)t_{run}} = 5 \qquad (1.17)$$

where the numerator and denominator are respectively the original completion time and the minimum time based on completely eliminating 80% of the process. This calculation suggests that when modest speedup is acceptable, optimization efforts should be focused on the code that occupies most of the run time. The search for those parts leads naturally to the following rule of thumb:

Pareto Principle: For many systems in which participants share resources, roughly 20% of the participants occupy 80% of the resources.

Specifically, experience indicates that for many codes, approximately 20% of the lines occupy 80% of the execution time. Though this situation might seem purely fortuitous, some form of Pareto rule always holds. In any system where participants (lines) share resources (execution time), there will always be a number k such that $k\%$ of the participates occupy $1-k\%$ of the resources so long as the number of participants is sufficiently large to admit a fraction equal to the chosen percentage. Two limiting cases are equal sharing ($k = 50\%$) and monopoly by one line ($k \to 100\%$). Hence, the Pareto Principle suggests most systems are skewed slightly towards monopoly. This very likely stems from the fact that, considering all possible resource allocations for N participants, there is only one that corresponds to an equitable distribution, whereas there are N possible monopolies.

Considering Amdahl's law vis-à-vis Pareto's rule leads to the design principle that guides this entire book: structure 80% of the code to reduce development time rather than run time. Focus run-time reduction efforts on the 20% of the code that occupies the largest share of the run time. Ultimately, this means design must be as analytical and empirical as the numerics and physics. Development time estimates must guide the design. Solution time estimates must guide the algorithm choice. Profiles of the run-time share occupied by each procedure must guide the optimization efforts. Rouson et al. (2008b) discussed a common case that is even far more extreme than Pareto's Principle: 79% of the run time is occupied by less than 1% of the code. The reason is that this short code segment calls an external library that does the heavy lifting. With increasing development of high-performance scientific libraries, this situation can be expected to become the norm.

The percentage of the code that must be highly optimized increases with the desired amount of speedup. When the optimization focuses on parallelization, the process of scaling up from running on a single execution unit to running on many thousands eventually takes the search for optimization into the darkest crevices of the code. If one starts from code that is already parallel, however, the Pareto-inspired rule-of-thumb holds: Targeting a fivefold increase in speed is as reasonable when going from 1 to 5 execution units as when going from 10,000 to 50,000.

May the Source Be With You

Write 80% of your source code with yourself and your fellow programmers in mind, that is, to make your life easier, not your computer's life easier. The CPU will never read your source code without an interpreter or compiler anyway. Arcane or quirky constructs and names written to impress on the CPU the urgency or seriousness of your task might well get lost in the translation.

1.6 How Design Informs Performance

Users who keep pace with the speed increases of leadership-class machines pace themselves against thousandfold speedup goals over decadal time periods. In going from the petaflops machines of the year 2010 to the exaflops platforms[4] projected for 2020, most of the speedup must come from distributing data and finding additional

[4] Petaflops and exaflops correspond to 10^{12} and 10^{15} floating-point calculations per second, respectively.

opportunities for parallel execution. Since current trends suggest that data distribution costs will dominate computing costs at the exascale, the chosen constructs must minimize the data distribution needs. A significant driver for data distribution is the need to synchronize memory when one execution unit modifies data that another unit needs. Over the past decade, the computer science community has issued many calls for functional programming as a solution that limits such communication: Side-effect-free procedures do not modify state that other procedures need.

Any strategy that promises exaflops must allow for a rapidly evolving landscape of hardware, programming languages, and parallel programming models. Hardware is likely to become heterogeneous with exotic combinations of processor types. Old languages will evolve while new languages are under development. The current parallel programming toolsets will themselves evolve whereas many of their ideas will ultimately be embedded into new and established languages. One way to mitigate the impact of this evolution on application programmers relies on establishing high-level, software abstractions that can remain reasonably static and self-consistent over time. Many of the most impactful application domains, for example, energy and climate sciences, will remain as high a priority a decade hence as a decade prior. In these physics-based fields, the tensor form of the governing equations will not change. Thus, tensor calculus abstractions hold great promise for achieving abstraction invariance.

Finally, candidate abstractions must be language-agnostic. OOD patterns thus provide a natural lexicon for expressing candidate abstractions. Developers can implement OOD patterns in any language that supports OOP, as Part II of the current text demonstrates in Fortran 2003 and C++.

All of the reasoning in this section suggests architecting scientific software by combining OOD patterns built around the concept of a parallel, tensor calculus based on side-effect-free, composable functions. This forms the central theme and ultimate aim of the current book, namely that the design aims of a book focused on expressiveness over performance ultimately align with performance points to a deep truth. Consider that the calculus abstractions employed throughout much of this book are modeled after the analytical expressions used in deriving closed-form, analytical solutions to PDEs. In writing analytical solutions, one often arrives all at once at the solution for the whole domain — or some unidirectional factor in the solution with methods such as the separation of variables or some globally varying term in the solution in the case of a series expansion. Thus, analytical solutions are intrinsically parallel. One expects the same to be true of software constructs patterned after continuous forms.

Cleanliness Is Next to Speediness

When the ultimate goal is to solve tensor PDEs, an approach intended to generate exceptionally clean code also illuminates a path to efficient parallel code. Functional programming limits communication. OOD patterns shield the application programmer from language dependency, while the use of high-level, tensor calculus constructs ensures the constancy of the abstractions over time.

1.7 Elements of Style

By definition, a high-level language is a human tongue meant for human eyes, so one of the ways to reduce a code's development time is to make it more easily readable for humans. To balance the emphasis on quantitative software descriptors in the remainder of Part I, this section offers a few qualitative tips to improve readability.

Rule 1.1. Write code that comments itself.

The suggested goal is to write code that reads like prose with the richness and structure of natural language grammar, including nouns, verbs, prepositions, and so on. Consider the following Fortran code:

```
type(gas)          :: air
real ,parameter :: viscosity=1.52E-05 ! meters/sec
if (.not.readFromDisk(air)) air = constructGas(viscosity)
```

and the corresponding C++:

```
gas                   air;
const float           viscosity=1.52E-05; // meters/sec
if (! readFromDisk(air))      air = constructGas(viscosity);
```

It should be clear that we have defined a data structure representing some abstraction of a gas, that the gas to be modeled is air, that the air's viscosity is constant at 1.52×10^{-5}m/s, that we will attempt to read the data structure from a disk, and that we will construct it based on its viscosity if the read is unsuccessful. As seen here, writing code that comments itself does not eliminate the need for comments but does minimize it.

A common exception to writing code prosaically arises when the notation in a given domain is sufficiently expressive that the code will be clearest in that notation. Inside procedures that solve, for example, the Navier-Stokes equations to model gas motion, one might use u or v for the gas velocity because those symbols have that meaning almost universally in the fluid dynamics literature. Many similar examples are discussed throughout this book. Nonetheless, the key point is to avoid the strong aversion most scientific programmers have toward using descriptive (and yes, lengthy) names for data structures. No matter how concise, convenient, or catchy it seems to use `VTP0` as the name of an array that holds velocities, temperatures, and pressures at time zero, such habits are to be avoided like illicit drugs. Just say no!

Rule 1.2. Name all constants.

Even worse than using cryptic names is the use of no name at all. Without an additional comment, the meaning of the numeric constant in the following code will be hard to decipher even by the programmer who wrote it if a long time has passed since its writing the following Fortran code:

```
if (.not.readFromDisk(air)) air = constructGas(1.52E-05)
```

and the corresponding C++:

```
if (! readFromDisk(air))     air = constructGas(1.52E-05);
```

The constant definitions in the previous examples serve dual roles as declarations and effectively as comments.

Rule 1.3. Make constants constant.

The only downside to deciding to declare and initialize the viscosity is that every time we declare a new data structure, we increase the possible data dependencies identified in section 1.2 as a key cause of increased programming complexity. Declaring the viscosity constant via the `parameter` or `const` keywords in Fortran and C++, respectively, gives the code read-only access to it, thereby reducing data dependencies.

Rule 1.4. Minimize global data.

Rule 1.5. Make global data constant.

Rule 1.6. Provide global type parameters, including precision specifiers.

Although global data sharing is to be discouraged, defining a minimal set of values with read-only access (i.e., constants) causes no harm and, in the case of type parameters, makes the code considerably more flexible. Other globally useful values include mathematical and physical constants. The `kind_parameters` module in Figure 1.4 illustrates. For example, if every module uses this module and every real type declaration takes the form:

```
real(rkind) :: x
```

then changing the entire software package from single- to double-precision proves trivial from the software perspective – numerical considerations aside.

Values that require coordination across multiple files also require global specification in order to avoid conflicts. These might include file names and unit numbers.

Rule 1.7. Provide global conditional debugging.

As proposed by Akin (2003), another useful global constant in Fortran is one that facilitates toggling between production mode and debugging mode as follows:

```
logical ,parameter :: debugging=.true.
```

One can use such a parameter to perform useful debugging tasks conditionally. One such task might be to print a message at the beginning and end of each routine. Making debugging a global constant allows most compilers to remove any code that depends on it when a high optimization level is employed. Lines such as:

```
if (debugging) print 'constructGas: starting'
...
if (debugging) print 'constructGas: finished'
```

would be removed during the dead-code removal phase of most optimizing compilers if `debugging` is `.false.` Hence, there is no run-time penalty during production runs. This practice proves most important in Chapter 2 and beyond, where we adopt a programming strategy that leads to calling sequences that can only be determined at runtime.

Rule 1.8. Declare intent for all procedure arguments.

Rule 1.9. Avoid side effects when possible, particularly in a `function`.

Rule 1.10. Make all derived type components `private`.

Each of these practices reduces data dependencies, thereby directly impacting bug search time in ways discussed throughout this chapter. Rule 1.9 also relates to the goal of writing code that comments itself. The notation for invoking procedures that return values resembles the notation for mathematical functions. Since mathematical functions do not modify their arguments, code reads more clearly when the same is true for function procedures.

Rule 1.11. Indent loops and procedure/module executable statements.

Rule 1.11 provides visual clues to code structure, indicating nesting levels in nested constructs.

Rule 1.12. Prevent implicit typing globally.

The `implicit none` statement in Figure 1.4 forces the programmer to explicitly declare the type of all variables used. Most Fortran programmers are aware of this rule. We mention it as a reminder to include it in every module.

Figure 1.4 also illustrates several other practices we find useful. First, we frequently align tokens in columns to expose semantic commonality. The alignment of the commas in lines 25-28 of the conduction module provides one example. We also attach commas to the tokens they precede, such as the intent specifiers in lines 25-28. This serves as a reminder to delete the comma if one deletes the token that necessitates the comma.

Second, as demonstrated in line 2 of Figure 1.4(b), we use the optional `only` construct to modify `use` statements. This explicitly notifies the reader of where to find definitions of entities used in a given module but not defined in that module.

Finally, line 2 of Figure 1.4(c) demonstrates the `iso_fortran_env` Fortran 2003 module that facilitates portable access to the runtime environment – specifically the `input_unit` file unit provides portable access to the pre-connected input stream known as standard input in Unix-like environments. We mention it here to advocate for using standard identifiers wherever possible.

Some of the previously mentioned rules warrant stricter adherence than others. For example, naming every constant can prove cumbersome. We occasionally skip it when we feel very confident that the meaning is clear. For example, the statement:

```
do file_unit=1,max_unit
```

would be a bit more clear (and more general) if rewritten as:

```
do file_unit=first_unit,last_unit
```

We nonetheless will use the first form when it seems reasonable "1" will be understood to be the first file unit and when its value is very unlikely to change.

Another rule we adhere to without complete consistency regards character case. Because Fortran is internally case insensitive, mixing upper and lower case can give a false impression. For example, a reader might think that entities `t` and `T` refer to different quantities – say, time and temperature – when the compiler considers them to be the same variable. We prefer lower case most of the time, but do occasionally switch case when we feel confident it clarifies more than confuses.

By contrast, we have never found a good reason to violate data privacy in production code, although it is sometimes useful temporarily while debugging. The point is to habituate oneself to writing robust code by default, rather than to impose an unnatural level of uniformity. As Ralph Waldo Emerson said, "A foolish consistency is the hobgoblin of small minds."

> **Love Thy Neighbor As Thy Source**
>
> Structure your code so that it comments itself and reads like prose. Missives that would never have impressed the CPU might win you the adoration of programmers who have to read your code. Those programmers include your future self.

EXERCISES

1. In some systems, cooling fans stay on after the rest of the system shuts down. To determine how long it takes to cool the fins of Section 1.2 down, consider the fin to be a lumped mass with a spatially uniform temperature, T, that obeys:

$$mc\frac{dT}{dt} = -\dot{Q}_{out} \tag{1.18}$$

where m is the fin mass, c is the specific heat, and $\dot{Q}_{out}$ is the heat transfer out of the fin.

   (a) Draw a flow chart using explicit Euler time advancement to read m, c, $\dot{Q}_{out}$, $T_{initial}$, and T_{final} from the keyboard, integrate the above equation until an input temperature is reached, and output the required cooling time.
   (b) Identify the sections of your flow chart that represent structured blocks.
   (c) Describe several options for decomposing your flow chart into procedures.

2. Structured blocks are defined in Section 1.4.
   (a) Count the structured blocks in Figure 1.3. (*Hint*: Blocks can be composed of blocks with varying levels granularity.)
   (b) Design a procedural decomposition of Figure 1.3, that is, define a set of procedure names, arguments, argument types, and return types (if any) that could accomplish the tasks defined by the figure.
   (c) Which of the procedures defined in part (b) are purely functional?

3. Repeat problem 2 for the flow chart constructed in problem 1(a).

4. Review the code in Figure 1.4.
   (a) Calculate the maximum data dependencies.
   (b) Cite specific lines that are data-independent and discuss how the actual number of dependencies relates to the maximum.

2 The Object-Oriented Way

"Believe those who are seeking the truth. Doubt those who find it."
Andre Gide

2.1 Nomenclature

Chapter 1 introduced the main pillars of object-orientation: encapsulation, information hiding, polymorphism, and inheritance. The current chapter provides more detailed definitions and demonstrations of these concepts in Fortran 2003 along with a complexity analysis. As noted in the preface, we primarily target three audiences: Fortran 95 programmers unfamiliar with OOP, Fortran 95 programmers who emulate OOP, and C++ programmers familiar with OOP. In reading this chapter, the first audience will learn the basic OOP concepts in a familiar syntax. The second audience will find that many of the emulation techniques that have been suggested in the literature can be converted quite easily to employ the intrinsic OOP constructs of Fortran 2003. The third audience will benefit from the exposure to OOP in a language other than their native tongue. All will hopefully benefit from the complexity analysis at the end of the chapter.

We hope that using the newest features of Fortran gives this book lasting value. Operating at the bleeding edge, however, presents some short-term limitations. As of January 2011, only two compilers implemented the full Fortran 2003 standard:

- The IBM XL Fortran compiler,
- The Cray Compiler Environment.

However, it appears that the Numerical Algorithms Group (NAG), Gnu Fortran (gfortran), and Intel compilers are advancing rapidly enough that they are likely to have the features required to compile the code in this book by the time of publication (Chivers and Sleightholme 2010).

Since Fortran is the primary language for Part I of this text, we default to the Fortran nomenclature in this chapter and the next one. Upon first usage of a particular term, we also include a parenthetical note containing the italicized name of the closest corresponding C++ construct. The relationship between the Fortran and C++ constructs, however, is rarely one of exact equality. Often, both could be used in

Table 2.1. *OOP nomenclature*

Fortran	C++	General
Derived type	Class	Abstract data type
Component	Data member	Attribute*
Class select type	Dynamic Polymorphism (emulated via `dynamic_cast`)	
Type-bound procedure	Virtual Member function	Method, operation*
Parent type	Base class	Parent class
Extended type	Subclass	Child class
Module	Namespace	Package
Generic interface	Function overloading	Static polymorphism
Final procedure	Destructor	
Defined operator	Overloaded operator	
Defined assignment	Overloaded assignment operator	
Deferred procedure binding	Pure virtual member function	Abstract method
Procedure interface	Function prototype	Procedure signature
Intrinsic type/procedure	Primitive type/procedure	Built-in type/procedure

*Booch et al. (1999)

additional ways that do not correspond with the common usage of the other. For example, we suggest below that Fortran 2003 derived types correspond to C++ classes, but Fortran derived types can be parameterized in the sense that parameters declared in the type definition can be used to set things like precision and array bounds directly in a type declaration without calling a constructor[1]. No analogous feature exists in C++. Likewise, C++ classes can be templatized, that is, types can be generic in the source and not resolved to actual types until compile-time. No analogous feature exists in Fortran, although some of the simpler features of templates can be emulated in Fortran as explained by Gray et al. (1999) and Akin (2003).

We focus on a capability subset common to Fortran and C++. When we draw comparisons between Fortran and C++ constructs, we imply only that the two largely correspond when used in the manner described herein. Table 2.1 summarizes the approximate correspondences along with some language-neutral terms generally employed for the same concepts. Most of the latter come from the Unified Modeling Language (UML), a schematic toolset for describing object-oriented software independently from any chosen implementation language (Booch et al. 1999). As demonstrated in the remainder of Parts I and II, and Appendix B using UML during the software analysis and design phase facilitates focusing on the high-level modeling concepts rather than specific implementations thereof.

[1] Section 2.2 defines and discusses the special type of procedure referred to in OOP terminology as a *constructor*.

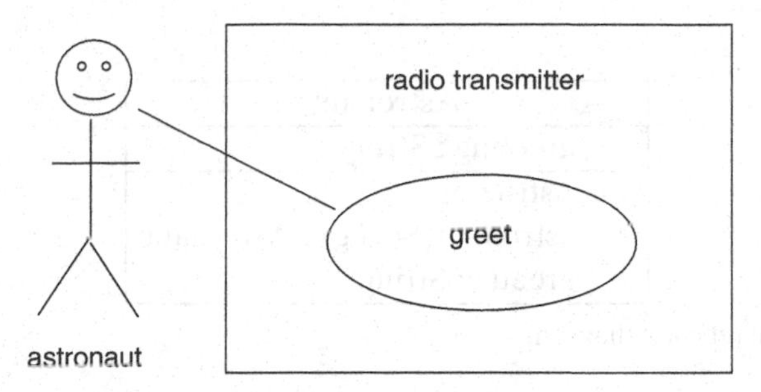

Figure 2.1. Spacecraft radio transmitter use case diagram.

A Rose By Any Other Nomenclature...

OOP concepts stand apart from any chosen programming language. While UML provides for programming-language-independent descriptions of OOP constructs, the constructs' implementations differ amongst languages. We focus on features common to Fortran 2003 and C++.

2.2 Object-Oriented Analysis and Design

Object-oriented analysis (OOA) and object-oriented design (OOD) comprise sequential steps in the modeling of software structure and behavior. In the *Unified Modeling Language User Guide*, Booch et al. (1999) described several techniques for modeling software abstractly. In the current text, we cover only those few aspects of OOA and OOD we have found useful in our work. In our experience, these correspond to the most frequently discussed techniques in computational science and engineering software development. These techniques have played critical roles in our discussions with research sponsors, in responding to journal referee reports, in presenting our work to colleagues, and in providing common idioms for communicating with codevelopers in multi-language projects.

A central activity in OOA involves determining the relevant set of actors and their actions. UML provides a standard schema for the graphical representation of the actors, their actions, and the system through which they interact. That schema entails drawing a UML *use case diagram*. As tradition dictates for programming discussions, we start with a program that simply prints "Hello, world!"

Constructing a use case diagram requires considering who might be the actor in such a model, what might be the system with which the actor interacts, and what uses the actor has for that system. The actor in a "Hello, world!" diagram might be an astronaut transmitting a message back to the earth via the spacecraft's radio transmitter, as shown in Figure 2.1. In this setting, the transmitter is the system and the usage case, or use case, is "greet the world," which describes the actor's goal in using the system. Use case diagrams depict each use case inside an oval and depict associations between the actors and the use cases by connecting the two with a line. Optionally, use cases can be encompassed by a box bearing a label for the system: the transmitter in the current context. As depicted, one always draws the actor external to the system.

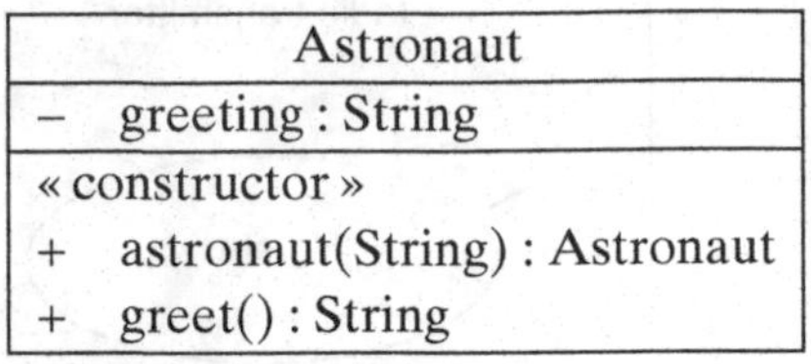

Astronaut
– greeting : String
« constructor »
+ astronaut(String) : Astronaut
+ greet() : String

Figure 2.2. Astronaut class diagram.

A more complete OOA might incorporate additional actors, such as a navigator, along with additional use cases such as "request landing location." OOA also considers various usage scenarios; however, one scenario suffices for the current, simple example. Given the simplicity of the current objective, we instead proceed to the OOD phase. OOD involves decomposing a problem domain into a set of useful classes, each of which is also termed an *abstract data type* (ADT). An ADT describes a collection of *objects* with a common state representation and functionality. Typically, an ADT embodies an *abstraction*, or abstract representation of whatever system it models. The term "abstract" implies the ADT retains only the features and behaviors required to represent the actual system in a particular context. One often regards an object as an anthropomorphic, or at least animated, entity that receives messages and collaboratively acts with other objects to accomplish some goal driven by the received message.

The OOD results in a specification of the ADTs that can be used to facilitate the given use cases. UML also provides a standard schema for diagramming a design's structure: the UML *class diagram*, so named because "class" is a common synonym for ADT. Due to differences in the meanings of the `class` keywords in Fortran and C++, we prefer the language-neutral "ADT" terminology; however, we use "class" interchangeably.

Figure 2.2 provides a UML class diagram for an astronaut ADT. A class diagram represents an ADT in a three-part box. The top part provides the class name. In Figure 2.2, the class is named "Astronaut."

The middle part of a class diagram describes the *attributes* – that is, data comprising the state of each instance of the class, each instance being termed an object. Names can be preceded by "+" or "–", indicating whether they are public or private, respectively. Public data can be referenced by code outside the module containing the class implementation. Private data can only be referenced inside the module. These can also be referred to as public or private scope. In this part of a class diagram, a datum's type follows its name and is separated from the name by a colon. In Figure 2.2, the state is represented by a private String "greeting."

The bottom part of a class diagram details the *operations* – that is, the methods. This part provides the names, scope, and return types of procedures implemented in the class. In OOP parlance, the procedure used to initialize, or *instantiate*, an object is termed a *constructor*. In UML, one customarily lists constructors first and precedes them by the stereotype name "constructor" (Deitel 2008). UML encloses stereotypes in guillemets (« and »). As indicated, the astronaut class implements a public constructor that takes a String and returns an astronaut instance. It also implements a public `greet()` method that returns a String.

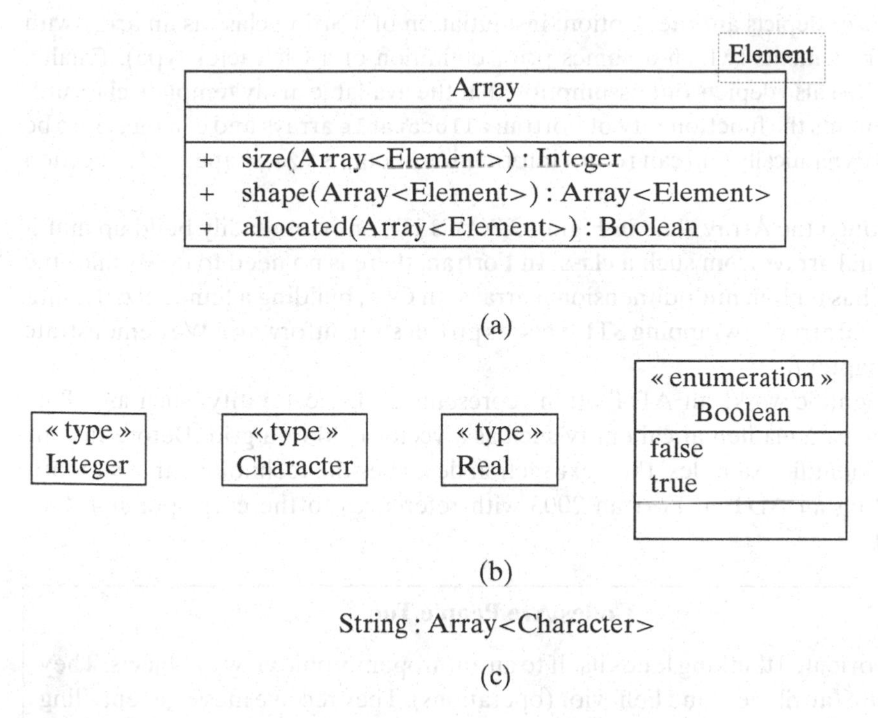

Figure 2.3. UML Array model: (a) Array template class, (b) supporting primitive classes, and (c) sample instantiation.

Figure 2.3 shows the attribute and operation portions of a UML diagram are optional, as is the listing of private information. The Integer, String, and Real types suppress all attributes and operations. The Array class diagram suppresses only attributes. The Boolean type suppresses only operations.

Although UML does not contain an intrinsic array type, any types defined by a UML model are considered UML model types. Henceforth, we assume in all of our UML diagrams the availability of a utility template class enabling the construction of arrays of various types. By definition, a template class is a class used to construct other classes. One often says such a class is *generic* or, in UML terminology, *parameterized*. In the template class of Figure 2.3(a), the parameter is "Element," which must represent another class. We have omitted the attributes of the Array class in Figure 2.3(a) under the assumption they are private. In most cases, they would include array descriptors holding the values returned by the methods depicted in the bottom box of in Figure 2.3(a).

We can also model intrinsic or, in UML terminology, *primitive* types as classes. The primitive types we use in our UML diagrams correspond to those in Figure 2.3(b). The Integer class name draws inspiration from the like-named Fortran type. The Boolean class provides an example of an enumeration of values: "true" and "false". It corresponds to the Fortran `logical` type.

The details of resolving the String type to an actual programming language type differ from language to language and even from implementation to implementation. In the code examples later in this chapter, we use allocatable deferred-length character variables, for which Fortran provides automatic dynamic memory management.

Figure 2.3(c) depicts another option: instantiation of a String class as an array with Character elements (which assumes prior definition of a Character type). Finally, Figure 2.3(a) also depicts our assumption that the available array template class utility implements the functionality of Fortran `allocatable` arrays and can therefore be allocated dynamically and can return information about its size, shape, and allocation status.

Although the Array class in Figure 2.3(a) is 1D, one can readily build up multidimensional arrays from such a class. In Fortran, there is no need to do so since the language has built-in multidimensional arrays. In C++, building a templatized multidimensional array by wrapping STL vectors proves straightforward. We demonstrate this in Chapter 8.

In scientific work, an ADT often represents a physical entity, such as a fluid or plasma, or a mathematical entity such as a vector field or a grid. Before turning to more scientific examples, the next section describes the technology that supports constructing an ADT in Fortran 2003 with references to the corresponding C++ constructs.

Codes Are People Too

Object-oriented thinking lends itself to an anthropomorphic view of objects. They have state (attributes) and behavior (operations). They receive messages entailing the operations invoked on them and the arguments passed to them. An operation's result conveys an object's response. A program thus consists of a set of objects collaborating to accomplish the user's goal. A UML use case diagram summarizes the actors in this drama along with their activities. A UML class diagram details the supporting class structure.

2.3 Encapsulation and Information Hiding

Encapsulation and information hiding provide the mechanisms for constructing an ADT by tightly coupling private, local data with a set of public and private procedures. Each of these procedures accesses the data, and the data is accessed by only these procedures. This two-way relationship creates the tight coupling. In Fortran, we accomplish encapsulation by defining a derived type (C++ *class*) with type-bound procedures (C++ *virtual member functions*) and wrapping each derived type in a Fortran `module` (C++ *namespace*). Each module encapsulates only one public derived type, though there may be other private ones included in the module. OOP's information-hiding philosophy dictates that the public derived type's components (C++ *data members*) each has the `private` attribute. Outside the module, access to the components is provided only via public type-bound procedures.

Figures 2.4(a) and 2.4(b) show structured and object-oriented "Hello, world!" programs, respectively. The structured program exhibits nothing noteworthy other than our choice to name the character string constant in keeping with Rule 1.2 from Section 1.7. The object-oriented program defines an astronaut class entailing a module that encapsulates an `astronaut` derived type with a component named `greeting` and a type-bound procedure named `greet()`. Following the information-hiding edict

(a)

```
1  program hello_world
2    implicit none
3    character(len=12),parameter :: message='Hello world!'
4    print *,message
5  end program
```

(b)

```
1  module astronaut_class
2    implicit none
3    private                 ! Hide everything by default
4    public :: astronaut  ! Expose type & constructor
5    type astronaut
6      private
7      character(:), allocatable :: greeting
8    contains
9      procedure :: greet ! Public by default
10   end type
11   interface astronaut  ! Map generic to actual name
12     procedure constructor
13   end interface
14 contains
15   function constructor(new_greeting) result(new_astronaut)
16     character(len=*), intent(in) :: new_greeting
17     type(astronaut) :: new_astronaut
18     new_astronaut%greeting = new_greeting
19   end function
20
21   function greet(this) result(message)
22     class(astronaut), intent(in) :: this
23     character(:), allocatable :: message
24     message = this%greeting
25   end function
26 end module
27
28 program oo_hello_world
29   use astronaut_class ,only : astronaut
30   type(astronaut) :: pilot
31   pilot = astronaut('Hello, world!')
32   print *, pilot%greet()
33 end program
```

Figure 2.4. "Hello, world!" programs: (a) structured and (b) object-oriented.

of OOP, line 3 defaults all module components to private scoping, while line 4 makes public the derived type name and the generic constructor name. The derived type component `greeting` is explicitly declared private. Since Fortran type-bound procedures are public by default, `greet()` also has public scope.

The main program `oo_hello_world` declares an `astronaut` named `pilot` and calls a like-named constructor that returns an `astronaut` instance. If the type components were public or had default initializations, we could have called the intrinsic structure constructor that Fortran 2003 provides with the same name as the type name. The result of that constructor could even be used to initialize `pilot` in its declaration on line 30. The information-hiding philosophy, however, recommends against exposing components, and Figure 2.4(b) omits any default initialization. The figure therefore includes a user-defined constructor.

The program then calls `greet()`, the invocation of which shows no arguments on the calling side. The argument received by `greet()` in the astronaut class implementation represents the object on which it is invoked: `pilot`. The Fortran 2003 standard refers to this argument as the *passed-object dummy argument*. In most OOP languages, including Java and C++, the type of that argument is assumed and need not be explicitly declared in the definition of the procedure. In these other languages, one refers to the corresponding entity by the keyword `this`, so we generally choose that to be its local name in Fortran as well. The `class` keyword employed in the declaration of `this` facilitates inheritance, which Section 2.5 discusses, and dynamic polymorphism, which Section 2.6 discusses.

A common model for understanding OOP views a type-bound procedure invocation as sending a message to the object on which it is invoked, that is, the passed-object dummy argument. The message conveys a request for action. Thus, one views line 32 in Figure 2.4(b) as the main program sending the request "greet" to the object pilot. That object responds by returning a character variable containing its greeting.

Figure 2.4(b) also leverages the 2003 standard's augmentation of Fortran's dynamic memory management capabilities. The `greeting` component of the `astronaut` type is an `allocatable` deferred-length character string. The semantics of `allocatable` entities obligates the compiler to automatically allocate sufficient space to store whatever character string gets assigned to `greeting` at line 18 and likewise for whatever string gets assigned to the `greet()` function result at line 24. Furthermore, the standard obligates the compiler to free any associated memory when the corresponding entities go out of scope. The compiler must also free (or reuse) any memory previously assigned to the left-hand-sides (LHSs) of these expressions.

This first foray into OOP demonstrates the overhead it adds to the software development process. Much of the additional overhead comes in the early phases of the process: the problem analysis and design that occur before fingers reach the keyboard. This extra effort manifests itself in discussions about the relevant actors, their goals, and their use of the system being modeled. For sufficiently small, one-off programming tasks, the overhead associated with OOA and OOD likely prove cost prohibitive in that the complexity of Figure 2.4(b) exceeds that of Figure 2.4(a) by virtually any straightforward measure: the number of lines of code, procedures, modules, language features, or programming constructs employed.

To the extent portions of Figure 2.4(b) might be more easily reused and extended, however, the additional overhead pays off many times over. In such cases, information hiding erects a partition between code inside the astronaut class and any outside code that relies on it. Specifically, data privacy opens up the possibility of completely changing the internal representation of the type components without affecting code external to the class. For example, suppose we wish to provide a default initialization for the `greeting` component in `pilot` to be printed in cases when the user forgets to instantiate objects. Because allocatable character string declarations cannot have default initializations, we might facilitate this by switching to the following fixed-width form:

```
type astronaut
  private
  character(len=27):: &
    greeting='Houston, we have a problem!'
contains
  procedure :: greet
end type
```

Making only this change would not force any changes to the main program. Specifically, the "astronaut()" and "greet()" procedure invocations at lines 31 and 32 could remain unchanged.

The general principle illustrated by the latter example can be articulated by considering the concept of a *class interface*, which comprises the collection of entities (procedure names, argument types, variables, and constants) a class makes public. Regarding public procedures, an actual interface might comprise one or more *interface blocks*, containing one or more *interface specifications* (C++ *function prototypes*). In Fortran, such interface specifications only need to be expressed in source code when dealing with *external procedures*. External procedures are those not encapsulated in a module. Those without a corresponding interface specification are said to have an *implicit interface*. The presence of an interface specification enables the compiler to check for argument type/rank/bounds mismatches between the calling code and the called code. Such type safety is provided automatically for *module procedures*. Module procedures are those contained in a `module`. Module procedures are said to have *explicit* (though confusingly tacit) *interfaces*.

In order to facilitate argument type checking, we encapsulate all procedures in this book inside modules. We therefore have no occasion to include interface specifications in programs except to link to C/C++ procedures in Chapter 11. We employ the term "interface" to refer to information that, in practice, we never isolate into one place separate from the executable code that implements that interface.[2]

Our class interface definition provides a succinct way to summarize one of the primary benefits of information hiding: Changes to privately scoped entities inside a class, including changes to the internal representation of its data, have no impact on code outside the class so long as the changes would require no modifications to the class interface. The power of this rule cannot be overstated. Section 2.7 provides an analysis of the impact this rule has on code complexity and robustness.

Friends Don't Let Friends Share Data

Encapsulating data in an ADT and hiding it from outside code gives developers the flexibility to alter implementations without impacting client code so long as the ADT's interface remains untouched.

[2] In Fortran 2008, it is possible to move the implementation of the type-bound procedures to a separate submodule, leaving only their interfaces in the module that defines the type. This allows complete documentation of the interface needed by users of the module without exposing details of how the services described in the interface are implemented.

2.4 Wrapping Legacy Software

Projects need not be object-oriented from their inception in order to benefit from OOA and OOD. In fact, a principle motivation behind OOP in Fortran 2003 stems from the prospect of seamlessly interfacing to the vast collection of Fortran programs already in use. These codes embody an indeterminately large number of people-years' worth of inherited knowledge. Reproducing this knowledge in new code, besides being inefficient, breaks down the trust the legacy codes have earned through years of use. This section addresses the process of wrapping legacy Fortran in modern, object-oriented Fortran, while Chapter 11 covers wrapping C++ with object-oriented Fortran.

Figure 2.5 delineates a strategy for modernizing legacy Fortran programs. This strategy has been adapted from one Norton and Decyk (2003) proposed. Whereas they discussed updating Fortran 77 codes to take advantage of Fortran 90/95 features, Figure 2.5 further contemplates transition to Fortran 2003 and potential inclusion in a CBSE-style framework. The first several steps might be considered steps in code *refactoring*, a term that most commonly refers to the process of tidying code. The first step addresses standard compliance. Because Fortran 90 retained all of Fortran 77 as a subset, Fortran 95 deleted only a small subset of Fortran 77, and Fortran 2003 contains all of Fortran 95, most standard-conforming Fortran 77 code also conforms to the later standards. The first step thus revolves primarily around removing compiler-specific extensions.

The second step in Figure 2.5 addresses obsolescent and deprecated features. The Fortran 95 standard equates obsolescence with redundancy, declaring features as obsolescent if Fortran 90 and 95 provided better alternatives. (Features that were already obsolescent by this definition in Fortran 77 were deleted in Fortran 95.) These included fixed-source form (now superseded by free-source form, which removes the significance of columns in source text), computed `go to` statements (superseded

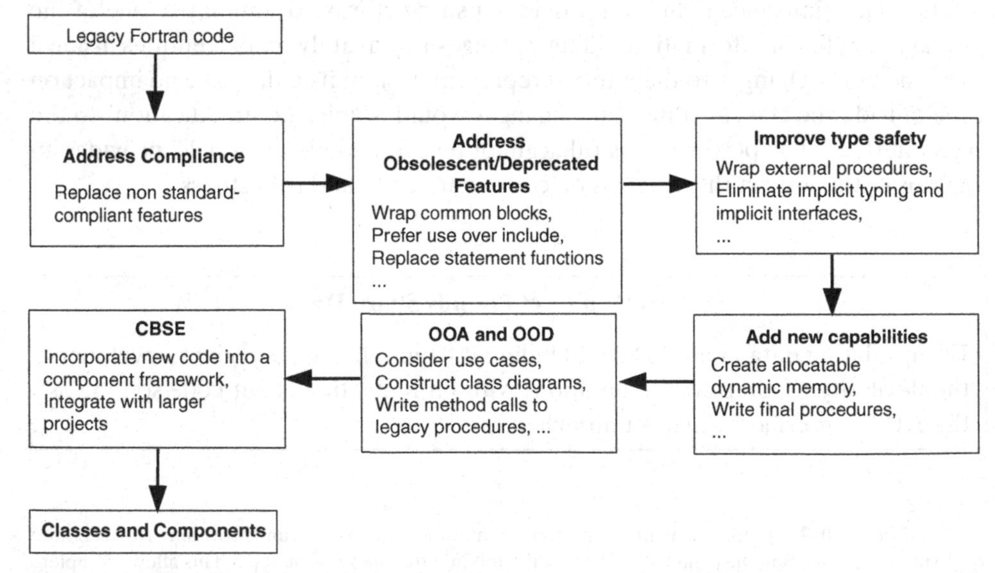

Figure 2.5. Steps for modernizing legacy Fortran programs (adapted from Norton and Decyk [2003])

by the `case` construct), statement functions (superseded by internal functions, i.e., functions preceded by a `contains` statement), and five other features.

In addition to the officially obsolescent features, Metcalf et al. (2004) deprecated several constructs and recommended they be declared obsolescent in a future standard. The features listed in Figure 2.5 fall into this category: `common` statements and `include` lines. Norton and Decyk (2003) described how Fortran 77 `common` blocks with static data can be reorganized to facilitate dynamic memory allocation. The process involves moving the declaration into a module, writing constructors, and replacing static data with allocatable entities. This adds considerable flexibility and functionality, including the methods in the Array class of Figure 2.3 and the ability to gracefully handle exceptions such as insufficient available memory.

Wrapping `include` lines in Fortran 90 `modules` also adds functionality and provides greater safety. A `module` can grant access to entities declared in the included code, but can also facilitate restricting access either via the `private` attribute inside the `module` or via an `only` clause in the `use` statement in client code. Wrapping included code in a `module` also prevents the accidental declarations that can occur when included legacy code contains `common` statements (see Metcalf et al. [2004]).

The third step in Figure 2.5 addresses type safety. Grouping external procedures into a set of `module` constructs gives them explicit interfaces and thereby precludes the argument type mismatches mentioned in Section 2.3. We prefer this approach to the construction of actual interface specifications. The lack of information redundancy between an actual procedure and an associated interface specifications eliminates the need to maintain consistency between the two during code revisions. Additional type safety derives from employing the `implicit none` statement to prevent implicit variable declarations.

A side benefit one incurs from the wrapping process will likely be greater type safety. Fortran 77 provided no way to explicitly declare a procedure interface. As a result, that version of the language uses a "sequence association" calling convention when passing arrays between procedures. There are many problems inherent in such calls due to the lack of type, rank, or bounds checking. Many legacy Fortran 77 codes took artful advantage of this situation. For example, as Norton and Decyk (2003) pointed out, many legacy Fotran 77 codes passed two-dimensional (2D) actual arguments that are received as 1D dummy arguments inside the called procedure. Modern wrappers can employ a generic name (via a generic interface) for such a procedure, resolving that name to the name of any one of multiple actual procedures that can accept the desired argument – the unsafe alternative being to make calls directly to an external legacy procedure and to thereby abandon type/rank/bounds checking on *all* arguments, not just the array in question.

In addition to being safer, wrapper interfaces can be simpler. Modern Fortran's intrinsic functions for inquiring about array properties obviate the need to pass these properties as arguments. Furthermore, the rich data structures and methods attendant to objects also shortens argument lists.

The fourth step in Figure 2.5 contemplates new capabilities. We strongly advocate liberal use of `allocatable` constructs for dynamic memory management. The language standards obligate compilers to free the memory associated with `allocatable` entities when they go out of scope, thus preventing memory leaks with any compliant compiler. Fortran 2003 greatly expanded the list of entities that can

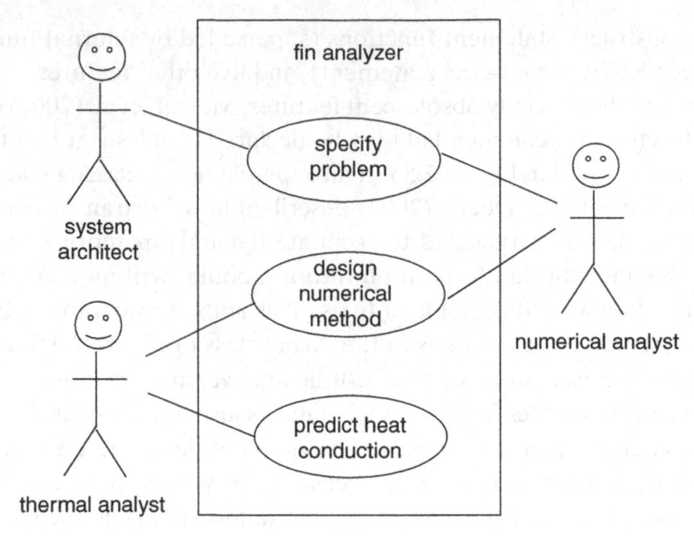

Figure 2.6. Use case diagram for a fin heat conduction analysis system.

have the `allocatable` attribute to include derived types and their components. We have found that this capability eliminates the vast majority of cases where we would otherwise use the `pointer` attribute. Nonetheles, when derived types do contain pointers, Fortran 2003 `final` procedures must be used to free associated memory when the pointer name goes out of scope.

The fifth step in Figure 2.5 turns to the chief subject of this book: object-orientation. To illustrate this process, we return to the structured heat fin conduction of Chapter 1 and consider how to wrap that code in an object-oriented "fin analyzer" framework. Figure 2.6 provides a use case for the usage of such a framework. The actors contemplated there include a system architect, a numerical analyst, and a thermal analyst. The architect uses the fin analyzer to specify the problem, providing the numerical analyst with a set of relevant physical parameters, including tolerances and boundary and initial conditions. The numerical analyst uses this specification to develop a mathematical model in the form of a discrete set of equations that can be solved on a digital computer. The thermal analyst uses the resulting mathematical model to predict the performance of various fins.

Having a use case diagram in hand enables us to think abstractly about the purpose of the software without tying us to a particular design or implementation. Starting from a given use case diagram, different software development strategies might generate designs and implementations bearing little resemblance to each other. Conventional practice would likely proceed directly to the structured program implementation in Figure 1.4. Another strategy might model classes after the actors as in the astronaut design of Section 2.3. A third might model classes after the activities. An OOD following the latter path might lead to the class interfaces of Figure 2.7, including a problem definition class, numerical differentiation class, and a heat conduction class. Chapter 3 presents a fourth approach that leads to a very different class decomposition.

The Problem class described in Figure 2.7(a) provides a constructor that takes an Integer file handle as its only argument. Although the implementation details

Problem
« constructor » + problem(file_handle : Integer) : Problem + boundary_vals() : Array<Real> + diffusivity() : Real + nodes() : Integer + time_step() : Real + spacing() : Real

(a)

Differentiator
« constructor » + differentiator(specification : Problem) : Differentiator + laplacian(T :Array<Real>,Tboundaries :Array<Real>) :Array<Real>

(b)

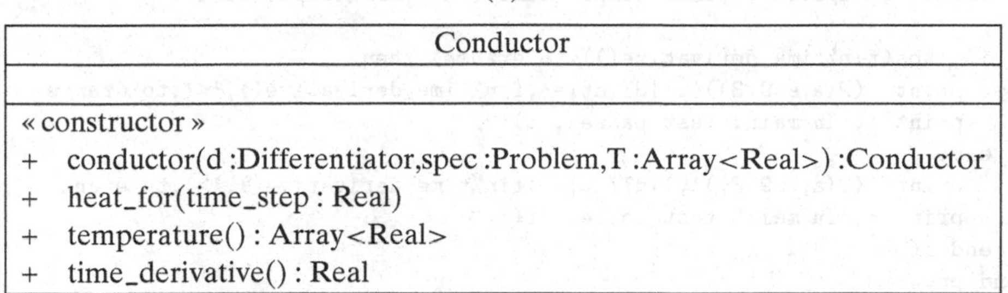

(c)

Figure 2.7. Object-oriented heat fin analysis package class interfaces: (a) problem specification class, (b) numerical differentiation class, and (c) heat conduction class.

remain unspecified by the interface, the corresponding file presumably contains the information required to initialize the private Problem attributes. The Problem class publishes operations that return physical parameters such as the boundary temperatures and the material diffusivity. Other operations return numerical parameters such as the number of nodes, the time step size, and the grid spacing.

The Differentiator class described in Figure 2.7(b) provides a constructor that takes a Problem argument and returns a Differentiator. It also provides a `.laplacian.` operation. That operation returns an array of second derivatives given arrays of internal temperatures and boundary temperatures.

The Conductor class described in Figure 2.7(c) provides a constructor that instantiates a Conductor from a Differentiator, a Problem, and a temperature array. Conductor also provides an operation that conducts heat for a specified time (one time step). A Conductor can also return an array of fin temperatures and the time derivative of the temperature at the chip boundary.

Having the class interfaces that result from an OOD provides an opportunity for *test-driven development*. This strategy advocates for writing the test that verifies a

(a)

```
program fin_test
  use iso_fortran_env ,only : input_unit
  use kind_parameters ,only : rkind
  use problem_class   ,only : problem
  use differentiator_class ,only : differentiator
  use conductor_class      ,only : conductor
  implicit none

  real(rkind), parameter :: tolerance=1.0E-06
  type(problem)    :: specification
  type(conductor) :: fin
  type(differentiator) :: finite_difference

  specification      = problem(input_unit)
  finite_difference  = differentiator(specification)
  fin = conductor(finite_difference,specification)
  print '(a,5g9.3)','initial temperature = ',fin%temperature()
  call fin%heat_for(specification%time_step())
  print '(a,5g9.3)','final temperature   = ',fin%temperature()

  if (abs(fin%time_derivative())<tolerance) then
    print '(2(a,es9.3))','|dT/dt|=',fin%time_derivative(),'<',tolerance
    print *,'In main: test passed. :)'
  else
    print '(2(a,es9.3))','|dT/dt|=',fin%time_derivative(),'>',tolerance
    print *,'In main: test failed. :('
  end if
end program
```

(b)

```
&physics alpha=1.,L_fin=1.,T_chip=1.,T_air=1. /
&numerics dt=0.01,nodes=3 /
```

(c)

```
initial temperature =  1.00      1.00      1.00      1.00      1.00
final temperature   =  1.00      1.00      1.00      1.00      1.00
|dT/dt|=0.000E+00<1.000E-06
 In main: test passed. :)
```

Figure 2.8. Test-driven development: (a) main program, (b) input, and (c) output for adiabatic test.

program's correctness before writing the program. To demonstrate, we consider the adiabatic case, wherein spatially constant temperatures preclude heat transfer and therefore correspond to a zero time derivative according to

$$T = \text{const} \Rightarrow \frac{\partial T}{\partial t} = \alpha \nabla^2 T = 0 \tag{2.1}$$

Figures 2.8(a)-(b) depict a main program and an input file that tests for this condition given a set of classes that implement the interfaces of Figure 2.7. Figure 2.8(c) shows the resulting output.

Figures 2.9–2.11 present implementations of the interfaces depicted in Figure 2.7. Each implementation wraps parts of the structured program in Figure 1.4. As discussed in the remainder this chapter, repackaging the structured program in this new object-oriented framework enhances its extensibility, flexibility, and stability. The enhanced extensibility derives at least in part from inheritance, the subject of Section 2.5. The increased flexibility derives in part from facilitating polymorphism, the subject of Section 2.6. Stability derives in part from limiting data dependencies, the subject of Section 2.7. Chapter 3 demonstrates how a very different OOD engenders even greater flexibility.

Cake From A Box Never Tasted So Good

Transitioning from structured programming to OOA, OOD, and OOP does not require mixing all your own ingredients and baking everything from scratch. Wrapping legacy software preserves the trust that software has earned from years of use. Appropriate wrapping increases type safety, simplifies procedure calls, and adds extensibility, flexibility, and robustness in ways covered in the remaining sections of this chapter.

Figure 2.9

```
1  module problem_class
2    use kind_parameters ,only: ikind,rkind
3    implicit none
4    private
5    public :: problem
6    type problem
7      private  ! Default values:
8      integer(ikind) :: num_nodes=4
9      real(rkind):: air_temp=1.,chip_temp=1.!boundary temperatures
10     real(rkind):: alpha=1.,length=1.      !physical parameters
11     real(rkind):: dt=0.1,dx               !numerical parameters
12   contains                !^ constructor computes default
13     procedure :: boundary_vals
14     procedure :: diffusivity
15     procedure :: nodes
16     procedure :: time_step
17     procedure :: spacing
18   end type
19   interface problem
20     procedure spec ! constructor
21   end interface
22 contains
23   type(problem) function spec(file)
24     use conduction_module &
25       ,only : read_physics,read_numerics,stable_dt
26     integer(ikind) :: file
27     integer(ikind) :: elements_default
28
29     elements_default = spec%num_nodes+1
30     spec%dx = spec%length/elements_default ! default element size
31
```

```
32      if (.not. read_physics(file                              &
33      ,spec%alpha,spec%length,spec%chip_temp,spec%air_temp)) &
34        print *,'In problem constructor: Using default physics spec.'
35
36      if (.not. read_numerics(file                         &
37        ,spec%dt,spec%num_nodes,spec%dx,spec%length)) &
38        print *,'In problem constructor: Using default numerics spec.'
39
40      spec%dt = min(spec%dt,stable_dt(spec%dx,spec%alpha))
41    end function
42    pure function boundary_vals(this)
43      class(problem) ,intent(in) :: this
44      integer(ikind) ,parameter :: end_points=2
45      real(rkind) ,dimension(end_points) :: boundary_vals
46      boundary_vals = (/this%chip_temp,this%air_temp/)
47    end function
48    pure real(rkind) function diffusivity(this)
49      class(problem) ,intent(in) :: this
50      diffusivity = this%alpha
51    end function
52    pure integer(ikind) function nodes(this)
53      class(problem) ,intent(in) :: this
54      nodes = this%num_nodes
55    end function
56    pure real(rkind) function time_step(this)
57      class(problem) ,intent(in) :: this
58      time_step = this%dt
59    end function
60    pure real(rkind) function spacing(this)
61      class(problem) ,intent(in) :: this
62      spacing = this%dx
63    end function
64  end module
```

Figure 2.9. Problem definition class implementation.

Figure 2.10

```
1   module differentiator_class
2     use kind_parameters ,only: ikind,rkind
3     implicit none
4     private                      ! Hide everything by default.
5     public :: differentiator ! Expose type/constructor/type-bound procs.
6     type differentiator
7       private
8       real(rkind),dimension(:,:),allocatable::diff_matrix
9     contains
10      procedure :: laplacian  ! return Laplacian
11      procedure :: lap_matrix ! return Laplacian matrix operator
12    end type
13    interface differentiator
14      procedure constructor
15    end interface
16  contains
17    type(differentiator) function constructor(spec)
18      use problem_class ,only : problem
```

```
19      use conduction_module ,only : differencing_stencil
20      type(problem) ,intent(in) :: spec
21      integer(ikind) ,parameter :: diagonals=3
22      allocate(constructor%diff_matrix(spec%nodes(),diagonals))
23      constructor%diff_matrix = &
24        differencing_stencil(spec%spacing(),spec%nodes())
25    end function
26    pure function laplacian(this,T,Tboundaries)
27      use conduction_module, only : differentiate
28      class(differentiator), intent(in) :: this
29      real(rkind) ,dimension(:) ,intent(in) :: T
30      real(rkind) ,dimension(:) ,allocatable :: laplacian
31      integer(ikind) ,parameter :: end_points=2
32      real(rkind), dimension(end_points), intent(in) :: Tboundaries
33      allocate(laplacian(size(T)))
34      laplacian = differentiate(& !compute derivative at internal points
35        this%diff_matrix,T,Tboundaries(1),Tboundaries(2) )
36    end function
37    pure function lap_matrix(this)
38      class(differentiator), intent(in) :: this
39      real(rkind) ,allocatable ,dimension(:,:) :: lap_matrix
40      lap_matrix = this%diff_matrix
41    end function
42  end module
```

Figure 2.10. Numerical differentiation class implementation.

Figure 2.11

```
1   module conductor_class
2     use kind_parameters ,only : ikind,rkind
3     use differentiator_class, only : differentiator
4     implicit none
5     private           ! Hide everything by default
6     public::conductor ! Expose type and type-bound procedures
7     integer(ikind), parameter :: end_points=2
8     type conductor
9       private
10      type(differentiator) :: diff
11      real(rkind) :: diffusivity
12      real(rkind) ,dimension(end_points) :: boundary
13      real(rkind) ,dimension(:), allocatable :: temp
14    contains                                 !^ internal temperatures
15      procedure :: heat_for
16      procedure :: temperature
17      procedure :: time_derivative
18    end type
19    interface conductor
20      procedure constructor
21    end interface
22  contains
23    type(conductor) function constructor(diff,prob,T_init)
24      use conduction_module ,only : read_physics
25      use problem_class     ,only : problem
26      type(differentiator) ,intent(in)    :: diff
27      type(problem)        ,intent(in)    :: prob
```

```
28       real(rkind), dimension(:) ,optional :: T_init
29       constructor%diff = diff
30       constructor%diffusivity = prob%diffusivity()
31       constructor%boundary = prob%boundary_vals()
32       if (present(T_init)) then
33         if (size(T_init)/=prob%nodes())stop 'In conductor: size mismatch.'
34         constructor%temp = T_init
35       else; allocate(constructor%temp(prob%nodes()))
36             constructor%temp = constructor%boundary(1)
37       end if
38     end function
39     function temperature(this)
40       class(conductor), intent(in) :: this
41       real(rkind), dimension(:), allocatable :: temperature
42       temperature = (/ this%boundary(1), this%temp, this%boundary(2) /)
43     end function
44     subroutine heat_for(this,dt)
45       class(conductor) ,intent(inout) :: this
46       real(rkind)      ,intent(in)    :: dt
47       real(rkind) ,dimension(:) ,allocatable :: T_xx !2nd derivative
48       allocate(T_xx(size(this%temp)))
49       T_xx = this%diff%laplacian(this%temp,this%boundary)
50       this%temp = this%temp + dt*this%diffusivity*T_xx
51     end subroutine
52     real(rkind) function time_derivative(this)
53       class(conductor)    :: this
54       real(rkind) ,dimension(:) ,allocatable :: T_xx_end
55       real(rkind)                            :: T_2
56       if (size(this%temp)>1) then
57         T_2=this%temp(2)
58       else; T_2=this%boundary(2)
59       end if
60       T_xx_end = this%diff%laplacian(this%temp,(/this%boundary(1),T_2/) )
61       time_derivative = this%diffusivity*T_xx_end(1)
62     end function
63   end module
```

Figure 2.11. Heat conductor class implementation.

2.5 Composition, Aggregation, and Inheritance

Central to OOP lies the goal of code reuse. Most reuse in OOP takes one of three forms: *composition*, *aggregation*, or *inheritance*. Each represents a relationship between two or more classes. Composition involves placing one or more instances of one class inside the definition of another class as a component of the second class. Composition can be thought of as a whole/part relationship.

In addition to conveying spatial information about the location of one object inside another, composition harbors temporal connotations, implying coincident lifetimes. The construction and destruction of the whole coincide with the construction and destruction of its parts. Hence, composition usually models relationships in which the part loses its usefulness in the absence of the whole.

Composition need not be universal. Although every Conductor defined by Figure 2.11 is composed of a Differentiator, not every Differentiator instance is part

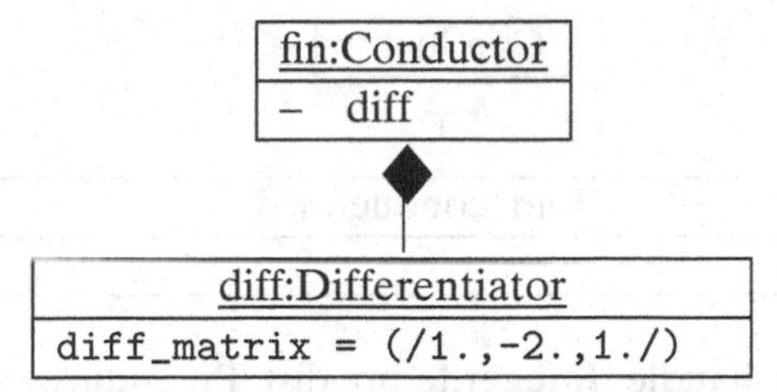

Figure 2.12. Conductor/Differentiator composite object diagram.

of a Conductor. The `finite_difference` object in the main program (`fin_test` in Figure 2.8) stands alone even though the `conductor()` constructor stores a copy of it named `diff` in `fin`. The compiler destroys `diff` when `fin` goes out of scope (at the program termination in this case).

Since not every Differentiator is part of a Conductor, their composite relationship might be best thought of as a relationship between two objects rather than between two classes. A UML *object diagram* renders an instantaneous snapshot of each object's state and its relationship to other objects. One portrays a relationship in UML as a line connecting objects in an object diagram (or classes in a class diagram). As Figure 2.12 shows, a software designer can indicate composition by adorning the connecting line with a closed diamond at the "whole" end of the whole/part relationship.

In object diagrams, one underlines the object and class names and separates them by a colon. Furthermore, the object state at the time of the snapshot can be indicated by assigning values to the object's attributes. The values assigned to `diff_matrix` in Figure 2.12 correspond to the simple case of a single internal node with unit grid spacing between that node and the boundaries.

When one class is composed of another, a software designer places a closed diamond at the "whole" end of the whole/part relationship as shown in Figure 2.12. If no symbol appears at either end (as in the bottom relationship in Figure 2.13), the diagram conveys no information about the nature of the relationship. The designer can also adorn both ends with a number, or number range, indicating the *multiplicity*, which is the allowable number of instances at either end. The absence of explicit multiplicities implies single instances.

Aggregation generalizes the composition concept by decoupling the lifetimes of the whole and the part. As with composition, the ends of the relationship connecting line can be adorned with a multiplicity indicating the number of possible instances at either end. Authors frequently refer to aggregation as a "has a" relationship in the sense that one class has an instance of another class. Since composition represents a specific instance of aggregation, another term for composition is "composite aggregation."

Inheritance denotes a very specific type of composite aggregation: one in which the whole supports the entire public interface of the part. Authors frequently refer to inheritance as an "is a" relationship in the sense that one class is a specific subcategory of another class because by default it retains the state and any nonoverridden behavior of the parent class. Fortran provides inheritance via *type extension*. When a child type extends a parent type, the compiler automatically inserts an instance of

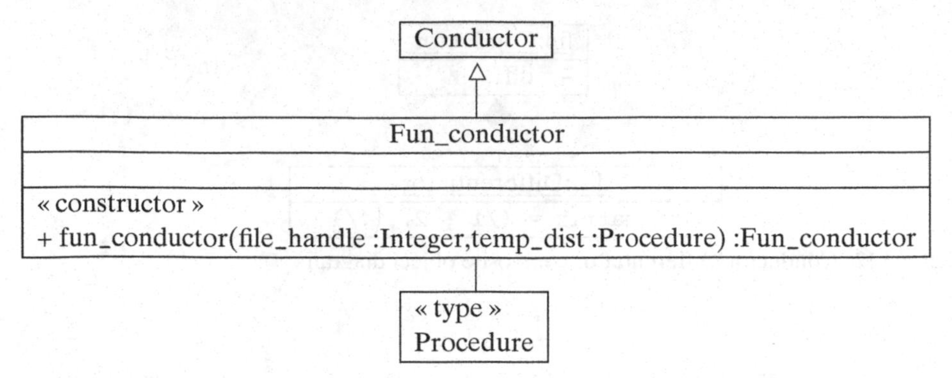

Figure 2.13. Function-initialized conductor inheritance hierarchy.

the parent type as an component in the child type. The child type can override any of the parent's methods by implementing them itself. Otherwise, the compiler handles any invocations of the parent's methods on the child by invoking these calls on the parent. The Fortran language standard refers to the child as an *extended type*.

Consider a developer who wants to construct a Conductor by passing a temperature distribution function. We designate this new class `fun_conductor` as a shorthand for "function-initialized conductor." Since the existing Conductor constructor in Figure 2.11 accepts an optional argument containing nodal temperature array, we can leverage this ability in the `fun_conductor()` constructor by accepting an argument that designates a temperature distribution function. The constructor could sample that function at the node locations and pass those samples to the parent type's `conductor()` constructor.

The open triangle in Figure 2.13 indicates that a Conductor generalizes a Fun_conductor. One usually models inheritance as generalization in UML. Fun_Conductor enhances the Conductor interface by defining a new constructor and simplifies the interface by hiding the Problem and Differentiator objects inside the `fun_conductor()` constructor. The test program in Figure 2.15 passes a file handle and temperature distribution function to the constructor, delegating the acquisition of all other construction-related data to the constructor.

Figure 2.13 declares the temperature distribution function to be of primitive type Procedure, which might represent any of several language mechanisms for passing a procedure argument. In C++, one might pass a function pointer. In Fortran, one might pass a procedure dummy argument (as in our implementation in Figure 2.14), a procedure pointer, or a C function pointer (see Chapter 11). An OOP purist might encapsulate the mechanism in a class, reserving the right to change the private implementation details.

```
1  module fun_conductor_class
2    use kind_parameters ,only : ikind,rkind
3    use conductor_class ,only : conductor
4    implicit none
5    private
6    public::fun_conductor
7    type ,extends(conductor) :: fun_conductor
```

```
8      private
9      real(rkind) :: dt
10   contains
11     procedure :: time_step
12   end type
13   interface fun_conductor
14     procedure constructor
15   end interface
16 contains
17    type(fun_conductor) function constructor(file,T_distribution)
18     use differentiator_class ,only : differentiator
19     use problem_class        ,only : problem
20     integer(ikind) ,intent(in)              :: file
21     type(differentiator)                    :: diff
22     type(problem)                           :: prob
23     real(rkind), dimension(:) ,allocatable :: T
24     real(rkind)                             :: dx
25     integer(ikind)                          :: i
26     interface
27       pure real(rkind) function T_distribution(x)
28         use kind_parameters ,only : rkind
29         real(rkind) ,intent(in) :: x
30       end function
31     end interface
32     prob = problem(file)
33     diff = differentiator(prob)
34     dx   = prob%spacing()
35     constructor%dt = prob%time_step()
36     allocate(T(prob%nodes()))
37     forall(i=1:prob%nodes()) T(i) = T_distribution(i*dx)
38     constructor%conductor = conductor(diff,prob,T)
39   end function
40   real(rkind) function time_step(this)
41     class(fun_conductor) :: this
42     time_step = this%dt
43   end function
44 end module
```

Figure 2.14. Function-initialized conductor implementation.

Figure 2.14 shows a sample implementation of a function-initialized conductor that inherits the type-bound procedures of the Conductor class implemented in Figure 2.11. Since linear temperature distributions also satisfy equation 2.1 and therefore define a steady-state solution, we can use a test analogous to the one used for the Conductor class. Figure 2.15 demonstrates such a test along with the input and output files.

The convenience of inheritance lies in the automated compiler support for applying the operations of the parent to the instances of the child. The type-bound procedures `temperature()` and `time_derivative()` invoked on `fin` in Figure 2.15(a) result in calls to the corresponding procedures in the Conductor implementation of Figure 2.11, even though the passed-object dummy argument

(a)

```
module initializer
  implicit none
  private
  public :: linear
contains
  pure real(rkind) function linear(x)
    use kind_parameters ,only : rkind
    real(rkind) ,intent(in) :: x
    linear = 1. - x
  end function
end module

program fun_fin_test
  use iso_fortran_env     ,only : input_unit
  use kind_parameters     ,only : rkind
  use fun_conductor_class ,only : fun_conductor
  use initializer         ,only : linear
  implicit none
  real(rkind) ,parameter :: tolerance=1.0E-06
  type(fun_conductor)    :: fin

  fin = fun_conductor(input_unit,linear)
  print '(a,5g9.2)','initial temperature = ',fin%temperature()
  call fin%heat_for(fin%time_step())
  print '(a,5g9.2)','final temperature   = ',fin%temperature()
  if (abs(fin%time_derivative())<tolerance) then
    print '(2(a,es9.3))','|dT/dt|=',fin%time_derivative(),'<',tolerance
    print *,'In main: test passed. :)'
  else
    print '(2(a,es9.3))','|dT/dt|=',fin%time_derivative(),'>',tolerance
    print *,'In main: test failed. :('
  end if
end program
```

(b)

```
&physics alpha=1.,L_fin=1.,T_chip=1.,T_air=0. /
&numerics dt=0.01,nodes=3 /
```

(c)

```
initial temperature =   1.0      0.75      0.50      0.25      0.0
final temperature   =   1.0      0.75      0.50      0.25      0.0
|dT/dt|=0.000E+00<1.000E-06
 In main: test passed. :)
```

Figure 2.15. Function-initialized conductor test: (a) main program, (b) input, and (c) output for steady-state test.

this those procedures receive is of type `fun_conductor` rather than the declared `conductor`. The next section describes this ability to pass objects of types other than declared type.

2.6 Static and Dynamic Polymorphism

Morpheus, the ancient Greek god of dreams, had the ability to take many different human forms when appearing in dreams. Similarly, polymorphism connotes the ability of a given construct to take many object forms or procedural forms. To facilitate taking different object forms, several objects might be able to respond to the same message – that is, the same method invocation – either by implementing their own version of the named method, by manually delegating to another object to respond, or by letting the compiler automatically delegate to an ancestor. To facilitate taking different procedural forms, several different procedures might have the same name, whether or not these procedures are class operations.

Polymorphism takes two forms: static and dynamic. Both appeared in code in the previous section. The current section highlights these occurrences and explains the related concepts.

Static polymorphism denotes situations in which the responder to a message can be determined at compile time, its identity henceforth fixed throughout execution. Many intrinsic functions are statically polymorphic. The expression `exp(A)` can refer to the exponential of a real scalar, a complex scalar, a real matrix, or a complex matrix, depending on the declared type of `A`. Each case invokes a different procedure and returns a result of a different type.

Thus far, our code examples have included only one constructor per derived type. Furthermore, our constructors are not type-bound procedures, so they are not inherited by extended types. For this reason, they cannot be invoked on children. We can use static polymorphism to enable the invocation of the parent Conductor constructor on instances of the child Fun_Conductor by adding a procedure to the `interface` statement in lines 13-15 of Figure 2.14 as follows:

```
interface fun_conductor
  procedure constructor, conductor_constructor
end interface
```

and adding a module procedure that delegates construction to the parent instance that the compiler automatically inserts as a component of the child:

```
type(fun_conductor) &
function conductor_constructor(diff,prob,T_init)
  use differentiator_class ,only : differentiator
  use problem_class        ,only : problem
  type(differentiator) ,intent(in)    :: diff
  type(problem)        ,intent(in)    :: prob
  real(rkind), dimension(:) ,optional :: T_init
  conductor_constructor%conductor = conductor(diff,prob,T_init)
end function
```

This effectively emulates inheritance of the parent constructor.

Dynamic polymorphism denotes situations in which the determination of which procedure will respond to a given message can be delayed until runtime. Consider again the Conductor type-bound procedures `temperature()` and `time_derivative()` invoked on `fin` in Figure 2.15. When writing these procedures in the Conductor class implementation of Figure 2.11, we originally planned for the passed-object dummy argument to be of type Conductor. We compiled and ran that code in the fin test of Figure 2.8(c). Dynamic polymorphism, as designated by the `class` keyword, enables the reuse of that compiled code, without recompilation, in the subsequent writing of the Fun_Conductor class. The compiler generates object code that can detect at runtime the actual argument passed and then dispatch code capable of invoking the specified method on the passed object. The underlying technology is also referred to as *dynamic dispatching*. It proves especially useful when the source code is unavailable, as might be the case in using a polymorphic library constructed using the techniques in Chapter 6.

Fortran links the concepts of inheritance and dynamic polymorphism by requiring programmers to replace the usual `type` keyword with its `class` counterpart in passed-object dummy argument declarations. The `class` keyword can be applied if and only if the passed-object dummy argument is extensible. The actual argument passed must then be of the type named in the `class` construct or any type that extends that type, including all generations of descendants. Such a chain of extended types comprises a *class hierarchy* or *inheritance hierarchy*.

2.7 OOP Complexity

Each of the techniques discussed so far reduce development time in some way. A little analysis helps determine which features of OOP offer the greatest benefit. It can be argued that inheritance and polymorphism attack problems that scale linearly with the size of the code (Rouson and Xiong 2004). Given a piece of code that is inherited across G generations of a parent/child class hierarchy, inheritance reduces the number of times this code must be replicated from G to 1. Given P common procedures to be performed on C classes of objects, polymorphism reduces the size of the protocol for communicating with those classes from CP to P. Note that the variables G, C, and P each appear with a unit exponent in the above expressions.

By contrast, encapsulation and information hiding attack a problem that scales quadratically with code size as shown in equation (1.16). Specifically, enforcing strict data privacy reduces the number of linkages in Figure 1.5, thereby limiting the number of lines one must visit to find the bug. Specifying argument `intent` limits the type of linkages by giving some procedures read-only access or write-only access to data. To reflect this in the debugging time estimate, we rewrite equation 1.16 as

$$t_{find} = \frac{rc}{2}\left(\frac{c}{2} - 1\right) t_{line} \tag{2.2}$$

where c is the number of lines with write access to the solution vector. In OOP, c corresponds to the number of lines in a class (assuming data privacy with one class per module in Fortran). Hence, significant reductions can only be accomplished if class size restrictions can be enforced. Although it might seem draconian to restrict class size across a project, Chapter 3 outlines a strategy that very naturally leads to roughly constant c across the classes.

2.8 More on Style

In our Fortran examples throughout this book, we typically avoid construct ending names such as the second occurrence of `foo` in the following:

```
interface foo
  procedure bar
end interface foo
```

Not using the ending name precludes the mistake of subsequently changing the first `foo` to `new_foo`, for example, without making the same change to the second `foo`. Though easily rectified, it seems best to avoid the problem in the first place when the second occurrence serves little useful purpose. An exception to this practice arises when the number of lines separating the beginning and ending of a construct likely exceeds one page on a screen or in print, as happens with some long procedures and main programs. In these instances, the second occurrence of the construct name provides a useful reminder of which construct is ending.

We make occasional use of the `forall` statement, as in Figures 1.4 and 2.14. We use this construct for two primary purposes. First, it occasionally proves more compact than a corresponding `do` loop because it has a single-statement form that does not necessitate an end statement. Second, it indicates opportunities for parallelism. Occurrences of `forall` can range over multiple array indices in a single construct and apply conditional masks as in:

```
forall(i=1,imax/2, j=1,jmax, c(j) > 0.  ) a(i,j) = b(i) + c(j)
```

where no assignment takes place for `j` values corresponding to positive `c(j)`. In addition to providing a compact syntax useful in a first pass through non-critical code, the semantics of `forall` require that it be possible for each of the resulting assignments to take place concurrently. Unfortunately, the attendant code speedup is typically moderate at best and negative at worst. Hence, the value `forall` adds in clarity, conciseness, and concurrency must be considered in the context of the given code's prominence in the overall performance profile.

We have chosen to write object constructors as standalone functions rather than type-bound procedures. We consider this approach the most natural for Fortran programmers because Fortran's intrinsic structure constructors take the same form. The intrinsic structure constructors – which are named after the type and have arguments named after the type components – can only be used when the derived type components are public or have default initialization.

Had we chosen to make the constructors type-bound procedures, the inheritance of parent constructors emulated in Section 2.6 would be handled by the compiler automatically. If the type-bound procedure constructor were to remain a function, however, it would have to return a polymorphic result to allow for extensibility. This would necessitate replacing the intrinsic assignment with a defined assignment. We deem such machinations overkill for the simple examples discussed so far.

A final alternative would be for the type-bound procedure to be a subroutine. With that option, however, we lose the resemblance to Fortran's intrinsic structure constructors. We also prefer the clarity of instantiating an object by assigning to it the result of a constructor named after the derived type. Nonetheless, the stated benefits must be weighed against the possibility that a compiler might make an unnecessary temporary version of the object before copying it into the left-hand-side object and destroying the temporary return instance. In many cases, optimizing compilers can eliminate such temporaries.

EXERCISES

1. Verify that the corresponding tests fail with nonlinear initial temperature distributions in (a) Figure 2.8(a) and (b) Figure 2.15(a).
2. In some systems, cooling fans stay on after the rest of the system shuts down. To determine how long it takes to cool the fins of Section 1.2 down, consider the fin to be a lumped mass with a spatially uniform temperature, that obeys $mc\frac{dT}{dt} = -\dot{Q}_{out}$, where m and c are the fin mass and specific heat and $\dot{Q}_{out}$ is the heat transfer rate.
   (a) Draw a UML class diagram for a lumped mass class capable of determining the time necessary to reach a final cooling temperature from a given starting temperature.
   (b) Write a main program that uses the above class to output the cooling time. List a few subclasses one might later desire to build from your lumped mass class.

3 Scientific OOP

"Software abstractions should resemble blackboard abstractions."

Kevin Long

3.1 Abstract Data Type Calculus

A desire for code reuse motivated most of Chapter 2. Using encapsulation and information hiding, we wrapped legacy, structured programs in an object-oriented superstructure without exposing the legacy interfaces. We used aggregation and composition to incorporate the resulting classes into components inside new classes. We also employed inheritance to erect class hierarchies. Finally, we employed dynamic polymorphism to enable child instances to respond to the type-bound procedure invocations written for their parents.

Much of the code in Chapters 1–2 proved amenable to reuse in modeling heat conduction, but few of the object-oriented abstractions presented would find any use for nonthermal calculations. Even to the extent the heat equation models other phenomena, such as Fickian diffusion, calling a procedure named `heat_for()` to solve the diffusion equation would obfuscate its purpose. This problem could be addressed with window dressing – creating a diffusion-oriented interface that delegates all procedure invocations to the `conductor` class. Nonetheless, neither the original `conductor` class nor its diffusion counterpart would likely prove useful in any simulation that solves a governing equation that is not formally equivalent to the heat equation. This begs the question: "When reusing codes, what classes might we best construct from them?"

Most developers would agree with the benefits of breaking a large problem into smaller problems, but choosing those smaller problems poses a quandary without a unique solution. If the primary goal is reuse, however, one can draw inspiration from the OOP proclivity for designing classes that mimic physical objects and mathematical constructs. When the entities being mimicked exhibit high levels of reusability in their respective domains, one naturally expects similar levels of reusability in their software counterparts.

In physics-based simulations, for example, fields and particles are nearly ubiquitous. Students first acquire the concept of a field as a function of space and time.

They get exposed to field instances starting with scalar fields such as pressure and temperature. These concepts get generalized to include vector fields such as electric and magnetic vector fields. Ultimately, further generalization leads to higher-order tensor fields such as stress fields in continuum mechanics and energy tensor fields in relativity.

The apparent reusability of the field concept makes it a prime candidate for encapsulation in software. A scalar field class constructed to model thermodynamic pressure could be reused to model absolute temperature or material density. Furthermore, one can imagine an ADT hierarchy in which scalar field objects become components of a vector field class.

A similar statement can be made for the particle concept. A neutral particle class used to model atmospheric aerosols might be reused to develop a charged particle class for studying electric precipitators. Inheritance might be employed to add a new charge component to a derived type, while adding a type-bound procedure that computes electric and magentic field forces on the particle.

Defining a set of algebraic, integral, and differential operators for application to fields enables compact representations of most physical laws. Fortran supports user-defined operators (C++ *overloaded operators*), the implementations of which can evaluate approximations to the corresponding mathematical operators. The results of expressions containing user-defined operators can be assigned to derived type instances either via a language-defined intrinsic assignment or by a defined assignment (C++ *overloaded assignment operator*). We refer to the evaluation of expressions involving the application of mathematical operators to ADT instances as *abstract data type calculus* or ADT calculus.

Supporting ADT calculus has been a recurring theme across numerous scientific software projects over the past decade, including the Sophus library at the University of Bergen[1] (Grant, Haveraaen, and Webster 2000), the Overture project at Lawrence Livermore National Laboratory[2] (Henshaw 2002), and the Sundance[3] and Morfeus[4] projects that originated at Sandia National Laboratories (Long 2004; Rouson et al. 2008). In many instances, these projects allow one to hide discrete numerical algorithms behind interfaces that present continuous ones. As noted on the Sundance home page, this lets software abstractions resemble blackboard abstractions. Users of these ADTs can write ordinary differential equations (ODEs) and PDEs formally in a manner that closely resembles their expression informally in blackboard discussions.

Table 3.1 depicts software abstractions that might correspond to various blackboard ones. The first row of the software abstraction column declares a variable `T`, its time derivative `dT_dt`, and its Laplacian `laplacian_T` to each be scalar fields. The second row provides a notation that might be used to define boundary conditions. The third row provides a software notation analogous to the subscript notation frequently used for partial derivatives – a time derivative in the case shown. The fourth row demonstrates forward Euler time advancement. The fifth row shows how the result of the `T%t()` operator might be calculated. In doing so, it also depicts the full

[1] http://www.ii.uib.no/saga/Sophus/
[2] https://computation.llnl.gov/casc/Overture/
[3] http://www.math.ttu.edu/~klong/Sundance/html/
[4] http://public.ca.sandia.gov/csit/research/scalable/morfeus.php

Table 3.1. *Translating blackboard abstractions into software abstractions*

Blackboard abstraction	Software abstraction
$T = T(x,y,z,t), \partial T/\partial t, \nabla^2 T$	`type(field) :: T,dT_dt,laplacian_T`
$T(x=0,y,z,t) = T_0$	`call T%boundary(x=0,T0)`
$T_t \equiv \partial T/\partial t$	`T%t()`
$T^{n+1} = T^n + \int_{t_n}^{t_n+\Delta t} T_t dt$	
$\approx T^n + T_t \Delta t$	`T = T + T%t()*dt`
$\partial T/\partial t = \alpha \nabla^2 T$	`dT_dt = alpha*laplacian(T)`
$\nabla^2 T \equiv +T_{xx} + T_{yy} + T_{zz}$	`laplacian_T = T%xx() + T%yy() + T%zz()`

expression of a PDE as it might appear inside an ADT that uses the `field` class to represent heat conduction physics. The final row demonstrates the evaluation of the PDE RHS by invoking the aforementioned subscript analogy. This expression would likely appear as the result of a function implemented inside the `field` class itself.

Figures 3.1–3.3 refactor the fin heat conduction problem using ADT calculus. Figure 3.1 offers a selection of tests of forward Euler quadrature:

$$T^{n+1} = T^n + \partial T/\partial t|_{t_n} \Delta t \tag{3.1}$$

$$= T^n + \alpha \nabla^2 T^n \Delta t \tag{3.2}$$

or backward Euler quadrature:

$$T^{n+1} = T^n + \partial T/\partial t|_{t_{n+1}} \Delta t \tag{3.3}$$

$$= T^n + \alpha \nabla^2 T^{n+1} \Delta t \tag{3.4}$$

$$= \left(1 - \alpha \Delta t \nabla^2\right)^{-1} T^n \tag{3.5}$$

where lines 27 and 30 in Figure 3.1 provide the corresponding software expressions.

In the case of implicit time advancement schemes such as backward Euler, the choice of software abstractions requires considerable thought to preserve clarity and generality without undue overhead in development time and runtime. One question regarding line 30 in Figure 3.1 concerns which terms to encapsulate as objects and which to represent using intrinsic data types. A preference for objects makes the code easier to maintain: Many details private to the employed classes can be changed without affecting the client code.

In our implementation, the " `*`" and " `-`" operators return `real` arrays, whereas the `.inverseTimes.` operator returns an `integrable_conductor` object. Practical concerns dominated our choice: The arrays contain more data than would fit in an `integrable_conductor`, whereas the result of `.inverseTimes.` contains the right amount of data for encapsulation in an `integrable_conductor`, as must be the case because it produces the RHS that ultimately gets assigned to `fin`.

The implementation of the `.inverseTimes.` operator invoked at line 30 also reflects an important efficiency concern: Rather than computing a matrix inverse and matrix-vector product, it performs Gaussian elimination and back substitution. Thus, the operator name, chosen by direct analogy to equation (3.5), need not correspond to the actual solution method so long as the two produce equivalent results. The choice stems from the associated operation counts, with Gaussian elimination and back substitution possessing a lower operation count than matrix inversion and matrix-vector multiplication (Strang 2003).

Figure 3.1

```
 1  program integrable_fin_test
 2    use iso_fortran_env              ,only : input_unit
 3    use kind_parameters              ,only : rkind,ikind
 4    use initializer                  ,only : linear
 5    use problem_class                ,only : problem
 6    use integrable_conductor_class ,only : integrable_conductor
 7    implicit none
 8    real(rkind) ,dimension(:,:) ,allocatable :: I ! identity matrix
 9    real(rkind) ,parameter     :: tolerance=1.0E-06
10    real(rkind)                :: dt=0.1           ! default time step
11    type(integrable_conductor):: fin              ! heat equation solver
12    integer(ikind)             :: scheme           ! quadrature choice
13    type(problem)              :: specs            ! problem specifications
14    namelist /test_suite/ scheme
15
16    enum ,bind(c)
17      enumerator forward_euler,backward_euler
18    end enum
19    if (.not. get_scheme_from(input_unit)) scheme=forward_euler ! default
20    specs = problem(input_unit)
21    fin   = integrable_conductor(specs,linear)
22    print '(a,5g9.2)','initial temperature = ',fin%temperature()
23
24    dt = specs%time_step()
25    select case (scheme)
26      case(forward_euler)
27        fin = fin + fin%t()*dt
28      case(backward_euler)
29        I = identity(fin%rhs_operator_size())
30        fin = ( I - fin%rhs_operator()*dt ) .inverseTimes. fin
31      case default; stop 'In main: no method specified.'
32    end select
33
34    print '(a,5g9.2)','final temperature   = ',fin%temperature()
35    if (abs(fin%time_derivative())<tolerance) then
36      print '(2(a,es9.3))','|dT/dt|=',fin%time_derivative(),'<',tolerance
37      print *,'In main: test passed. :)'
38    else
39      print '(2(a,es9.3))','|dT/dt|=',fin%time_derivative(),'>',tolerance
40      print *,'In main: test failed. :('
41    end if
42  contains
43    logical function get_scheme_from(file)
44      integer(ikind) ,intent(in) :: file
45      integer(ikind) ,parameter  :: found=0
```

```
46      integer(ikind)           :: namelist_status
47      read(file,nml=test_suite,iostat=namelist_status)
48      if (namelist_status == found) then
49        get_scheme_from = .true.
50      else
51        get_scheme_from = .false.
52      end if
53      rewind(file) ! future reads start from the file head
54    end function
55
56     pure function identity(n)
57        integer(ikind) ,intent(in) :: n
58        integer(ikind)             :: i
59        real(rkind) ,dimension(:,:) ,allocatable :: identity
60        allocate(identity(n,n))
61        identity = 0.
62        forall(i=1:n) identity(i,i) = 1.
63    end function
64  end program
```

Figure 3.1. Integrable conductor test.

The structure of line 30 reflects a desire to decouple the time integration code from the specific PDE solved. A more equation-specific form might be:

```
real(rkind) :: alpha
alpha= specs%diffusivity()
fin= (I - alpha*fin%laplacian_operator()*dt) .inverseTimes. fin
```

where we assume the existence of a suitably defined type-bound procedure that returns a matrix-form finite difference approximation to the Laplacian operator. The above implementation would tie the time advancement code and the `integrable_conductor` interface to the heat equation. For a similar reason, the main program obtains its own copy of the time step from the `problem` object, keeping the physics model free of any of the details of the quadrature scheme used for time advancement.

Handling derived type components inside defined operators requires discipline to perform only those operations on each component that are consistent with the meaning of the operators. When writing a time-integration formula, this suggests each component must satisfy a differential equation. For `alpha` in Figure 3.2, the effective differential equation is $d\alpha/dt = 0$. For `stencil` in Figure 3.3, we instead choose to make it a module variable. Giving it module scope enables its use throughout the Field class without incorporating it into the `field` derived type. We also give `stencil` the `protected` attribute so only procedures inside the enclosing module can modify it. We recommend this practice for all module variables that lack the `private` or `parameter` attributes. We designate `stencil` `protected` in case it becomes `public` in a future revision.

Whereas Chapter 2 presented the use of UML diagrams in OOA and OOD, these diagrams also find use in reverse engineering existing codes. This need arises when handed legacy software without design documents. It also arises with new software for which the development path appears sufficiently straightforward to proceed to OOP without going through OOA and OOD — as might occur in ADT calculus, given the close correspondence between the program syntax and the high-level mathematical expressions being modeled.

Even in the above cases, UML diagrams can illuminate subtle aspects of the software structure. Figure 3.4 depicts two new types of relationships in a class diagram reverse engineered from the code in Figures 3.2–3.3. Whereas the composition and inheritance relationships in Chapter 2 could be discerned simply by looking at the source-code definitions of the corresponding derived type definitions, the relationships in Figure 3.4 require deeper inspection of the code. The dashed arrows drawn to the Problem class denote dependencies that result from argument passing. The solid line drawn between the Differentiator and Field classes indicates the use of an object that is neither aggregated explicitly nor inherited implicitly: It has module scope.

UML diagrams can also explain software behavior. Important behavioral aspects of ADT calculus relate to the depth of the resulting call trees and the fluidity with which the trees change with the addition or rearrangement of terms in the generating formulae. The UML sequence diagrams in Figure 3.5 depict the calling sequences for the forward and backward Euler time-integration formulas at lines 27 and 30 of Figure 3.1. Following the message-passing interpretation of OOP, one can read a sequence diagram from top to bottom as a series of messages passed to the objects in a simulation.

The evaluation of the forward Euler expression `fin + fin\%t()*dt` involves first sending the message "`t()`" to `fin` instructing it to differentiate itself with respect to time. This results in `fin` sending the message "`xx()`" to its `field` component `T_field`, which in turn creates a new, temporary Field object. This new Field initializes its allocatable array component by sending the message "`laplacian()`" to the Differentiator object `stencil`. Subsequently, the temporary Field invokes the division operator on itself – represented as a loopback in the sequence diagram. Finally, `fin` relays its multiplication and addition messages `*` and `+` to `T_field` to complete formation of the RHS of the forward Euler expression.

Figure 3.5(b) diagrams the sequence of calls in the evaluation of the backward Euler expression `(I-fin\%rhs_operator()) .inverseTimes. fin`. Upon receiving the message `rhs_operator()`, `fin` queries its `T_field` component for its second-order finite difference matrix. `T_field` in turn requests the corresponding matrix from `stencil`. Because the compiler handles the ensuing intrinsic array subtraction operation, the sequence diagram does not show this step. Finally `fin` invokes the `.inverseTimes.` operator on itself by relaying the same message to its `T_field` component.

The relationship between scientific research and ADT calculus is not unique: ADT calculus could find use outside science (in financial mathematics for example) and other programming paradigms find use in science. Nonetheless, it appears to be the one major, modern programming idiom that originated amongst scientific programmers. Thus, when we refer to Scientific OOP (SOOP), we refer to ADT calculus. As a novel paradigm, SOOP warrants its own analysis and scrutiny. The next section provides such analysis and scrutiny.

Coming Soon to a Screen Near You!

With ADT calculus, your screen becomes your blackboard. Writing a high-level language resembles using a symbolic runtime environment such as Mathematica, Maple, or Mathcad. The techniques described in this chapter demonstrate how to support the natural syntax of these packages in your own source code.

Figure 3.2

```
1  module integrable_conductor_class
2    use kind_parameters ,only : ikind ,rkind
3    use field_class     ,only : field, distribution
4    implicit none
5    private
6    public :: integrable_conductor
7    type integrable_conductor
8      private
9      type(field) :: T_field ! temperatures
10     real(rkind) :: alpha   ! thermal diffusivity
11   contains
12     procedure                :: t
13     procedure                :: rhs_operator
14     procedure                :: rhs_operator_size
15     procedure                :: temperature
16     procedure                :: time_derivative
17     procedure ,private       :: product
18     generic                  :: operator(*) => product
19     procedure ,private       :: total
20     generic                  :: operator(+) => total
21     procedure ,private ,pass(rhs) :: inverseTimes
22     generic                  :: operator(.inverseTimes.) => inverseTimes
23   end type
24   interface integrable_conductor
25     procedure constructor
26   end interface
27 contains
28   type(integrable_conductor) function constructor(spec,T_distribution)
29     use problem_class ,only : problem
30     type(problem) :: spec ! problem specification
31     procedure(distribution) T_distribution
32     constructor%T_field = field(spec,T_distribution)
33     constructor%alpha   = spec%diffusivity()
34   end function
35
36   pure integer(ikind) function rhs_operator_size(this)
37     class(integrable_conductor) ,intent(in) :: this
38     rhs_operator_size = this%T_field%field_size()
39   end function
40
41   function rhs_operator(this)
42     class(integrable_conductor) ,intent(in)  :: this
43     real(rkind) ,dimension(:,:) ,allocatable :: rhs_operator
44     rhs_operator = this%T_field%xx_matrix()
45   end function
```

```
46
47    function temperature(this)
48      class(integrable_conductor) ,intent(in)  :: this
49      real(rkind) ,dimension(:) ,allocatable :: temperature
50      temperature = this%T_field%nodal_values()
51    end function
52
53    pure type(integrable_conductor) function product(this,factor)
54      class(integrable_conductor) ,intent(in) :: this
55      real(rkind)                  ,intent(in) :: factor
56      product%T_field = this%T_field * factor
57      product%alpha   = this%alpha   * factor
58    end function
59
60    pure type(integrable_conductor) function total(lhs,rhs)
61      class(integrable_conductor) ,intent(in) :: lhs
62      type(integrable_conductor)  ,intent(in) :: rhs
63      total%T_field = lhs%T_field + rhs%T_field
64      total%alpha   = lhs%alpha   + rhs%alpha
65    end function
66
67    pure function t(this) result(d_dt)
68      class(integrable_conductor) ,intent(in) :: this
69      type(integrable_conductor)              :: d_dt
70      d_dt%T_field = this%T_field%xx()*this%alpha
71      d_dt%alpha   = 0.0_rkind
72    end function
73
74    pure real(rkind) function time_derivative(this)
75      class(integrable_conductor) ,intent(in) :: this
76      time_derivative = this%alpha*this%T_field%xx_boundary()
77    end function
78
79    type(integrable_conductor) function inverseTimes(lhs,rhs)
80      real(rkind) ,dimension(:,:) ,allocatable ,intent(in) :: lhs
81      class(integrable_conductor)            ,intent(in) :: rhs
82      inverseTimes%T_field = lhs .inverseTimes. rhs%T_field
83      inverseTimes%alpha = rhs%alpha
84    end function
85  end module integrable_conductor_class
```

Figure 3.2. Integrable conductor class implementation.

Figure 3.3

```
1   module field_class
2     use kind_parameters      ,only : ikind, rkind
3     use differentiator_class ,only : differentiator
4     implicit none
5     private
6     abstract interface
7       pure function distribution (x) result(ret)
8           import
9           real(rkind), intent(in) :: x
10          real(rkind) ret
```

```
11        end function
12      end interface
13      type field
14        private
15        real(rkind) ,allocatable ,dimension(:) :: node ! internal points
16      contains
17        procedure               :: nodal_values
18        procedure               :: field_size
19        procedure ,private      :: product
20        generic                 :: operator(*) => product
21        procedure ,private      :: ratio
22        generic                 :: operator(/) => ratio
23        procedure ,private      :: total
24        generic                 :: operator(+) => total
25        procedure ,private ,pass(rhs) :: inverseTimes
26        generic                 :: operator(.inverseTimes.) => inverseTimes
27        procedure               :: xx          ! 2nd-order spatial derivative
28        procedure               :: xx_boundary ! 2nd-order boundary derivative
29        procedure               :: xx_matrix   ! matrix derivative operator
30      end type
31      interface field
32        procedure constructor
33      end interface
34      public :: field, distribution
35      type(differentiator) ,protected    :: stencil
36      integer(ikind)       ,parameter    :: end_points=2
37      real(rkind) ,dimension(end_points) :: boundary
38    contains
39      type(field) function constructor(spec,sample)
40        use problem_class ,only : problem
41        type(problem) :: spec
42        procedure (distribution) sample
43        real(rkind)    :: dx
44        integer(ikind) :: i
45        stencil = differentiator(spec)
46        dx      = spec%spacing()
47        allocate(constructor%node(spec%nodes()))
48        forall(i=1:spec%nodes()) constructor%node(i) = sample(i*dx)
49        boundary = spec%boundary_vals()
50      end function
51
52      pure type(field) function xx(this)
53        class(field) ,intent(in) :: this
54        xx%node = stencil%laplacian(this%node,(/boundary(1),boundary(2)/))
55      end function
56
57      pure real(rkind) function xx_boundary(this)
58        class(field) ,intent(in) :: this
59        type(field)              :: this_xx
60        this_xx = this%xx()
61        xx_boundary = this_xx%node(1)
62      end function
63
64      pure function xx_matrix(this)
65        class(field) ,intent(in) :: this
```

```
66     real(rkind) ,dimension(:,:) ,allocatable :: xx_matrix
67     xx_matrix = stencil%lap_matrix()
68   end function
69
70   pure integer(ikind) function field_size(this)
71     class(field) ,intent(in) :: this
72     field_size = size(this%node)
73   end function
74
75   pure function nodal_values(this)
76     class(field) ,intent(in) :: this
77     real(rkind) ,allocatable ,dimension(:) :: nodal_values
78     nodal_values = this%node
79   end function
80
81   pure type(field) function ratio(this,denominator)
82     class(field) ,intent(in) :: this
83     real(rkind)  ,intent(in) :: denominator
84     ratio%node = this%node/denominator
85   end function
86
87   pure type(field) function product(this,factor)
88     class(field) ,intent(in) :: this
89     real(rkind)  ,intent(in) :: factor
90     product%node = this%node*factor
91   end function
92
93   pure type(field) function total(lhs,rhs)
94     class(field) ,intent(in) :: lhs
95     type(field)  ,intent(in) :: rhs
96     total%node = lhs%node + rhs%node
97   end function
98
99   function inverseTimes(lhs,rhs)
100    use linear_solve_module ,only : gaussian_elimination ! See Appendix A
101    real(rkind) ,dimension(:,:) ,allocatable ,intent(in) :: lhs
102    class(field)                             ,intent(in) :: rhs
103    type(field)                 ,allocatable :: inverseTimes
104    allocate(inverseTimes)
105    inverseTimes%node = gaussian_elimination(lhs,rhs%node)
106  end function
107 end module field_class
```

Figure 3.3. Field class implementation.

3.2 Analysis-Driven Design

One expects programs that employ ADT calculus to exhibit a degree of elegance in some sense, which naturally generates two fundamental questions: to what degree and in what sense? This section presents steps toward answering these questions.

The stated aim requires developing quantitative, analytical ways to describe scientific software architecture. Attempts to do so in the broader software engineering community have met with some controversy (Fenton and Ohlsson 2000; Schach 2002; Shepperd and Ince 1994). Much of the controversy swirls around the

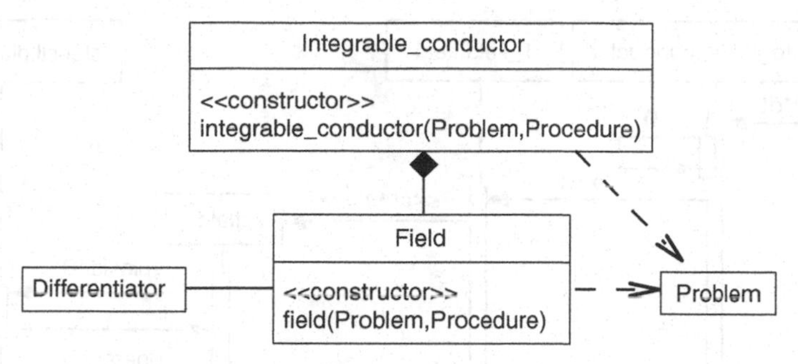

Figure 3.4. SOOP heat conduction class diagram.

validity and utility of measurable properties of software, or *design metrics*, when employed as

1. Fault rate predictors,
2. Programmer productivity evaluators, and
3. Development time estimators.

For example, in programs with modular architectures, one might expect fault rates to increase monotonically with module size; yet, as noted in Section 1.2, Fenton and Ohlsson detected nearly constant fault rates independent of module size in their study of two releases of a major telecommunications software project. In fact, the only significant departure from this behavior occurred at the smallest module size, where the fault rates dropped precipitously in one release and spiked upward in the subsequent release. They hypothesized that the latter behavior stemmed from the increased ratio of interface content to body content for the smallest modules, so that complexity in the body had been pushed into the interface.

As tools for evaluating productivity, design metrics suffer from potential manipulation by programmers and misinterpretation by their managers. To use the simplest example, were one to use the much maligned source lines of code (SLOC) as the measure of productivity, even novice programmers could manipulate the result by inserting dead code or splitting steps that could easily be combined. On the manager's side, a tendency to task the more experienced developers with the more difficult problems might yield the lowest SLOC for the most adept people.

The use of design metrics to estimate development time provokes possibly the most pessimistic backlash. For example, Lewis (2001) argued that neither a program's size nor its complexity can be objectively estimated *a priori* and likewise for its development time and the absolute productivity of its developer. Lewis backed these claims with algorithmic complexity theory. The basic argument regarding program size and complexity follows: For a string s, the associated algorithmic complexity, or Kolmogorov complexity, $K(s)$, defined as the shortest program that produces s, is provably noncomputable. Taking s to be the desired program output suggests that the smallest program (including input data) that produces this output cannot be estimated.

Alternatively, easily derived upper bounds on the minimum program size, such as the size of a program that simply prints the output string, do not prove useful. The lack of useful, computable bounds on program size implies a similar statement for development time because it depends on program size as well as a similar statement

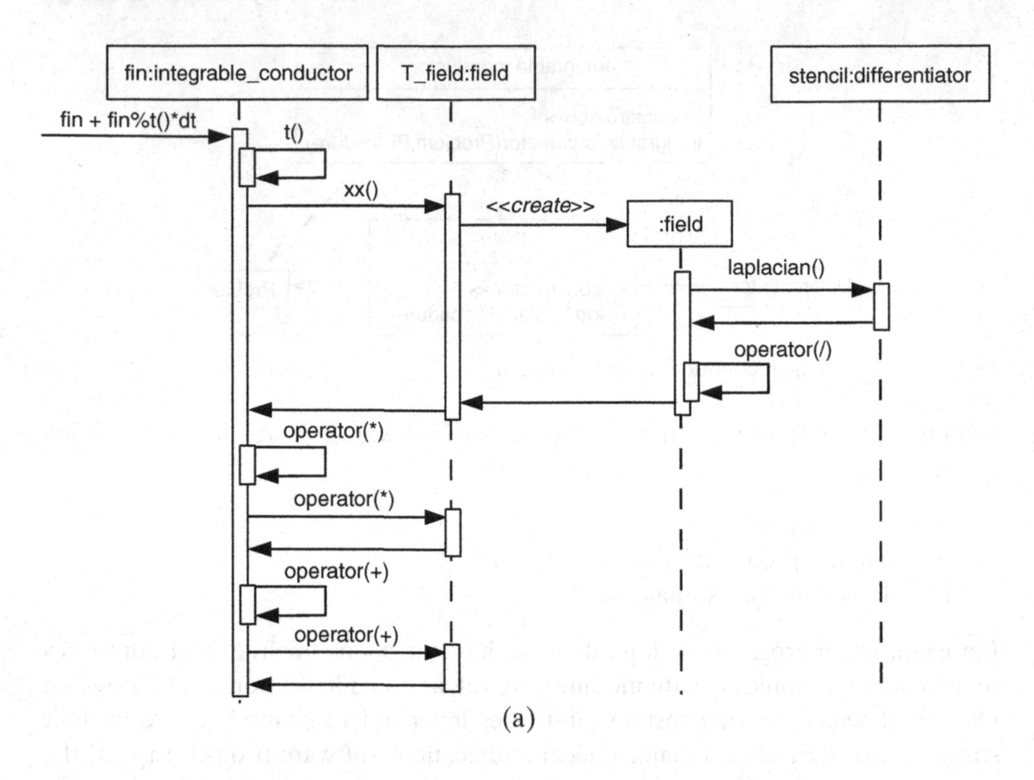

(a)

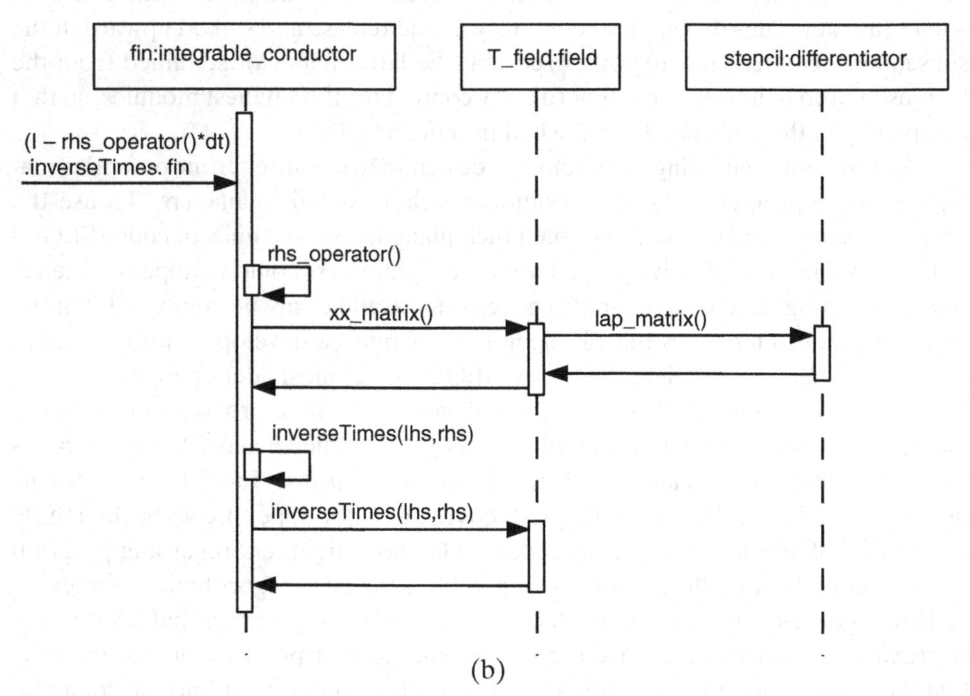

(b)

Figure 3.5. Sequence diagrams for the two quadrature schemes in Figure 3.1: (a) forward Euler and (b) backward Euler.

for programmer productivity as measured by program size divided by development time. Ultimately, Lewis concluded that even approximate estimations cannot be constructed theoretically and must therefore be judged purely on empirical grounds.

None of the above precludes calculating algorithmic complexity, or other formal properties of software, within restricted domains wherein certain program properties can be guaranteed. For example, Chaitin (1996) demonstrated "how to run algorithmic information theory on a computer" by restricting the program to being expressible in so-called "pure LISP." In this context, one assumes programs to be self-delimiting – that is, to contain information about their lengths. In such a system, one can construct successively sharper lower bounds on a program's algorithmic complexity.

Somewhat ironically, Kirk and Jenkins (2004) provided empirical evidence of the usefulness of the very entity, algorithmic complexity, that Lewis employed in arguments against theoretical estimates. They took the concatenation of a compressed piece of source code and the compression software that produced it as an upper bound on $K(s)$. Furthermore, they used code obfuscators to estimate the worst-case complexity and compared this with the complexity added by the developer during a revision cycle. Obfuscators alter source code in ways that make it more difficult for humans to understand and more difficult to reverse engineer. Using the compression software bzip2[5], they estimated the information content of obfuscated files at varying levels of obfuscation. Based on their results, they proposed analogies between their information measures and thermodynamic entropy. They further suggested that these analogies could lead to studying material-like phase transitions as software becomes more brittle through stronger coupling between modules.

This section builds on the complexity discussions in Chapters 1 and 2 by exploring three avenues for analyzing ADT calculus: object-oriented design metrics, computational complexity theory, and information theory. Each subsection analyzes the problem of time advancement in the context of the forward and backward Euler evaluation lines in Figure 3.1. Subsection 3.2.1 employs two design metrics that appear frequently in the OOP literature: coupling and cohesion. Subsection 3.2.2 explores scaling arguments and a more realistic bug search time estimate than that considered in the complexity sections of Chapters 1–2. Finally, subsection 3.2.3 develops two information-theoretic arguments that quantify another development-time process: developer communications.

3.2.1 Design Metrics

The proponents of design metrics aim to improve software engineering practices by encouraging code designs that are coherent, flexible, and robust. In many regards, metrics share these purposes with ADT calculus. Not surprisingly then, the implementation of ADT calculus tends to foster trends in certain metrics. Specifically, ADT calculus implementations tend to exhibit high *cohesion* within classes and low *coupling* between classes.

The term "cohesion" describes the extent to which a class's procedures relate to each other in purpose. Stevens et al. (1974) ranked various types of cohesion from

[5] http://www.bzip.org

weakest to strongest. Coincidental cohesion, the weakest form, occurs when parts of a system have no significant relationship to each other. Functions in classical, procedural mathematics libraries exhibit coincidental cohesion in that most procedure pairs have in common only that they evaluate mathematical functions. Functional cohesion, the strongest form, occurs when each part of a system contributes to a single task.

In the discrete time-advancement schemes, each of the operators and type-bound procedures contributes to time advancement. This guarantees functional cohesion at least in a minimal ADT that implements only the requisite calculus.

The term "coupling" describes the extent to which different ADTs depend on each other. As with cohesion, one can rank types of coupling from loosely coupled to tightly coupled. Authors typically rank data coupling as the loosest form (Pressman 2001; Schach 2002). Data coupling occurs when one part of a system depends on data from another part. Control coupling, a moderately tighter form, arises when the logical flow of execution in one part depends on flags passed in from another part. Content coupling, which typically ranks as the tightest form of coupling (Pressman; Schach), occurs when one violates an abstraction by providing one ADT with direct access to the state or control information inside another ADT. An especially egregious form of content coupling can occur when it is possible to branch directly into the middle of a procedure inside one ADT from a line inside another.

Because Fortran requires defined operators to be free of side effects (all arguments must have the `intent(in)` property), a developer can write the time advancement expressions in Figure 3.1 without any information about the state or flow of control inside the objects and operators employed. ADT calculus thus allows for strict adherence to data privacy, which precludes content coupling.

Furthermore, depending on the design of the ADT calculus, one can write discrete time advancement expressions in which nearly all the procedure arguments are instances of the class being advanced. This is the case for the forward Euler algorithm in Figure 3.1. The arguments arise implicitly when the compiler resolves the statement

```
fin = fin + fin%t()*dt
```

into a procedure call of the form

```
call assign(fin,total(fin,product(t(fin),dt)))
```

where `total()`, `product()` and `t()` are the defined addition, multiplication, and time differentiation procedures, respectively, in Figures 3.2–3.3 and where `assign()` could be a suitably defined assignment procedure.[6] Thus, in the absence of any flags internal to the derived type itself, no control coupling arises when employing an ADT calculus. This leaves only the loosest form of coupling: data coupling. At least in terms of the top-level semantics, even data coupling is kept to a minimum given that all operators are either unary or binary, so at most two arguments can be passed.

[6] We explicitly reference the defined assignment here for clarity. In practice, Fortran's intrinsic assignment suffices unless the derived type contains a pointer component, in which case one needs to define an assignment if one desires a "deep copy" of all the RHS data into the LHS object. Otherwise, the intrinsic assignment executes a "shallow copy" in which only a pointer assignment occurs, resulting in the LHS and RHS pointer components sharing a target.

Whereas cohesion is essentially qualitative, coupling is more easily quantified. Following Martin (2002), we define an ADT's afferent couplings (*Ca*) as the number of other ADTs that depend on it, whereas its efferent couplings (*Ce*) are the number of ADTs on which it depends. Its instability

$$I \equiv \frac{Ce}{Ce + Ca} \tag{3.6}$$

vanishes when modifications to other ADTs do not impact the ADT in question (a completely stable situation) and approaches unity when changes to *any* other ADT can potentially affect the one in question (a completely unstable situation). In an exercise at the end of this chapter, we ask the reader to calculate the instability of the packages in our various fin models. Chapter 8 discusses the instability metric in the much richer context of multiphysics modeling.

Building up the instability metric from the afferent and efferent couplings represents a subtle shift from measuring properties of software to measuring properties of the software development process. Although one could argue that the couplings themselves have little importance in static code, it would be hard to downplay their potential impact on the code revision process. The instability metric captures that potential impact. The next two subsections also focus on development-time processes: bug searches and developer communications.

3.2.2 Complexity Theory

Chapter 1 analyzed a bug search algorithm, estimating that the average number of lines searched to find all bugs in a traditional, procedural code grows quadratically in λ for code with the λ lines operating on a globally shared data set. The bug search algorithm considered in Chapter 1 has the virtue of being easy to analyze but the vice of not resembling what most programmers do. The following bisection method takes us a step closer to how developers actually debug code:

1. Bracket the suspected offending code with statements that verify the satisfaction of necessary constraints. An appropriate set of constraints will be satisfied at the beginning of the code segment (as indicated by the "true" conditions in Figure 3.6) but not satisfied at the end (as indicated by the "false" conditions in Figure 3.6).
2. Bisect the code segment and insert constraint checks at the midpoint.
3. Form the next segment from the code between the midpoint and the end point at which the result is the opposite of the midpoint result. In Figure 3.6, for example, the verification passes at the first midpoint, so the next segment lies between the midpoint and the end point.

This process can be repeated until the segment is reduced to a single line of code. That line is the first place where the constraint checks fail.

The bisection search draws inspiration from the analogous method for finding a root of a function of a single variable. In that setting, a necessary and sufficient condition for convergence to a particular root is the existence of exactly one sign change in the initially bracketed interval. That sign change indicates the presence of a root.

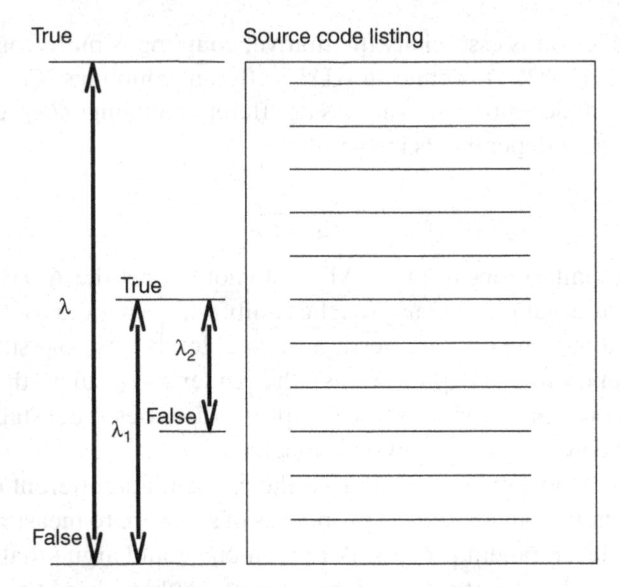

Figure 3.6. Bisection method bug search.

The corresponding condition for the bisection bug search is the presence of exactly one place where the constraint statements change from truth to falsehood. A sufficiently tight bracketing of the bug in a chronological listing of executed lines would likely suffice. The number of lines λ_n bracketed at the nth iteration of a bisection search provides as a measure of the maximum distance between the lines being tested (the end points of the nth segment) and the bug. Because the bisection methods cuts the size of the interval in half at each iteration, the maximum distance from the bug at iteration n is

$$\lambda_n = \lambda_0/2^n \tag{3.7}$$

where λ_0 represents the initialize segment size.

The search converges when the code segment reaches a single line, which happens after

$$n = \log_2 \lambda_0 \tag{3.8}$$

iterations. At the beginning of the bisection search ($n = 0$), two lines are checked, those being the first and last lines of the suspect code. At each subsequent iteration, one new line is checked, that being the midpoint line. Thus, the search for all $r\lambda$ bugs terminates after

$$\lambda_{\text{searched}} = r\lambda(2 + \log_2 \lambda_0) \leq r\lambda(2 + \log_2 \lambda) \tag{3.9}$$

lines, or asymptotically $O(\lambda \log_2 \lambda)$ lines maximum.

Following the arguments in section 2.7, the number of unique lines tested in the course of the search is very likely bounded by the size of the class that contains the problematic data: $\lambda_0 \leq c$ for a class with c lines sharing access to private data. The authors' experiences suggest this inequality holds in ADT calculus. Although occasional excursions outside the class in question aid in understanding the flow of control and ensure that the client code supplies correct data to the class, the excursions prove brief and straightforward due to the simplicity of the expressions that generate the call tree and due to these software expressions' close resemblance to the blackboard expressions from which they derive. Hence

$$\lambda_{\text{searched}} \leq rc(2 + \log_2 c) \tag{3.10}$$

where the inequality must be considered approximate.

An even more exciting feature of ADT calculus appears when we factor c into the product of the number of procedures p in the suspect class and the class's average procedural line density $\rho \equiv c/p$. Two reasons motivate this factoring. First, reviewing the forward Euler expression

```
fin = fin + fin%t()*dt
```

suggests that the number of procedures one must implement to advance any ADT over time would be the same as those defined for `fin`. These procedures number three to four: the addition operator (+), the time differentiation procedure (`t()`), the multiplication operator (*), and, if necessary, a defined assignment (=). Chapter 8 describes a strategy for building a multiphysics package in which one can time-advance a coupled group of single-physics ADTs using a single expression analogous to the forward Euler expression for `fin`. With ADT calculus, the same three procedures must be defined for each single-physics abstraction. Thus, p remains constant as the size of the software project grows to incorporate new physics.

Second, the algorithm implemented inside the aforementioned procedures is often the same for each ADT. The * operator multiplies an object's state vector by a constant, while the + operator adds the state vectors of two objects, and = copies an object's state vector. Because the algorithm is the same for each class, the number of lines required to express the algorithm will be very nearly the same. Hence, ρ remains very nearly constant across the project. This leads to rewriting equation (3.10) as

$$\lambda_{\text{searched}} \leq r\rho p[2 + \log_2(\rho p)] \tag{3.11}$$

and to the assertion that the maximum number of lines searched remains very nearly constant independent of the size of the overall project.

The above analysis focuses on testing for the presence of a bug during program execution. This would appear to neglect the possibility of errors in the input data. However, sufficiently restrictive validation constraints should detect such errors during the two initial tests that start the bisection search – namely the one on the first executed line.

3.2.3 Information Theory

In his seminal paper on information theory, Shannon (1948) quantified the information content in telecommunication signals. It seems reasonable then to build on his foundation to describe the content in other types of communications such as communications between developers. Of interest here is the amount of information developers communicate in supplying new classes that extend the capability of a multiphysics framework as described in the previous subsection. In OOP, the minimum information each developer must provide is the interface to the class that developer contributes. In the time-integration problem, that interface contains a derived type along with the aforementioned mathematical operators, time differentiator, and defined assignment.

Shannon reasoned as follows about the set of all possible messages that can be transmitted between two locations (two developers in the present case):

> If the number of messages in the set is finite then this number or any monotonic function of this number can be regarded as a measure of the information produced when one message is chosen from the set, all choices being equally likely.

The underlying notion here is that communication hinges on novelty. If only one message is possible, the transmission always contains that message and no new information arrives. Imagine an analog electrical transmission line on which the voltage remains fixed. Such a line cannot be used to transmit information. It follows that the degree to which the voltage can vary, and likewise the number of possible states into which the line can be placed, the more novel is the choice of a particular voltage, the more information can be conveyed.

Shannon went on to choose the logarithm as the monotonic function because it satisfies several conditions that match our intuitive understanding of information. For example, consider a digital transmission line that can transmit one of two messages, say 0 or 1. Adding two identical new lines would cube the number of states available to the collective system: There would be 2^3 possible messages. Intuitively, however, one expects the amount of information to triple in going from one line to three. The logarithm rescales the number of states to match our intuition: $\log_2 2^3 = 3\log_2 2 = 3\cdot 1$, where the base of the logarithm determines the size of one unit of information. In digital systems, the natural unit is the binary digit, or "bit."

More generally, Shannon defined information entropy, a measure of information content, as

$$H \equiv -\sum_{i=1}^{N} p_i \log p_i \tag{3.12}$$

where p_i is the frequency of occurrence of the ith token (e.g., a character or keyword) in some large sample representative of the domain of interest, and where N is the number of unique tokens. In Shannon's entropy, $\log p_i$ represents the amount of information communicated in each occurrence of the ith token. In probability theory, summing the product of a quantity and its probability over the entire space of probabilities provides its *expected value* or *mean*. Taking all choices (tokens) as equally likely leads to $p_i = 1/N$ and therefore

$$H = \log N \tag{3.13}$$

which returns us to the simpler definition: the logarithm of the number of possible messages.

To relate these concepts to source code, consider that the information received over time on a transmission line can also be spread over space in a program source file. Thus one draws an analogy between the probability of occurrence of a token in a signal and its probability of occurrence in a program. One can measure the probability by calculating the token's frequency of occurrence in a signal (or program file) sufficiently long to provide accurate statistics.

Considering the information entropy of an ADT calculus elucidates how the calculus simplifies developer communications. First, consider that the class interfaces comprise the minimum amount of information the developer of each class must communicate to users of that class. In ADT calculus, wherein each class that supports the calculus must implement the same operators, the class interfaces exhibit a high degree of repetition. To see this, imagine extending our heat conduction solver to include one-dimensional convection via the PDE

$$\frac{\partial T}{\partial t} = -\frac{\partial}{\partial x}(Tu^*) + \alpha\frac{\partial^2 T}{\partial x^2} \tag{3.14}$$

where $u^* \equiv u/(\rho c)$ is a constant flow speed divided by the material density and specific heat capacity. To enable forward and backward Euler time advancement, we might build the new type as follows:

```
module integrable_fluid_ADT
  implicit none
  private
  public :: integrable_fluid
  real(rkind) ,parameter :: u_star =1.0_rkind
  type ,extends(integrable_conductor) :: integrable_fluid
    private
  contains
    procedure  :: t
    procedure  :: rhs_operator
    procedure ,private ,pass(rhs) ::        inverseTimes
    generic :: operator(.inverseTimes.) => inverseTimes
  end type
  interface integrable_fluid
    procedure constructor
  end interface
contains
  ! implementation omitted
end module
```

where we have overridden three of the parent type's type-bound procedures: `t`, `rhs_operator`, and `inverseTimes`. Two compute the new time derivative and the RHS operator matrix. A third solves the linear system, allowing for a more sophisticated Gaussian elimination process, for example, one with pivoting, due to potential changes in the matrix properties. The new type inherits the parent type's remaining type-bound procedures.

The character/keyword patterns that recur in the `integrable_conductor` and `integrable_fluid` types would occur in any ADT designed for forward and backward Euler time advancement. Considering each character as a token yields high p_i and low H when adding `integrable_fluid` type to the package. This slow information growth reflects in very succinct human communications between programmers. The test writer simply informs the developers of the required operators. The developers provide interfaces supporting those operators.

Taking larger groups of characters as the tokens yields even lower H. At one extreme, the entire listing of `integrable_conductor` type-bound procedures from the `contains` to the `end type` could be one token. This token would recur across all ADTs that support the desired calculus without inheriting any procedure bindings. The resulting low H values would indicate the possibility of very high file compression, the significance of which we consider next.

The dependence on the token set demonstrates the nonabsolute nature of information entropy H. Kolmogorov complexity K provides a more general quantification of information in that it does not depend on a probability model (which depends on the choice of tokens). Although the cost of this generality lies in the noncomputability of K, estimating K by compressing files links K to H, which is computable. Shannon's *relative entropy* provides the link:

> The ratio of the entropy of a source to the maximum value it could have while still restricted to the same symbols will be called its relative entropy. This is the maximum compression possible when we encode into the same alphabet.

Thus, the best available compression software yields the tightest estimate to the Kolmogorov complexity as well as the best estimate of the relative entropy.[7]

Equation (3.12) is formally analogous to the thermodynamic entropy of statistical mechanics, wherein it describes the information one obtains by determining the microscopic state of a collection of matter (as determined by the quantum states of its constituent parts) when it is in a particular macroscopic state. The thermodynamic limit treats all microscopic states as equally likely so $H = k \ln N$, where the Boltzmann constant k facilitates converting the logarithm to the natural base and thereby sets the unit of information.

Kirk and Jenkins (2004) took the thermodynamic analogy further. They estimated the entropy of Java byte code. They also showed that other properties of the compressed code behave like other thermodynamic variables such as volume, temperature, and pressure. They plotted these variables against each other for an ensemble of student projects. We will not pursue an empirical study of source code entropy in the current text, but an exercise at the end of Chapter 8 presents a very simple, quantifiable entropy measure.

The Audacity of Hope

Pessimism abounds regarding the utility, and even the possibility, of reasoning quantitatively about software properties. Yet heuristic arguments with varying degrees of sophistication suggest properties of the software development process are amenable to analysis. OOD metrics suggest ADT calculus makes packages stable under revision. Complexity theory suggests it renders bug search times nearly independent of project size. Information theory suggests it reduces the required developer communications by limiting the growth in the interface information content incurred when adding new abstractions.

[7] Encapsulation and information hiding also reduce interface content in any OOP project due to the shorter argument lists they facilitate. SOOP augments these strategy by limiting most argument lists to one or two for the unary and binary operators, respectively, of an ADT calculus.

3.3 Still More on Style

The codes in Figures 3.1–3.2 present several new constructs along with new structural and behavioral features. Lines 16-18 in Figure 3.1 present this book's first use of one of the Fortran 2003 features designed to promote interoperability with the C programming language. The `enum` construct assigns distinct integer values to the names listed in the `enumerator` clause. The `bind(c)` attribute ensures that the type kind parameter is interoperable with C in the sense that the bit representation of the values corresponds to that which the companion C compiler uses for an enumeration. Chapter 11 details several additional, useful C interoperability features in Fortran 2003.

Given the depth of the call trees in ADT calculus, one must take care to avoid inefficiencies associated with excessive function invocations. One straightforward strategy involves collecting large amounts of coarsely partitioned data at the lowest levels of abstraction (corresponding to the deepest invocations in the call tree) and pushing computationally intensive calculations down to those same abstractions. With this strategy, the computational work accomplished inside the most deeply nested procedures greatly exceeds the cost of the function invocations required to get there.

As we detail next, the chief penalties in terms of execution speed and memory utilization relate to the separation of operations that might be more efficient if combined. Consider once more the forward Euler expression:

```
fin = fin + fin%t()*dt
```

The aforementioned predilection for coarse-grained data implies the presence of a sizeable array component inside the time-differentiation result. Considering most modern processors' ability to perform a multiplication/addition pair in a single clock cycle, separating the scaling of this result by the factor `dt` from the addition of the resulting scaled array and the array component inside `fin` represents a potential doubling of the number of clock cycles required to evaluate the forward Euler RHS expression. On a closely related note, if these array components exceed the size of a processor's memory cache, looping through an entire array means retiring results of the multiplication to main memory before reloading them for the addition.

Surmounting these performance hurdles requires combining operators. This can happen in three ways. First, in some instances, optimizing compilers can combine the operators during their interprocedural optimization stage. Second, programmers can define combined operators explicitly in the source code. A combined forward Euler operator might take the form

```
fin = fin%plus_times(fin%t(),dt)
```

where `plus_times` scales the `fin\%t()` result's array component by `dt` and adds the scaled array it to the corresponding `fin` array component in a single loop. A third approach available in C++ is to write *expression templates* that teach the compiler how to resolve the original forward Euler syntax into an invocation of a combined procedure such as `plus_times`. As always, the choice of strategy should be informed by empirical profiles of the relative runtime share of various processes in the simulation. On many problems of modest size, the gains will not justify the loss in semantic elegance and the attendant sacrificing of interface clarity.

Optimizing compilers are very likely to be efficient with intrinsic assignment of RHS objects to the LHS ones. Whereas naive compiler implementations might form the RHS result's array component before copying it into the LHS, more sophisticated compilers often recognize the opportunity to overwrite the LHS in the process of constructing the RHS result. Doing so, however, might require separating the multiplication and addition steps, because overwriting the `fin` component during the multiplication phase (or likewise during the combined addition/multiplication phases) generates the wrong result. This subtle interplay between memory management and operation count underscores one of the reasons such optimizations are best left to the compiler wherever possible.

The issue of creating a temporary object to hold the RHS result represents one example of the larger issue of how to handle temporaries created as the result of all of the operators and type-bound procedures in an expression. The compiler knows it will overwrite the LHS of an assignment, but one cannot safely expect similar insight regarding the fate of other type-bound procedure results. One strategy involves writing all ADT calculus procedures such that they return intrinsic types. In the forward Euler expression, having `fin\%t`, `operator(*)`, and `operator(+)` all return `real` arrays, exposes a great deal of information to the compiler that can be used for the kinds of optimizations that might otherwise require combined operators. This has the drawback, however, of limiting the programmer to the intrinsic data structures. There would be no way to write a procedure that works only with the nonzero elements of sparse arrays, for example. Chapter 10 outlines a strategy that enables users to eliminate most temporaries while retaining control over the data abstractions.

A common theme in object-oriented software design (the theme of Part II of the current text) is a preference for aggregation over inheritance. Both technologies involve an instance of one object serving as a component of another. With aggregation, the programmer explicitly imposes this structure. With inheritance, the compiler creates this structure automatically. In many instances, however, employing inheritance resembles using a machete to slice butter. Much of the power is wasted if the two types do not share a significant portion of their interfaces and the writer of the child type must override most or all of the type-bound procedures. An even worse case occurs when some of the procedures in the parent type's interface have no reasonable meaning in the child type's interface.

This dilemma exposes the ambiguities of the "has a" and "is a" metaphors for aggregation and inheritance relationships, respectively. These metaphors are interchangeable in theory. Nonetheless, aggregation often feels more natural in practice. To wit, saying "an `integrable_conductor` *is a* `field` that satisfies the heat equation" is essentially equivalent to saying "an `integrable_conductor` *has a* `field` that satisfies the heat equation." Yet several of the procedures bound to the `field` type would need different interfaces for clarity in the `integrable_conductor`. It makes little sense to talk about a conductor's second derivative. Furthermore, some of the procedures exposed in the `field` interface would prove completely superfluous in the `integrable_conductor` interface. For example, the ADT calculus the `integrable_conductor` supports for time advancement does not require a division operator (/).

Figure 3.3 marks this book's first use of the keyword `abstract`. It plays two roles in Fortran 2003. In an `abstract interface` such as at line 6 in Figure 3.3, it enables the declaration of a procedure signature, comprising the procedure arguments' attributes, types, and, if it is a function, its return type and attributes. In doing so, the abstract interface facilitates the use of the dummy procedure argument at line 42 in Figure 3.3. At runtime, one can pass as an actual argument the name of any procedure with matching argument attributes, types, and return type. As compared to the Fortran-95-style procedure argument in Figure 2.14, this new form avoids error-prone, duplicate interface declarations. Even an external procedure could be substituted for the declared one.

Section 4.2.2 describes the second usage of `abstract`, wherein it serves as a type attribute. An abstract type defines a specification to which its extended types must conform. One can never instantiate an abstract type, and its type-bound procedures might or might not have default implementations. In lieu of procedure implementations, abstract types often have *deferred bindings* that specify an abstract interface to which any extended types' type-bound procedures must conform.

A final salient aspect of the sample code is the aforementioned usage of variables with module scope at lines 35-37 in Figure 3.3. An alternative would be to include the relevant variables in the type definition, which would have the chief drawback of requiring that the variables be initialized or computed in every procedure that constructs a new `field` object. These definitions or calculations prove superfluous when all fields share a common state (as with C++ *static variables*) that does not change with time. One must take care to remember, however, that module variables must become type components when different instances need different values — as might be the case with the `stencil` variables for so-called staggered grid calculations, wherein different variables might be sampled at different spatial locations.

3.4 The Tao of SOOP

The Chinese symbol for tao most often translates as "way" or "path." Part I of the current text attempts turn the focus of the scientific programmer from the numerical algorithms and their runtime cost toward their high-level expression in software and its development costs. This largely represents a redirection of developers' attention from the properties of a program (such as its accuracy and speed) to the properties of the programming (such as the complexity of the associated bug search process and the requirement that process places on developers to clearly communicate their designs to each other). In other words, Part I shifts attention from the product to the path.

The Merriam-Webster dictionary offers the following alternative definition[8]:

> the art or skill of doing something in harmony with the essential nature of the thing

For many hundreds of years, the essence of scientific discourse could be thought of metaphorically as a collective discussion at the blackboard or on a handwritten page.

[8] http://www.merriam-webster.com/dictionary/tao

The advent of scientific computing, however, has largely turned this open dialogue into sorcery. Individual researchers or small research groups develop codes that are largely indecipherable to the uninitiated. This has ramifications both for the openness of the scientific dialogue and the ability to independently verify codes and validate their output.

Via ADT calculus, SOOP empowers researchers to build software constructs that express the essential nature of the mathematical constructs they represent. In many respects, the syntax of ADT calculus resembles many commercial, symbolic computation engines such as Mathematica, Mathcad, or Maple. Although some of these packages evolved from open versions, their modern incarnations are proprietary. In this regard, ADT calculus aims to bring the nonprofessional programmer – the person who sees scientific insights as their primary product rather than software – behind the curtain to see how their code can more closely resemble the natural syntax that professional programmers provide them via the aforementioned packages.

EXERCISES

1. The Crank-Nicolson time advancement algorithm applied to the heat equation takes the form

$$T^{n+1} = \left(1 - \frac{\alpha \Delta t}{2} \nabla^2\right)^{-1} \left(T^n + \frac{\alpha \Delta t}{2} \nabla^2 T^n\right) \tag{3.15}$$

   Modify Figure 3.1 to include a third enumeration named `crank_nicolson`. Add a `case(crank_nicolson)` to the `select case` construct in the same figure to include a software expression of the form

```
fin = ( I - fin%rhs_operator()*(alpha*dt/2.) &
      .inverseTimes. ( fin + fin%t()*(alpha*dt/2.)
```

   where `I` is an appropriately sized identity matrix and `fin%rhs_operator()` returns a finite difference matrix operator approximating a 1D Laplacian operator. Test the new scheme by defining a new, nonlinear initial condition and advancing to steady state until the the magnitude of the value returned by `fin%time_derivative()` falls below `tolerance`.
2. Draw a UML sequence diagram to describe the sequence of execution corresponding to the above Crank-Nicolson code.
3. Use the Fortran 2003 `pass` attribute in the `product` type-bound procedure binding in Figure 3.2 to enable the products

```
fin%rhs_operator()*(dt/(alpha*2.))
```

   and

```
fin%t()*(alpha*dt/2.)
```

   to be written in the reversed order

```
(alpha*dt/2.)*fin%rhs_operator()
```

   and

```
(alpha*dt/2.)*fin%t()
```

respectively.

4. Add a `debugging` parameter of `logical` type to the `kind_parameters` module (or, for clarity, create a new module) and modify each module in the integrable conductor model to give it access to this parameter via a `use` statement. Insert lines of the form

```
if (debugging) print *,'In (procedure name): start'
```

and

```
if (debugging) print *,'In (procedure name): end'
```

at the beginning of each procedure throughout the model.

(a) Set `debugging=.true.` and print the calling sequence in the integrable conductor model.

(b) Print the calling sequence again after reordering the terms in the forward Euler expression, for example

```
fin = fin%t()*dt + fin
```

Try other rearrangements of this expression.

5. Calculate the afferent couplings (Ca), efferent couplings (Ce), and instability (I) of all classes in the fin heat conduction models in Chapters 2–3. Which approach appears to be the most stable in terms of the average ADT stability across the design?

PART II

SOOP TO NUTS AND BOLTS

4 Design Patterns Basics

> "There is one timeless way of building."
>
> Christopher Alexander

4.1 Essentials

Whereas code reuse played an important role in Part I of this text, design reuse plays an equally important role in Part II. The effort put into thinking abstractly about software structure and behavior pays off in high-level designs that prove useful independent of the application and implementation language. Patterns comprise reusable elements of successful designs.

The software community typically uses the terms "design patterns" and "object-oriented design patterns" interchangeably. This stems from the expressiveness of OOP languages in describing the relationships and interactions between ADTs. Patterns can improve a code's structure and readability and reduce its development costs by encouraging reuse.

Software design patterns comprise four elements (Gamma et al. 1995):

1. **The pattern name**: a handle that describes a design problem, its solution, and consequences in a word or two.
2. **The problem**: a description of when to apply the pattern and within what context.
3. **The solution**: the elements that constitute the design, the relationships between these elements, their responsibilities, and their collaborations.
4. **The consequences**: the results and trade-offs of applying the pattern.

Although there have been suggestions to include additional information in identifying a pattern, for example, sample code and known uses to validate the pattern as a proven solution, authors generally agree that elements 2-4 enumerate the three essential factors in each pattern. Alexander et al. (1977) referred to a slightly different "three-part rule" as "a relation between a certain context, a problem, and a solution."

The current text takes a soup-to-nuts approach, providing the three essential elements followed by complete, compilable Fortran and C++ examples. We choose simplicity and clarity over speed and semantic power. Sophisticated programmers might exploit more advanced language features to enhance our code's generality, speed it up, or reduce its resource utilization. We encourage readers to revise and extend our examples. Publishing versions that employ more advanced programming techniques, for example, generic programming, or that address the performance needs of larger scale applications, might launch a much needed dialogue about scientific software design.

Demonstrating patterns in code completes a cycle. All patterns start out as working designs that developers reused across well-engineered projects without conscious recognition of their universality. The code came first. Gamma et al. (1995) distilled recurring themes out of the resulting source code after observing those designs' positive impact on projects. Thus, going from ubiquitous truths to simple examples brings us full circle: Software begets patterns beget software. An additional, unique step the current text adds to the cycle involves theoretical analysis along the lines of Section 3.2.

Pattern recognition requires perspective. A frequently used program construction in one language might be an intrinsic facility in another. For instance, as pointed out by Gamma et al. (1995) and demonstrated by Decyk et al. (1997b–1998), one might construct patterns for inheritance, encapsulation, or polymorphism in a procedural language, whereas these exist in object-oriented languages. Thus, the patterns one sees at a particular level of abstraction reflect the programming capabilities supported at that level.

At the source code level, intrinsic language constructs represent patterns supported by the creators of the language. Thus, whereas patterns enable us to see the forest in the trees, such a view treats trees as an intrinsic construct. Were one to attempt to construct trees from scratch using designer genes, the "tree" concept itself would represent a biological design pattern.

Patterns comprise tried-and-true constructions well worn by the time of their inclusion in a pattern canon, or *pattern language*. The subsequent chapters of Part II present a pattern menagerie for which we make none of the comprehensiveness claims that might justify calling it a pattern language. Rather, Part II whets the reader's appetite for patterns with a few we have found broadly useful.

It's Easy Being Green

OOD patterns emerge like the independent recurrence of a solution to a common biological problem in multiple evolutionary tree branches. In this sense, they resemble organic growth in nature. Moreover, as the construction of natural beings, software bears consideration as a product of nature. Copying time-honored designs that naturally arise in one environment eases the creation of designs that feel natural in another.

4.2 Foundations

This section sketches the evolution of patterns research from its origins in building architecture through the identification of patterns in software architecture to their recent penetration into scientific software architecture.

4.2.1 Building Architecture

As the quote at the beginning of this chapter suggests, patterns hail from time immemorial. Their modern identification and labeling began with architecture professor Christopher Alexander[1] and his collaborators at the Center or Environmental Structure three-volume series: Volume 1, *The Timeless Way of Building* (Alexander 1979); Volume 2, *A Pattern Language: Towns, Buildings, Construction* (Alexander et al. 1977); and Volume 3, *The Oregon Experiment* (Alexander, Silverstein, Angel, Ishikawa, and Abrams 1975).

Opposite to the numbering of the volumes, the chronological order of publication mirrors the progression outlined in Section 4.1: Practice precedes patterns precedes theory. *The Oregon Experiment* describes an architectural planning project conducted for the University of Oregon. As such, it shows an actual implementation of the patterns and theory in the other two volumes. *A Pattern Language* describes more than 250 archetypal patterns – patterns with lasting value, as likely to be used centuries hence as centuries prior. *The Timeless Way of Building* (hereafter, *Timeless Way*) puts forth a theoretical framework and fleshes out and contextualizes the pattern concept.

Timeless Way is at once poetic and philosophical. Alexander intersperses short, italicized paragraphs with the body of the text. He refers to these as "headlines." He recommends one method of skimming the book by reading each chapter's start and finish along with the headlines in between. Read in this manner, the expressiveness and brevity of the headlines sometimes give them the feeling of the stanzas of a poem.

The book's emphasis on "the way" evokes Taoism. Its occasional use of seeming contradictions feels reminiscent of Zen Buddhism, wherein *koans* provoke the reader to reconcile paradoxes as in the famous koan: "Two hands clap and there is a sound; what is the sound of one hand?"[2] Likewise, in order to purge ourselves of the constrictions imposed by other design methods, Alexander urges the reader to learn a new discipline only to give it up:

> To purge ourselves of these illusions, to become free of all the artificial images of order which distort the nature that is in us, we must first learn a discipline which teaches us the true relationship between ourselves and our surroundings.
>
> Thence, once this discipline has done its work, and pricked the bubbles of illusion which we cling to now, we will be ready to give up the discipline, and act as nature does.

This is immediately reminiscent of another Zen Buddhist concept: *shoshin*, or beginner's mind, an openness to possibilities that might seem closed off to an expert schooled in the ways of the field.

Timeless Way concerns itself with a process for releasing one's innate understanding of an unnamed quality that breathes life into buildings and towns and makes them whole. That process starts with observing patterns of events that occur in these places. As these patterns unfold over time, so do they in space. Their spatial characteristics define characteristic geometries that designers and inhabitants can weave together into a complete building or town. This dialogue between designers and inhabitants is

[1] The first recipient of a Ph.D. in architecture from Harvard University, Alexander was also awarded the first Gold Medal for Research by the American Institute of Architects and is now Professor Emeritus of Architecture at the University of California, Berkeley.

[2] Oral tradition attributes this koan to Hakuin Ekaku, 1686–1769.

one of the hallmarks of the timeless way. It is a dialogue spoken in pattern languages. Because each person possesses a pattern language unique to the events and cycles of their own life, facilitating mutual understanding necessitates constructing a common pattern language.

Several aspects of Alexander's writing strike chords that resonate across many schools of philosophical and scientific thought, Eastern and Western, ancient and modern. In stating, "To seek the timeless way, we must first know the quality without a name," he acknowledges that fundamental truths evade precise articulation in everyday language. A second universal theme is Alexander's useage of dialectics: He maintains that vibrant architecture can only emerge from an interplay of ideas between its designers and its intended inhabitants. Pattern languages give the interested parties a set of design elements along with rules for combining those elements to extract designs from a combinatorial space of possibilities. Languages allow infinite variation in many of the ultimate details while still maintaining thematic consistency. For instance, *Timeless Way* outlines a pattern language for stone houses in the South of Italy and sketches several plans that could be instantiated from the language.

In addition to identifying the fundamental nature of patterns and their importance, Alexander postulates that the distinction between living and dead patterns lies in the extent to which everyone shares an understanding of them. In cultures where the patterns are alive, all people share a basic understanding of building. Such was the case in agricultural societies. Farmers felt comfortable building barns. Home owners felt comfortable laying out their homes. While this might seem farfetched in modern times, Alexander points out that as recently as fifty years ago in Japan, for example, every child learned how to lay out a house.

By contrast, the specialization required in industrial societies tends to kill patterns:

> Even within any one profession, professional jealousy keeps people from sharing their pattern languages. Architects, like chefs, jealously guard their recipes so that they can maintain a unique style to sell.
>
> The languages start out by being specialized, and hidden from the people; and then within the specialties, the languages become more private still, and hidden from one another and fragmented.

In such settings, so much wisdom fails to propagate down through the generations that even the specialists begin to make obvious mistakes. Alexander gives examples from his campus where two seminar rooms lack the geometry and lighting to foster vibrant, comfortable discussions. The architects who designed the room forgot several basic rules of thumb:

- Rooms need natural light from two directions.
- Windows too low create silhouettes of occupants as viewed by other occupants.
- A reflective display surface orthogonal to an adjacent window reflects the window's view into its viewers' eyes.

One modern response to this situation involves centralized control over urban design. Centralization, however, fails to generate vibrant buildings and towns due to its insufficient responsiveness to people's needs. It exacerbates the very problem it seeks to solve.

Timeless Way defines architectural design patterns in terms of the context within which they occur, a conflict that arises in that context, and the way in which the pattern resolves the conflict. A sample context→conflict→resolution triplet reads "communal room→conflict between privacy and community→ alcove opening off communal room." This three-part logic imparts a degree of empiricism: One can verify whether the context exists, whether the conflict arises, and whether the pattern resolves the conflict. Alexander asserts that this empiricism moves the statement that a pattern is "alive" from the realm of taste to the realm of objective verifiability. Alexander's final step in defining a pattern is naming it. Thus, the aforementioned triplet becomes the ALCOVE pattern.

Whereas patterns share their empirical nature with physics, they share another quality with computer science: invariance. As Chapter 10 describes, the formal methods subdiscipline of computer science deals in part with logical statements that must be true throughout the execution of an algorithm. These are invariants of the algorithm. The search for patterns represents a search for invariant aspects of designs that can be observed across all valid implementations of the pattern.

Timeless Way advocates precision in specifying patterns. An ENTRANCE TRANSITION pattern describes the junction at which the indoors meets the outdoors. Examples of precision in defining this pattern include stipulations that doors be 20 feet from the street, that windows facing the transition region not be visible from the street, that the character of surfaces differ from that in the adjoining spaces, and that it include a glimpse of something entirely hidden from the street. These stipulations imbue patterns with a property many consider to be among the most fundamental in all of science: falsifiability. Their precision opens them up to challenge.

The book also acknowledges the elusive nature of even some precisely defined concepts. In one sense, an entrance transition never exists in its own right but merely describes a relationship between two other entities: indoors and outdoors. Specifically, it describes their intersection. In fact, the transition described might even occur fully indoors. This points up a paradox Alexander identifies in the goal of making designs whole. Wholeness implies limits and self-containment, and yet patterns might exist solely as relationships or as characteristics of a particular part of a design, say, the part of the indoors closest to the outdoors.

Lest patterns appear overly formal and sterile, *Timeless Way* connects them intimately with the realm of social science. The ultimate test of a pattern comes in the way it makes occupants feel. This exposes one of the reasons patterns vary among cultures. Nonetheless, within the appropriate culture or subculture, Alexander reports a remarkably high degree of unanimity about the feelings patterns evoke in people.

Not every concept from *Timeless Way* maps neatly or obviously to software. This seems particularly true in its discussion of construction. Nonetheless, contemplating the relationship between the two fields generates useful insights. Consider Alexander's comparison of the emergence of a design to the growth of an organism from an undifferentiated clump of cells. Rather than stitching together preformed parts, the organism comprises a single, whole entity throughout its maturation. Different organs arise in constant contact with each other. Though infinitely more complex, this process seems reminiscent of a technique the current authors employ in program construction: We replicate a skeleton ADT multiple times before differentiating it for the specific role it will ultimately play. In Fortran, such a skeleton might take the form:

```
module skeleton_ADT       ! Manage entity scope/lifetime
  use    ,only :          ! Surgical use of module entities
  implicit none           ! Prevent implicit typing
  private                 ! Hide everything by default
  public :: skeleton      ! Expose type/constructor/methods
  type skeleton
    private               ! Hide data
    ! derived type components
  contains
    procedure ::          ! Public methods
    final :: destructor ! Free pointer-allocated memory
  end type
  interface skeleton      ! Generic constructor name
    procedure constructor
  end interface
contains
  type(skeleton) function constructor()
    ! return new instance
  end function
  elemental subroutine destructor(this)
    type(skeleton) ,intent(inout) :: this
    ! deallocate any allocated pointer components
  end subroutine
end module skeleton_ADT
```

Such a skeleton is not viable in the sense that it cannot be compiled. It simply provides a scaffolding containing the minimum number of elements required for a well-designed ADT. These include the enforcement of explicit typing, default privacy, modular scoping, and memory management. What it does not contain is the equivalent of DNA molecules driving a process of self-organization. Only the programmer can drive that process.

The `skeleton` ADT contains this book's first `elemental` procedure. Fortran requires rank and parameter matches for automatic invocation of final subroutines. If a program instantiates an array of `skeleton` objects, `destructor` will only be called if a final subroutine explicitly accepts an array argument of matching rank or if `destructor` is `elemental` and therefore supports all ranks.

Timeless Way concludes by providing "the kernel of the way." True to form, it does so by pointing out a paradox: By enforcing a discipline based on reconnecting your design to its living purposes, a pattern language frees you from its very own strictures. Once you have mastered it, you no longer need it. You can create beautiful designs by relying on your innate skills now that you have learned to trust them. Alexander refers to this state of mind as "egoless."

Moving from *Timeless Way* to the second volume, *A Pattern Language*, one finds fewer specifics of direct utility in software. Yet the structure and manner in which one uses the book carries over. In *A Pattern Language*, Alexander et al. (1977) catalog

253 patterns and order them from the largest to the smallest. Each intermediate pattern helps complete larger patterns that come before it. Each is itself completed by smaller patterns that come after it. The progression of patterns goes from regions and towns to clusters of buildings, buildings themselves, building subunits, and construction details.

Each pattern comprises, in order:

- A picture of an archetypal example of the pattern.
- A paragraph that contextualizes the pattern within larger patterns.
- A one- to two-sentence headline summarizing the pattern's essence.
- The body of the pattern, detailing its empirical background, evidence of its validity as a solution, and its range of manifestations.
- The solution, including the physical and social relationships required to solve the problem.
- A diagram.
- A list of smaller patterns required to complete or enhance the current pattern.

Thus, *A Pattern Language* fleshes out each pattern rather completely, providing much of the information required of a trusted manual for design and construction.

Alexander and his colleagues suggest a methodology for reading and using the book in an actual project. Scanning the master sequence of patterns, one finds and reads the largest pattern describing the overall project scope. One then marks on the master list those smaller patterns delineated at the end of this largest pattern. Next, one reads the smaller patterns and progressively marks the even smaller patterns they list. This cascade generates a network of patterns. To prevent unmanageable growth of this network, the authors suggest skipping the smaller patterns that seem of doubtful value. However, they suggest inserting your own patterns where no appropriate one exists in the language.

Alexander et al. (1977) tag their patterns in a way that distinguishes their certainty in each pattern's capturing of an invariant – that is a property common to all possible solutions of the problem the pattern addresses. In one set, they believe they have found an inescapable feature of all well-formed solutions. In a second set, they believe the pattern gets us closer to this inescapable set but that other possible solutions almost certainly exist. Finally, in a third set, they express certainty that they have not found an invariant. They also suggest modifying patterns to suit your purposes. The resulting pattern network represents a pattern language unique to your project.

Interestingly, the frontmatter in *A Pattern Language* makes the aforementioned connection to poetry concrete. In a section entitled THE POETRY OF THE LANGUAGE, Alexander and colleagues distinguish poetry from prose by the density of interpretations each writing style affords. Prosaic English most often conveys one, straightforward meaning. English poems, by contrast, harbor manifold interlocking messages even in single sentences or words. The authors argue that a building that strings together loosely connected patterns resembles prose whereas one that overlaps multiple patterns in space captures many interpretations in its nooks. Such density engenders profundity. As the authors state, compressing many patterns into the smallest possible space encourages economy and leads to "buildings which are poems."

The third volume, *The Oregon Experiment*, illustrates the application of the theory expounded by *Timeless Way* and the structural template laid out in *A Pattern Language* to a specific architectural project at the University of Oregon. It would at first seem that the more concrete nature of building design and construction would lend itself to fewer direct comparisons with software design and construction. Yet the chief force driving the adoption of patterns in one setting mirrors that in the other: managing growth. *The Oregon Experiment* arose from the crisis brought about by rapid growth over a single decade from a century-long size of a few thousand students to a student body of 15,000. Recognizing that computer programs experienced a contemporaneous period of rapid increase in scale, leading to what is often termed "the software crisis," it starts to seem less farfetched that the work of Alexander and colleagues ultimately influenced software design.

The Oregon Experiment comprises six chapters, each expounding a single principle the authors believe "will probably have to be followed in all communities where people seek comparably human and organic results." The six principles comprise organic order, participation, piecemeal growth, patterns, diagnosis, and coordination. The authors set the principle of organic order up in opposition to the totalitarian rule imposed by traditional master plans. Such plans attempt to map out the organization of a site envisioned for the distant future. Organic order relies instead on a pattern language adopted by thousands of local projects meeting immediate needs. A critical driver of organic order is a planning board small enough to keep meetings focused and balanced enough between users and decision makers such that real, informed decision making happens in the room.

With the case for user participation already detailed in the first two volumes, *The Oregon Experiment* focuses on a specific example of user participation in designing a set of new buildings for the School of Music. The embodiment of user participation takes the form of a user design group. This group drafts pattern-inspired design schematics before architects enter the project.

Piecemeal growth harbors two important implications. First, it acknowledges the need for constant repair of existing structures as part of the growth process. Second, it suggests new pieces be added slowly, in small increments over time. The authors suggest enforcing piecemeal growth by distributing a capital budget evenly across budget categories. With the same amount of funds allocated for small projects in aggregate as for large projects in aggregate, there will be many more small projects than large ones.

Of the 250 patterns in *A Pattern Language*, *The Oregon Experiment* cites 200 relevant to the university. Of these, they find 160 of appropriate scope for individual user groups to employ – those dealing with building rooms and gardens, for example – while 37 are of sufficiently large scope they require university-wide, multi-project agreement to complete. These include BIKE PATHS AND RACKS and PROMENADE patterns, for example. They also identify 18 domain-specific patterns of special interest to universities, for example, CLASSROOM DISTRIBUTION and REAL LEARNING IN CAFES patterns. In keeping with the empirical philosophy of *Timeless Way*, the authors suggest planning boards adopt new patterns or revise old patterns only when the patterns have experimental or observational support.

Diagnosis refers to an annual process of reevaluating which spaces are alive and which are dead. As with the timeless nature of patterns, diagnosis has deep roots. Alexander and colleagues provide evidence of annual surveys of 13th-century Siena, Italy. Laypersons, as opposed to building specialists, populated the survey committee. The committee drafted plans for specific projects aimed at improving important public works.

Coordination happens when the planning board considers newly proposed projects with explicitly articulated relationships to the currently adopted patterns and diagnosis. The board prioritizes the submitted projects, judges them by the extent of their conformance to the current patterns and diagnosis, and partitions the list so that each project competes only with those of a similar size.

As these concepts apply to scientific software, the organic growth and user participation principles suggest scientific software projects might most often succeed when the code designers are embedded in the organizations that will ultimately use the code. Piecemeal growth acknowledges the importance of ongoing maintenance and incremental expansion. The necessity for domain-specific patterns suggests tailoring some abstractions to the scientific domain of interest. The need for diagnosis encourages gathering users annually to inform the global view of the health of various parts of the software. We have observed these practices across several of the best-run scientific software projects. The coordination mechanism, however, frequently differs depending on the extent to which funding is central or distributed.

4.2.2 Software Architecture

Although University of Oregon alumnus Kent Beck and coauthor Ward Cunningham introduced the design pattern concept to the computer science community in 1987 (Beck and Cunningham 1987), the idea first gained a substantial following with the publication of *Design Patterns: Elements of Reusable Object-Oriented Software* by Gamma et al. (1995), often referred to as the Gang of Four or GoF. The first part of the GoF book defines "design pattern" in the context of software and demonstrates the utility of patterns via a case study. The second part catalogs general design patterns of three varieties: creational, structural, and behavioral. Creational patterns deal with object construction. Structural patterns describe the composition of collections of classes with the chief compositional mechanism being inheritance. Behavioral patterns describe class or object interactions and responsibilities.

Much as Alexander and his colleagues did, the GoF employ a literary analogy for their work, noting:

> Novelists and playwrights rarely design their plots from scratch. Instead, they follow patterns like "Tragically Flawed Hero" (Macbeth, Hamlet, etc.) or "The Romantic Novel" (countless romance novels).

As they point out, object-oriented software designers follow patterns as well. Much like common literary themes, different patterns arise in different genres or domains. The GoF explained that their book only captures a fraction of the likely knowledge of expert designers. They made clear they were not attempting to present patterns for concurrency or distributed programming, for example; nor did they cover patterns specific to particular application domains such as user interfaces, device drivers, or databases.

Inspired by Alexander, the GoF draw analogies between software architecture and building architecture, comparing objects and interfaces to walls and doors. Expanding on the three-part rule, the GoF include several additional features of the patterns they present, including:

- Pattern Name and Classification: the name captures the essence of the pattern and becomes part of the design lexicon, while the classification specifies the pattern's role (creational, structural or behavioral) and scope (whether they determine static, compile-time class relationships or dynamic, run-time object relationships):
- Intent: what a pattern does, its rationale, and the problem it solves.
- Also Known As: pattern aliases.
- Motivation: an archetypal scenario.
- Applicability: situational context.
- Structure: schematic descriptions of class relationships and interactions.
- Participants: classes or objects involved.
- Collaborations: the roles of the participating classes.
- Consequences: design trade-offs.
- Implementation: issues that arise in constructing the solution.
- Sample Code: illustrative code fragments.
- Known Uses: examples from real systems.
- Related Patterns: differences from, and relationships to, other patterns.

While we do not adopt the GoF's complete structure for describing patterns, we include various elements of this structure in our pattern descriptions.

The GoF identified the key drivers that complicate OOD: encapsulation, granularity, dependency, flexibility, performance, evolution, and reusability. They aptly pointed out that these drivers often oppose each other, sometimes in complicated multidimensional ways. The desire to reduce dependencies between abstractions complicates encapsulation. Performance demands often discourage high levels of granularity in scientific applications where efficiency necessitates long loops over large data sets. Frequent reuse of a given class early in the evolution of a package might inhibit the package's flexibility in later generations.

Two guiding principles reverberate throughout the GoF patterns:

1. Favor object composition over class inheritance.
2. Program to an interface, not an implementation.

Section 3.3 discusses the first principle. Illustrating the second principle requires introducing a new concept: an abstract class. An abstract class defines the interface that its child classes must implement. No instances of an abstract class can be instantiated. In its purest form, an abstract class contains no data and defers the implementation of all its methods to its child classes. Any class that is nonabstract is termed *concrete*.

In Fortran and C++, the language mechanisms typically employed to relate abstract classes to concrete ones match those employed to relate parent classes to their children: type extension in Fortran and inheritance in C++. Whereas a concrete child class inherits from its parent class any state and behavior that it does not specifically override, no such state and behavior exist if the parent class encapsulates no data and provides no default procedure definitions. This is the case for pure abstract classes

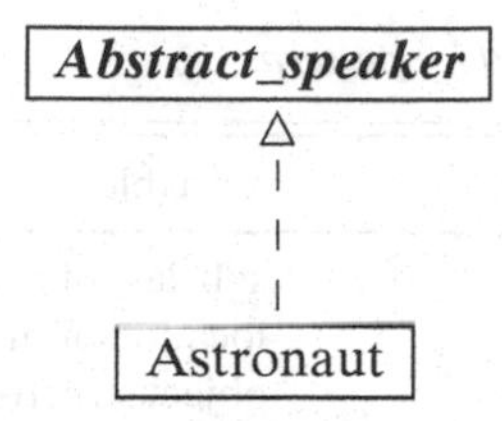

Figure 4.1. Abstract class realization: Concrete Astronaut realizes, or implements, an Abstract_speaker interface.

(C++ *pure virtual classes*). In such a case, one more naturally refers to the child class as *implementing* or *realizing* the interface defined by the abstract parent. UML models a realization relationship with a dashed line adorned by an open triangle at the interface end, as demonstrated in Figure 4.1. Furthermore, UML denotes abstract class names in bold italics.

Fortran abstract types facilitate creating abstract classes. Figure 4.2 demonstrates the definition of an abstract type containing no components, and specifying a deferred binding `speak`, leaving the implementation of the corresponding method to extended types. It constrains those implementations to conform to the abstract interface `talk`. Figure 4.2 further depicts the extension of the `abstract_speaker` type to create an `astronaut` with the same purpose as the like-named type in Figure 2.4.

The construction of an abstract type as a base for extension in Figure 4.2(a) does not enhance the concrete type implemented in Figure 4.2(b). It enhances the `say_something()` client code in Figure 4.2(c), enabling it to be written using the abstract type with knowledge of neither what concrete type actually gets passed at runtime nor what procedure actually gets invoked on that type. At runtime, any concrete type that extends `abstract_speaker` can be passed to `say_something`. This flexibility is a hallmark of design patterns and explains why abstract class construction plays a recurring role across many patterns.

Neither Fortran nor C++ formally defines class interfaces in a way that decouples this concept from a class's implementation. In both languages, however, one can approximate the class interface concept with abstract classes. The GoF advocate that all extensions of an abstract class merely define or redefine the abstract methods of their parent abstract class without defining any new methods. All child classes so defined share an identical interface. This enables the writing of client code that depends only on the interface, not on the implementation.

As the GoF pointed out, one must instantiate concrete classes eventually. Creational patterns do so by associating an interface with its implementation during instantiation, thereby abstracting the object creation. We describe a creational pattern, FACTORY, in Chapter 9. Other GoF patterns we have found useful in our research include the behavioral patterns STRATEGY and TEMPLATE METHOD, and the structural pattern FACADE. Each of these is the subject of, or inspiration for, a section in Part II of the current text. FACTORY METHOD encapsulates object construction for a family of subclasses. STRATEGY represents a family of interchangeable algorithms. FACADE presents a unified interface to a complicated subsystem. TEMPLATE METHOD specifies an algorithm in skeleton form, leaving child classes to implement many of the algorithm's steps.

Table 4.1. *Design variables supported by patterns*

Pattern	Variable
Factory Method	subclass of instantiated object
Proxy	location or method of accessing an object
Composite	object structure and composition
Facade	interface to a subsystem
Iterator	how an aggregate's elements are accessed
Mediator	how and which objects interact with each other
Strategy	an algorithm
Template Method	steps of an algorithm

The essence of a pattern often lies in the flexibility it facilitates by enabling variation in design elements. The GoF tabulated the design variations supported by each of their patterns. Table 4.1 excerpts the list of GoF patterns that appear in or inspire patterns in the current text.

The GoF outline how this flexibility enhances software applications, toolkits, and frameworks. One might define an application as any collection of code that includes a main program. Applications benefit from the increased internal reuse, maintainability, and extensibility that design patterns foster. The GoF define toolkits as useful classes that user codes call, that is, the OOP counterpart to conventional subroutine libraries. Toolkits benefit from the loose coupling that design patterns encourage. Such loose coupling makes toolkits more reusable across disparate applications. The GoF define frameworks as domain-specific system architectures that call user code. Frameworks benefit from the increased consistency and clarity exhibited by pattern-based project.

Part I of the current text demonstrates application design and construction, focusing primarily on heat conduction. Refactoring that application in terms of design patterns would allow for varying the numerical methods or the objects that encapsulate those methods or those objects' properties, including their external view and internal structure.

Part II corresponds closely to toolkit design and construction. The designs presented herein can be reused across a plethora of applications. As a natural analog to library development, for example, the designs in Chapters 6–7 could be reused to build a numerical quadrature library.

Part III outlines framework development. Chapter 11 outlines the development of Fortran interfaces to a C++ solver framework. Chapter 12 sketches the design of the Morfeus multiphysics Fortran framework.

Figure 4.2(a)

```
1  module speaker_class
2    implicit none
3    private
4    public :: say_something
5
6    type ,abstract ,public :: speaker
7    contains
```

```
 8      procedure(talk) ,deferred :: speak
 9    end type
10
11    abstract interface
12      function talk(this) result(message)
13        import :: speaker
14        class(speaker) ,intent(in) :: this
15        character(:) ,allocatable  :: message
16      end function
17    end interface
18
19  contains
20
21    subroutine say_something(somebody)
22      class(speaker) :: somebody
23      print *,somebody%speak()
24    end subroutine
25
26  end module
```

Figure 4.2(b)

```
 1  module astronaut_class
 2    use speaker_class ,only : speaker
 3    implicit none
 4    private
 5    public :: astronaut
 6
 7    type ,extends(speaker) :: astronaut
 8      private
 9      character(:) ,allocatable :: greeting
10    contains
11      procedure :: speak => greet
12    end type
13
14    interface astronaut  ! Map generic to actual name
15      procedure constructor
16    end interface
17  contains
18    function constructor(new_greeting) result(new_astronaut)
19      character(len=*), intent(in) :: new_greeting
20      type(astronaut) :: new_astronaut
21      new_astronaut%greeting = new_greeting
22    end function
23
24    function greet(this) result(message)
25      class(astronaut) ,intent(in) :: this
26      character(:) ,allocatable :: message
27      message = this%greeting
28    end function
29  end module
```

Figure 4.2(c)

```
 1  program abstract_hello_world
 2    use astronaut_class ,only : astronaut
```

```
3    use speaker_class ,only : say_something
4    implicit none
5    type(astronaut) :: pilot
6    pilot = astronaut('Hello, world!')
7    call say_something(pilot)
8  end program
```

Figure 4.2. Abstract class: (a) abstract speaker, (b) extended type, and (c) driver.

4.2.3 Scientific Software Architecture

Object-oriented design patterns apparently first entered in the scientific software literature early in the 21st century (Blilie 2002). Gardner and Manduchi (2007) published the first book-form coverage of applying OOD patterns in scientific research.[3] Their book covers a case study on the design and construction of an open-source "waveform browser" for use in data acquisition system on plasma fusion experiments.

Gardner and Manduchi implement their designs in Java to leverage the Java Virtual Machine's platform-independence and Java's explicit support for parallelism. Part 1 of their book outlines a first implementation of a waveform browser dubbed PreEScope. Part 2 refactors PreEScope into a pattern-based EScope, itself a simplified version of the 30,000-line, public-domain jScope that is in production use at several plasma fusion laboratories.

Gardner and Manduchi design a FACADE that provides a unified interface to two subsystems in EScope's graphical user interface (GUI). One subsystem handles EScope's display and control. The other interacts with the data server. The facade has two parts: an abstract facade class that client code uses, and a concrete facade class that interacts with the subsystems. The concrete facade fulfills requests the client submits via the abstract facade.

Gardner and Manduchi employ a MEDIATOR to communicate information between classes within the subsystems. Doing so eliminates the need to trace complicated linkages between classes to understand the communication patterns. All information passes through the mediator.

Gardner and Manduchi use the TEMPLATE pattern to express in a parent superclass those parts of an algorithm that are common to a collection of subclasses. The superclass delegates the subclass-specific portions of the algorithm to the subclasses. They use a TEMPLATE to separate the classes that deal with drawing various parts of graphs: the waveform, the axes, the labels, and any messages.

Gardner and Manduchi discuss a collection of creational patterns under the rubric "Factory patterns." These include FACTORY METHOD, ABSTRACT FACTORY, BUILDER, PROTOTYPE, and SINGLETON. Chapter 9 of the current text describes how

[3] Given the importance of parallel programming in high-performance scientific computing, it is tempting to give precedence to *Patterns for Parallel Programming* (Mattson, Sanders, and Massingill 2005). Although most of their examples are scientific, they aim to capitalize on the industry-wide trend toward parallelism and to impact programming practice beyond science. Furthermore, they do not limit themselves to a specific programming paradigm. As such, their use of the term "pattern" refers more broadly to recurring themes in program design independent of whether those programs are object-oriented.

to use a FACTORY METHOD to add flexibility to the spatial differencing scheme incorporated into our heat equation solvers in Part I.

Each of the previous mentioned patterns covered by Gardner and Manduchi has an analog in the current text. The ABSTRACT CALCULUS pattern of Chapter 6 functions as a domain-specific facade. Our implementations of the STRATEGY pattern in Chapter 7 effectively embed TEMPLATE patterns. Our PUPPETEER pattern plays the role of a MEDIATOR.

Other GoF patterns that Gardner and Manduchi cover include DECORATOR, ADAPTER, OBSERVER, ADAPTER, STATE, CHAIN OF RESPONSIBILITY, and ABSTRACT FACTORY. They also propose one new pattern as a variation on a GoF pattern useful in their application domain: ARTICULATED FACADE. Finally, they touch on concurrent programming patterns that arose after the GoF's seminal text.

In terms of presenting design patterns in scientific programming, the book by Gardner and Manduchi sits closer to the current book than any other of which we are aware. The application they study, however, differs considerably from that of prime interest herein: multiphysics simulation. Moreover, their language selection precludes the core technology we feature: ADT calculus. Java neither allows the overloading of existing operators, as Fortran and C++ do, nor the definition of new operators, as Fortran allows. Although this makes Java unsuitable for current purposes, the language has much to recommend it. Java stands as the only one of the three languages that was object-oriented, parallel, and distributed from conception. It therefore conforms much more closely with OOP philosphy. Nearly every data entity in Java is an object, and even the simplest "Hello, world!" program is inherently object-oriented. Furthermore, the multithreading capabilities of Java long preceded the expected entry of similar facilities in the upcoming Fortran 2008 and C++0x standards. The network socket model intrinsic to Java lacks standard counterparts for distributed programming in Fortran and C++. Finally, the garbage collection for all dynamic memory eliminates one of the most vexing challenges encountered by C++ programmers and, to some extent, by Fortran programmers.

There's a There There

Writing scientific software traditionally comprised no more than expressing algorithms in code. Algorithms describe a code's temporal structure, saying nothing about its spatial structure. Without a notion of how to locate oneself spatially relative to other parts of the code,"there's no there there" in the words of author Gertrude Stein. Thinking architecturally leads to representations of the code layout in terms of class and object relationships. Design patterns provide a common language for describing the resulting spatial structures along with behavioral and creational structures. This school of thought took root in the building architecture community before making the leap to software architects and finally to scientific software architecture.

4.3 Canonical Contexts

When the remaining chapters of Part II refer to "the problem," they generally refer to a dilemma that arises in designing software to embody a mathematical model. From a design pattern viewpoint, the mathematical problem statement itself is simply the context within which the software design problem occurs. Defining a set of canonical contexts here facilitates focusing the rest of Part II more tightly on the software problems, their solutions, and their consequences.

A desirable canon would contain contexts that are simple enough to avoid distracting from the software issues yet complex enough to serve as a proxy for the contexts commonly encountered in scientific research. The chosen ones involve solving coupled ordinary differential equations, coupled integro-differential equations, and a partial differential equation. All exhibit nonlinearity. In contexts involving temporal dependence, the time integration algorithms applied in these contexts range from straightforward explicit Runge-Kutta schemes to multistep, semi-implicit, and fully implicit schemes. In contexts involving spatial dependence, the finite difference methods employed likewise range from the explicit and straightforward to the implicit and more complicated. Three important complications missing from our canonical contexts are addressed in Part III of this text: complicated geometry and boundary conditions and runtime scalability.

Section 4.3.1 provides the context for Chapters 6–8. Section 4.3.2 provides the context for Chapter 5. Section 4.3.3 provides the context for Chapter 9.

4.3.1 The Lorenz Equations: A Chaotic Dynamical System

The software complexity issues that patterns address become most acute as a package scales up to large numbers of classes. In scientific simulation code, this increase in scale often involves adding new physics. Starting from a classical, Navier-Stokes turbulence simulation package, one might add particle tracking to study multiphase flow, magnetic field transport to study magnetohydrodynamics (MHD), scalar fields to study molecular diffusion, or quantum vortices to study superfluidity. Each of these topics forms a case study in Chapter 12 in Part III.

In most instances, the resulting multiphysics package greatly exceeds the size desirable for a sample pattern demonstration. We therefore choose a very simple mathematical model that can be formulated in a way reminiscent of a multiphyics model. We defer discussion of true multiphysics solvers to Part III. The current section aims to describe the mathematical motivation and to provide some basic insights into its dynamics.

One can describe a broad swath of multiphysics phenomena with equations of the form:

$$\frac{\partial}{\partial t}\vec{U}(\vec{x},t) = \vec{F}(\vec{U}(\vec{x},t)), \;\; \vec{x} \in \Omega, \;\; t \in (0,T] \tag{4.1}$$

where $\vec{U} \equiv \{U_1, U_2, ..., U_n\}^T$ is the problem state vector; $\vec{x}$ and t are coordinates in the space-time domain $\Omega \times (0,T]$; and $\vec{F} \equiv \{F_1, F_2, ..., F_n\}^T$ is a vector-valued operator that couples the state vector components via a set of governing ordinary-, partial-, or integro-differential equations. Closing the equation set requires specifying appropriate boundary and initial conditions:

$$\vec{B}(\vec{U}) = \vec{C}(\vec{x},t), \;\; \vec{x} \in \Gamma \tag{4.2}$$

$$\vec{U}(\vec{x},0) = \vec{U}_0(\vec{x}), \;\; \vec{x} \in \Omega \tag{4.3}$$

where Γ bounds Ω, $\vec{B}(\vec{U})$ typically represents linear or nonlinear combinations of $\vec{U}$ and its derivatives, and where $\vec{C}$ specifies the values of those combinations on Γ.

A common step in solving equation (4.1) involves rendering its right-hand side discrete by projecting the solution onto a finite set of trial-basis functions or by replacing all spatial differential operators in $\vec{F}$ by finite difference operators and all spatial integral operators in $\vec{F}$ by numerical quadratures. Often, one also integrates equation (4.1) against a finite set of test functions. Either process can render the spatial variation of the solution discrete while retaining its continuous dependence on time. One commonly refers to such schema as semidiscrete approaches. The resulting equations take the form:

$$\frac{d}{dt}\vec{V(t)} = \vec{R}(\vec{V}(t)) \;\; t \in (0,T] \tag{4.4}$$

where the right-hand side vector function $\vec{R}$ contains linear and nonlinear discrete operators and where $\vec{V} \equiv \{V_1, V_2, ..., V_m\}^T$ might represent $m = np$ samples of the n elements of $\vec{U}$ on a p-point grid laid over $\Omega \cup \Gamma$.

Alternatively, $\vec{V}$ could represent expansion coefficients, for example, Fourier coefficients, obtained from projecting the solution onto the aforementioned space of trial functions, such as complex exponentials. In any case, we can now assume that the boundary conditions have been incorporated into equation (4.4) and no longer need to be specified separately as in equation (4.2). This allows us to focus on equation (4.4) as a self-contained, continuous dynamical system. For most of Part II, we ignore the original spatial dependence in equation (4.1) and use a specific dynamical system as our prototypical multiphysics model: the Lorenz system (Lorenz 1963):

$$\frac{d}{dt}\left\{\begin{array}{c} v_1 \\ v_2 \\ v_3 \end{array}\right\} = \left\{\begin{array}{c} \sigma\,(v_2 - v_1) \\ v_1\,(\rho - v_3) - v_2 \\ v_1 v_2 - \beta v_3 \end{array}\right\}. \tag{4.5}$$

We return to the issue of spatial dependence in Chapters 9 and 12.

The Lorenz equations retain some of the challenges posed by more complicated equation sets: coupling terms, nonlinearities, and complex behavior. In certain regions of the σ-ρ-β parameter space, the Lorenz equations exhibit chaotic dynamics. In subsequent chapters, we select the chaos-inducing values $\sigma = 10$, $\rho = 28$, and $\beta = 8/3$. This provides a stringent test for codes in that the extreme sensitivity to initial conditions characteristic of chaos magnifies numerical inaccuracies over time. One often refers to such extreme sensitivity to minor perturbations as the *butterfly effect*, wherein a butterfly in India flaps its wings and causes a hurricane in Florida.

The path by which Lorenz arrived at equation system (4.5) corresponds to a simplified semidiscretization of a set of PDEs governing forced, dissipative fluid flow in a horizontal layer of uniform depth with a fixed temperature difference between its top and bottom. All motion is assumed to be in horizontal planes. Lorenz's discretization process involved Fourier approximations to the solution of the original PDE. The dependent variables v_1, v_2, and v_3 are the Fourier coefficients retained after truncating the Fourier series to three terms. In particular, v_1 measures the strength of the resulting convective motions, v_2 varies proportionately with the overall temperature difference, and v_3 varies proportionately with the vertical temperature profile

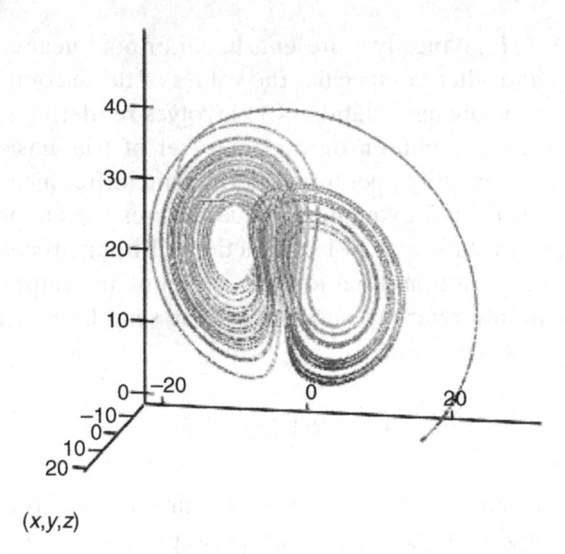

Figure 4.3. Phase-space trajectory of the Lorenz equation system solution.

from linearity. The σ parameter characterizes the relative magnitudes of the diffusion of momentum to that of thermal energy, whereas ρ determines the stability of buoyancy-driven motions and β relates to the wavelength of the Fourier modes retained in the discretization. When $\sigma > \beta + 1$, solutions are unstable at sufficiently high ρ values (> 24.74). Such behavior indicates turbulence in the original flow problem.

Explicit Euler time advancement of the Lorenz equations takes the form:

$$\vec{V}^{n+1} = \vec{V}^n + \Delta t \vec{R}\left(\vec{V}^n\right) \tag{4.6}$$

where $\vec{V}^n$ and $\vec{V}^{n+1}$ are the solution vectors at times t_n and $t_{n+1} \equiv t_n + \Delta t$, respectively, and where $\vec{R}$ is the RHS of equation (4.5). Figure 4.3 depicts the behavior of an explicit Euler time integration of the Lorenz system in phase space – that is, in v_1-v_2-v_3 space. The butterfly-shaped geometrical object toward which the initial solution trajectory evolves is a *strange attractor* (so called because it possesses a fractional, or *fractal*, spatial dimension) characteristic of a variety of forced, nonlinear, dissipative systems. In the current case, the forcing stems from the energy input required to maintain the overall thermal gradient.

4.3.2 Quantum Vortex Dynamics in a Superfluid

The Lorenz system solution variables depend only on time. Among problems in which the solution depends also on space, some of the most difficult involve time-varying topologies. In such problems, both the geometry and connectivity of the domain evolve over time. A tangle of quantum vortices in superfluid liquid helium provides one of the simplest prototypes of such a problem because it can be accurately modeled with one-dimensional, curvilinear objects subject to deformation, tear, and reconnection.

Helium occurs naturally in two isotopes: ^{4}He and ^{3}He. Liquid ^{4}He exists in two phases: He I and He II. He I is an ordinary liquid. He II, a superfluid, forms below a critical temperature of approximately 2.17 K. Below this temperature, He II exhibits exotic behavior. This behavior includes the formation of swirling vortices, the circulation rate of which must be an integer multiple of the quantum of circulation: $\kappa \equiv h/m = 9.97 \times \text{cm}^2/\text{s}$, where h is the Planck constant and m is the mass of a helium atom.

At the core of each quantum vortex lies an evacuated region approximately 1-Angstrom across. When only the macroscopic dynamics of He II are of interest, these thin cores can be approximated very accurately as 1D filaments. They induce a superfluid velocity $\mathbf{v}(\mathbf{r},\mathbf{t})$ throughout the surrounding helium described by the Biot-Savart law:

$$\mathbf{v} = \frac{\kappa}{4\pi} \int \frac{(\mathbf{S}-\mathbf{r}) \times d\mathbf{S}}{\|\mathbf{S}-\mathbf{r}\|^3} \tag{4.7}$$

where $\mathbf{S} = \mathbf{S}(\xi, t)$ traces the vortex filament in 3D space as a function of arc-length ξ and time t, $\times$ denotes a vector product, and the integration is performed along all vortex filaments in the simulation. The superfluid velocity is thought to arise from the motion of atoms at the ground energy level. The remaining excited atoms move with a macroscopic velocity $\mathbf{u}$ that can be accurately described by classical mechanics (see section 12.2.1).

In the absence of any background superfluid motion imposed by boundary or initial conditions, the equation of motion for a point $\mathbf{S}$ on a filament is:

$$\frac{d\mathbf{S}}{dt} = \mathbf{v} - \alpha \mathbf{S}' \times (\mathbf{u}-\mathbf{v}) - \alpha' \mathbf{S}' \times [\mathbf{S}' \times (\mathbf{u}-\mathbf{v})] \tag{4.8}$$

where $\mathbf{S}' \equiv d\mathbf{S}/d\xi$ and where α and α' are temperature-dependent constants. The first term on the RHS is the total velocity induced by other points on the same filament and other filaments. At absolute-zero temperature, with all the atoms in the ground state, this would be the only term on the RHS. The first RHS term causes a circular loop, such as that of Figure 4.4, to propagate in a direction orthogonal to the plane of the loop. The second and third RHS terms model the effect of mutual friction

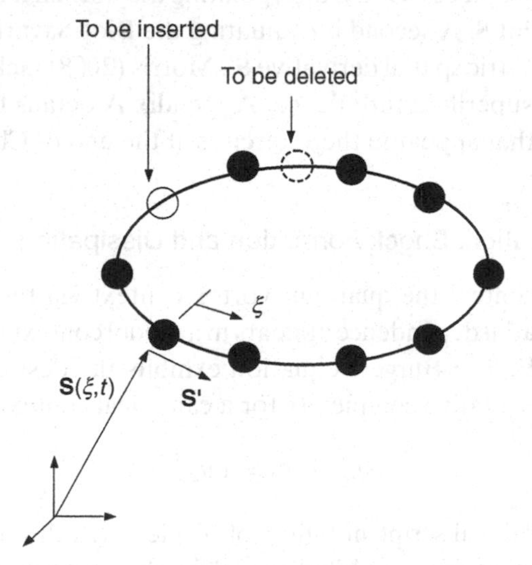

Figure 4.4. Quantum vortex filament discretization and adaptive remeshing.

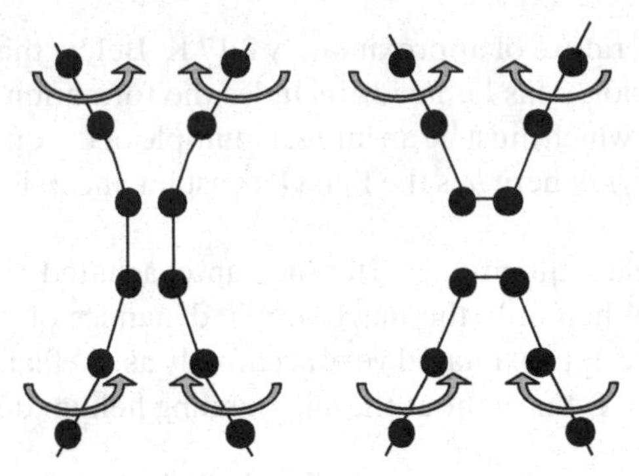

Figure 4.5. Quantum vortex filament before (left) and after (right) reconnection. Curved arrows indicate direction of circulation.

between the superfluid component and the excited atoms, which comprise a normal fluid occupying the same volume. For a circular loop, the second term causes the vortex to shrink as it propagates. For the same loop, the third term, which is typically the smallest in magnitude, imposes a drag that opposes the induced velocity.

One renders the continuous vortex core discrete by placing a series of mesh points along each filament as in Figure 4.4. Writing equations (4.7)–(4.8) for each mesh point yields a coupled set of nonlinear, integro-differential equations. The motions of these mesh points determines the filament motion and shape.

Two processes change the connectivity of the points. First, as the distance between mesh points changes, new points must be inserted and existing points removed to balance resolution requirements against computational cost and complexity (Morris 2008). Second, when two filaments approach one another, their interaction tends to draw them closer, resulting in antiparallel vortices (Schwarz 1985, 1988). Koplik and Levine (1993) showed that the filaments join and reconnect whenever two antiparallel vortex filaments approach within a few core diameters of each other, as demonstrated in Figure 4.5.

Broadly speaking, numerical solution of the discretized vortex equation system breaks down into three pieces. One is interpolating the normal fluid velocity $\mathbf{u}$ at each superfluid vortex point $\mathbf{S}$. A second is evaluating the Biot-Savart integral. A third is evaluating the parametric spatial derivative $\mathbf{S}'$. Morris (2008) tackled these problems in a dissertation on superfluid turbulence. Appendix A details those aspects of her numerical methods that appear in the Exercises at the end of Chapter 5.

4.3.3 Burgers' Equation: Shock Formation and Dissipation

Spatial dependence enters the quantum vortex context via the line integral (4.7). More commonly, spatial dependence appears in a model context in the form of spatial derivatives in a PDE. The Burgers equation exhibits the desired balance between simplicity and representative complexity for a canonical context:

$$u_t + uu_x = \nu u_{xx} \tag{4.9}$$

which we write in the subscript notation of Table 3.1. Although Burgers (1948) introduced this equation as a simplified model for fluid turbulence, in which setting

u plays a role analogous to velocity and ν to kinematic viscosity, the equation also crops up in settings ranging from superconductivity to cosmology (see Canuto et al. 2006).

Despite its apparent simplicity, solving the Burgers equation presents sufficiently vexing challenges to warrant flexibility: A numerical method that works in one regime across a given range of the parameter *nu* for a given set of initial and boundary conditions might not work well in other regimes. Great value accrues from ensuring the ability to swap numerical methods within a single solver framework. Furthermore, the dynamics can change sufficiently even within a single solution time interval to warrant dynamic adaptivity. In particular, the nonlinear second term on the LHS of equation (4.9) causes solutions to develop steep gradients characteristic of the nearly discontinuous shock waves that occur in settings as widely disparate as supernovae and rocket plumes. These might necessitate locally adaptive, shock-capturing schemes in the vicinity of the steep gradients (Berger and Colella 1989), while less demanding discretizations could be used elsewhere in space and time as the RHS term dissipates these gradients.

The perils of approximating the Burgers equation arise partly from oscillating instabilities related to bifurcations in the behavior of the discretized versions of the equation. For small ν, solutions near these bifurcations display a rich variety of behaviors. These include limit cycles (asymptotic approaches periodicity in time) and strange attractors (Maario et al. 2007).

An additional attraction to including the Burgers equation in our canon stems from its unusual status as one of the few nonlinear PDEs with known exact solutions. Hopf (1950) and Cole (1951) demonstrated that transforming the Burgers equation according to:

$$u = -2\nu\phi_x/\phi \tag{4.10}$$

converts it into the linear heat equation:

$$\phi_t = \nu\phi_{xx}. \tag{4.11}$$

which admits solutions of the form:

$$\phi(x,t) = \frac{1}{\sqrt{4\pi\nu}} \sum_{n=-\infty}^{\infty} e^{-(x-2\pi n)^2/(4\nu t)} \tag{4.12}$$

in the case of periodic boundary conditions. Figure 4.6 plots an initial condition $u(x,0) = 10\sin(x)$ along with exact and approximate periodic solutions on the interval $[0,2\pi)$. The numerical solution integrates the conservative form:

$$u_t = \nu u_{xx} - (u^2/2)_x \tag{4.13}$$

so called because discretizations of the RHS conserve the integral of $u^2/2$ over the domain when $\nu = 0$. One solution in the figure employs second-order, Runge-Kutta time advancement and sixth-order accurate Padé scheme for spatial differencing (Moin 2001). Appendix A summarizes these two algorithms, the advantage of using the high order finite difference scheme is to achieve spectral-like accuracy in space. Another solution plotted in the same figure, obtained with a second-order central difference scheme for spatial discretization, is also ploted in Figure 4.6 for

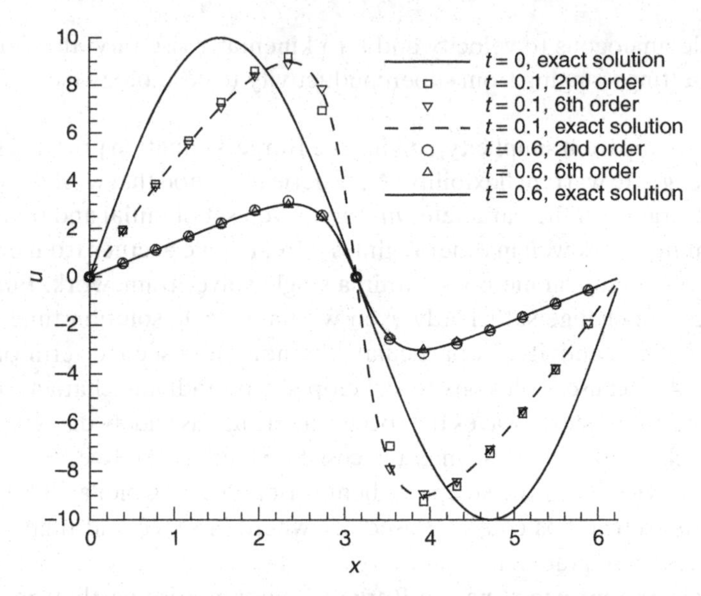

Figure 4.6. Solutions of the Burgers equation with initial condition $u = 10\sin x$ and periodic boundary conditions.

comparison. With only 16 grid points, the sixth-order Padé scheme leads to more accurate results than the second-order central difference scheme.

Yes We Canon

Simple pattern demonstrations run the risk of missing important complications found in actual scientific codes. Complicated ones run the risk of losing the forest in the trees. We have constructed a small canon of relatively simple problems that exhibit some of the chief complexities found in more complicated scientific simulations.

EXERCISES

1. Write an abstract `vortex_filament` derived type in Fortran with a deferred `remesh` binding that implements an abstract interface.
2. Write a pure virtual `VortexFilament` class derived type in C++ with a pure virtual `remesh` member function.
3. List recurring themes you see in buildings you like. Are there analogies you can make to software architecture?

5 The Object Pattern

"Memory is a crazy woman [who] hoards colored rags and throws away food."
Austin O'Malley

5.1 The Problem

Large software development efforts typically require a degree of consistency across the project to ensure that each developer follows practices consistent with the critical goals of the project. In high-performance computing (HPC), for example, Amdahl's law (Chapter 1) suggests that scaling up to the tens or hundreds of thousands of processor cores available on leadership-class machines requires that every dusty corner of the code make efficient use of the available cores. Otherwise, whichever fraction of the code speeds up more slowly with increasing numbers of cores eventually determines the overall speedup of the code.

Another form of consistency proves useful when one desires some universal way to reference objects in a project. Doing so facilitates manipulating an object without knowledge of its identity. The manipulated object could equally well be an instance of any class in the project.

In HPC, communicating efficiently between local memories on distributed processors represents one of the most challenging problems. One might desire to ensure consistency in communication practices across the project. In these contexts, two broad requirements drive the desire to impose a degree of uniformity across a design: one stemming from a need for consistent functionality, and the other stemming from a need for consistent referencing.

Opposing these forces is the desire to avoid overconstraining the design. In the worst-case scenario, imposing too much uniformity stifles creativity and freezes counterproductive elements into the design. Continuing with the HPC example, were one to mandate that all communication take place through a shared-memory protocol, for example, porting the code to a distributed-memory platform would be difficult or impossible. In a more innocuous scenario, one might force all classes to implement methods that nominally support a protocol that proves superfluous for those classes that carry out embarrassingly parallel tasks — that is, tasks that involve little or no communication. Writing superfluous methods wastes the time of the writers and potentially sets up unfounded expectations on the part of clients.

> **Tie Me Up–don't Tie Me Down**
>
> Classes at the base of an inheritance hierarchy can ensure that those logically further up the hierarchy implement mission-critical capabilities. They can also do harm when those capabilities prove superfluous for certain ADTs. The best universal base classes contain minimal or no functionality and of sufficiently broad utility that they might be candidates for incorporation into the language.

5.2 The Solution

In a procedural programming project, one might achieve the desired consistency in functionality by writing utilities that perform tasks that are common within the given application domain. The project leader could require that all developers use these utilities whenever applicable. In HPC, for example, the Basic Linear Algebra Communication Subprograms (BLACS) serve this role for ScaLAPACK (Blackford et al. 1997).

In an OOP project, one might construct classes that provide methods that perform common functions on encapsulated data. Continuing with the HPC example, some methods might communicate data between processors. This communication might even be hidden from client code. Developers could build atop these classes by accepting their instances as arguments, by extending them via inheritance or type extension, or by aggregating them into new classes. The Trilinos[1] solver framework's Epetra package takes this approach (Heroux et al. 2005). When built atop Trilinos, most user code never needs to communicate information explicitly, leaving that detail to Trilinos.

Fulfilling the expressed desire for consistency in referencing proves somewhat trickier and potentially language-dependent. Since the application context in the current chapter does not present a need for universal referencing, we defer the related discussion to subsequent chapters. Chapter 6 discusses the necessary language infrastructure: the `select type` construct in Fortran and the `dynamic_cast` operator in C++. Chapter 11 presents an example of using this infrastructure in an actual open-source software package.

The OBJECT pattern takes its name and inspiration from a native Java construct: the `Object` class. In Java, `Object` is a universal base class that every Java class extends. By definition then, every class inherits any `Object` functionality that it does not explicitly override. This satisfies the need for common functionality at least insofar as the minimum functionality desired by the language designers. Furthermore, any instance of a Java class can always be treated as an `Object` instance. This satisfies the need for a universal reference without need for exception handling.

Hence, the OBJECT pattern defines a base class hierarchy and stipulates that every class in a given project extends the class at the top of that hierarchy. As with our discussion of a two-part FACADE in section 4.2.3, this pattern potentially maps to multiple

[1] http://trilinos.sandia.gov

classes. Several reasons motivate implementing OBJECT as a unidimensional, hierarchical collection of classes rather than a single class. First, when the desired common functionality involves unrelated methods, collecting them into one class results in the coincidental cohesion, the weakest and least desirable form (see Section 3.2.1). A more understandable and maintainable design results from separating concerns.

Second, any deferred bindings in the parent type (base class) hierarchy must be implemented in all concrete classes in the project. The OBJECT pattern should therefore be extremely stable. Only deferred bindings expected to survive for the life of the project should be included. Nonetheless, if some deferred bindings seem likely to be removed later, placing these deeper in the ancestral hierarchy inoculates the child classes from subsequent deletion of the relevant ancestor, because the classes only reference the ultimate descendant – that is, ultimate from viewpoint of the OBJECT hierarchy. Declarations of instances of this ultimate descendant elsewhere in the package need not be changed when more distant ancestral relatives are removed.

Since efficient use of HPC platforms usually requires a sophisticated understanding of the underlying architecture and the chosen parallel programming paradigm, we defer further discussion of HPC to Part III. There we exploit the C interoperability features of Fortran 2003 to construct a mixed Fortran/C++, object-oriented, scalable ADT calculus. That calculus circumvents many of the details required for more traditional approaches ensuring runtime scalability.

Here we focus on an example involving the bane of every programmer whose project requires considerable use of dynamic memory allocation: memory leaks. By introducing `allocatable` entities, Fortran relieves the programmer from thinking about leaks in a large fraction of the cases where they could occur in C++. C++ has a long-standing reputation of difficult memory management relating to exposing low-level entities directly to the programmer – specifically, exposing machine addresses via pointers. Many partial solutions exist (standard container classes, management via lifetime of stack allocated objects, reference counting, garbage collectors, etc.), but the current language standard does not provide any general, portable, or efficient solution to the problem of managing object lifetimes. Section 5.2.2 presents an invasive reference-counting scheme, but in the absence of a language-provided solution, this approach will almost certainly prevent a compiler from eliminating temporaries and a host of other optimizations.

Another approach puts each developer in charge of managing memory but gives them tools that assist them in doing so. Although the use of `allocatable` entities in Fortran obviates the need for developers to explicitly free memory in the vast majority of circumstances, at least three scenarios warrant giving programmers this responsibility. First, as recently 2006, three of five compilers tested exhibited memory leaks resulting from not freeing memory in the manner in which the Fortran 2003 standard obligates them (Rouson et al. 2006). Until widespread availability of leak-free compilers becomes the norm, developers will often need to adopt defensive postures. Second, as recently as December 2009, only two of seven compilers surveyed supported the final subroutines the standard provides for programmers to instruct the compiler as to how to free memory automatically when a pointer component goes out of scope (Chivers and Sleightholme 2010). Third, programmers might want to explicitly free memory before the associated entity goes out of scope.

The flexibility and power of `allocatable` entities for automatic dynamic memory management makes the need for pointers (and hence final subroutines) quite rare in Fortran. The quantum vortex filament discretization of Chapter 4, however, maps quite nicely onto a linked list data structure with pointer components. Figure 4.4 naturally evokes notions of translating each mesh point into a node in the list and each line segment into a pointer or bidirectional pair of pointers. We use this example in the remainder of this section.

5.2.1 Fortran Implementation

Markus (2003) identified several scenarios wherein memory leaks can occur in Fortran code in settings that resonate with the central theme of the current book: evaluating expressions built up from defined operators and defined assignments. Rouson et al. (2005) and Markus independently proposed nearly identical memory management algorithms that eliminated leaks in most cases. In those cases, the simplicity of their strategy makes it attractive. Chapter 10 uses it to motivate formal design constraints.

Stewart (2003) proposed a more complicated approach that eliminates leaks in several situations not covered by Markus and Rouson and colleagues. He considered a derived type `Rmat` containing a pointer component used in a manner analogous to that shown in Figure 5.1. The first leak occurs at line 11 in 5.1(b), which implicitly invokes the type-bound procedure `total` in 5.1(a). The leaks at lines 17–18 appear similar but have subtle differences discussed in Chapter 10.

The `total` procedure allocates memory for a result that becomes the `rhs` argument of the assignment procedure `assign`. Herein lies the dilemma: Once `assign` terminates and control reverts to the `main` program, `rhs` goes out of scope – that is `main` contains no variable name associated with the memory referenced by `rhs`, the contents of which have been copied into `lhs`, which `main` refers to as `c`. More importantly, `main` contains no variable name associated with the component `rhs%matrix`. Hence, while the compiler will deallocate `rhs`, the language semantics do not provide for deallocation of `rhs%matrix`. With a standard-compliant Fortran 2003 compiler, one can remedy this situation most simply by replacing the `pointer` component with an `allocatable` component, obligating the compiler to free the memory. Alternatively, one could define a final subroutine that invokes the `deallocate` intrinsic on the pointer component. Otherwise, the memory's fate rests with the programmer.

In order to ensure that each developer has access to memory management tools and implements certain supporting functionality, one might place the tools in a universally extended parent type hierarchy. Figure 5.2 demonstrates this strategy using the procedures `SetTemp`, `GuardTemp`, and `CleanTemp` adapted from Stewart (2003). The hierarchy in this example includes only one abstract derived type: `hermetic`. The `hermetic` type satisfies the aforementioned desire to encourage certain universal practices. If every class extends `hermetic`, it could also satisfy the desire for universal referencing. However, it is usually better to separate these two concerns into different types. Chapter 11 discusses an example in which the the need for universal referencing arises and a second type in the universal parent hierarchy provides for it.

Following Markus, one marks all function results as temporary. As employed in ADT calculus by Rouson et al., this covers any intermediate result an operator allocates and passes to an operator of lower precedence or to a defined assignment. As implemented by Stewart, `SetTemp` marks a result temporary by allocating its integer pointer component if necessary and by setting that component's target to 1. The use of an integer pointer component rather than an integer component facilitates complying with the Fortran standard requirement that the dummy arguments of defined operators have the `intent(in)` attribute. The application of this attribute to a derived type flows down to the derived type components. In the case of a pointer component, `intent(in)` prevents changing with *which* target a pointer is associated, but places no restriction on changing the values stored in the target. This allows for incrementing setting and incrementing the integer as described next.

In Stewart's scheme, each subroutine or function that receives an argument of type `Rmat` that could be temporary, it guards it from premature deallocation by passing it to `GuardTemp`. If it is temporary, `GuardTemp` increments the tally stored in the integer pointer component. Stewart refers to this tally as a *depth count* due to his interest in the deeply nested function call trees characteristic of defined-operator expressions.

The final step in Stewart's approach requires calling `CleanTemp` just before exiting each procedure that receives a possibly temporary `Rmat` argument. If `CleanTemp` receives a temporary object, it checks whether the depth count exceeds 1. If so, it decrements the count. Then if the count it equals 1, `CleanTemp` frees the memory associated with the argument. In Stewart's concrete `Rmat` type, `CleanTemp` deallocates the memory itself. Our OBJECT implementation in Figure 5.2 defines a `hermetic` type. Since this is an abstract type, its only utilities derives from extending it. Furthermore, because the private scoping of its integer pointer component applies at the module level and because all extended types are implemented in separate modules, `hermetic` publishes several accessors, that is, methods that control access to its private state. It also publishes the deferred binding `force_finalization`. `CleanTemp` invokes this binding, which concrete extended types must implement. With a compiler that supports type finalization, `force_finalization` might simply invoke the type's final subroutine. Otherwise, it would free the memory itself. We discuss next an interesting hybrid of these two approaches.

The `vortex` type of Figure 5.3 demonstrates the beginnings of a solver for the quantum vortex dynamics context of Section 4.3.2. The `vortex` type extends `hermetic`. Its `ring` constructor creates a circular linked list of vortex points. Much like the ring discretization of Figure 4.4, the list connects tail to head.

The `vortex` implementation also provides a `finalize` final subroutine and a `manual_finalizer` implementation of the `force_finalization` binding. Both subroutines exhibit subtle, elegant behaviors associated with Fortran type finalization. The circular referencing of the linked list returned by `ring` precludes the simple null-terminated traversal of the linked list common to linear linked list implementations. Accordingly, the first step in `finalize` involves artificially converting the list to a linear form by breaking a link at line 47 in Figure 5.3. This happens immediately after storing a reference to the node targeted by that link. After these two steps, `finalize` passes the linked node to `deallocate` if the link is associated with a target.

The Fortran standard obligates compilers to finalize an object whenever a programmer deallocates it as line 48 does.[2] This makes `finalize` inherently recursive as indicated by the `recursive` attribute on line 43. This artifice neatly provides for automatic traversal and destruction of the entire linked list in the three executable lines 46–48.

The language semantics harbor equally subtle and elegant implications for `manual_finalizer`. When one invokes this procedure to free memory in advance of when a compiler might otherwise do so automatically via the final subroutine, it is critical to ensure that the two procedures do not interact in harmful ways. This issue vanishes with compilers that do not support final subroutines. For these, the programmer must remove the `final` binding at line 15 manually or automatically, for example via Fortran's conditional compilation facility or C's preprocessor. In such cases, `manual_finalizer` breaks the chain in the same manner as `finalizer` and traverses the vortex, deallocating each node until it returns to the broken link.

When the compiler supports final subroutines, the `deallocate` command at line 58 implicitly invokes `finalizer`, after which the destruction process continues recursively and automatically, making only one pass through the `do while` block that begins at line 56. The automatically invoked final subroutine and the explicitly invoked manual destruction procedure peacefully coexist.

Figure 5.4 demonstrates vortex construction, output, and destruction. Exercises at the end of this chapter guide the reader in writing code for point insertion, removal, and time advancement via ADT calculus. Line 5 in Figure 5.4 marks the first Fortran pointer assignment in this text. Were the constructor result produced by the RHS of that line assigned to a LHS object lacking the pointer attribute, an intrinsic assignment would copy the RHS into the LHS and then deallocate the RHS. This would start the aforementioned finalization cascade and destroy the entire ring! The pointer assignment prevents this.

Figure 5.1(a)

```
1  module Rmat_module
2    implicit none
3    type Rmat
4      real ,dimension(:,:) ,pointer :: matrix=>null()
5    contains
6      procedure :: total
7      procedure :: assign
8      generic   :: operator(+) => total
9      generic   :: assignment(=) => assign
10   end type
11 contains
12   type(Rmat) function total(lhs,rhs)
13     class(Rmat) ,intent(in) :: lhs,rhs
14     allocate(total%matrix(size(lhs%matrix,1),size(lhs%matrix,2)))
15     total%matrix = lhs%matrix + rhs%matrix
```

[2] Although the standard does not specify whether finalization or deallocation comes first, it seems reasonable to assume compilers will finalize first to avoid breaking final subroutines that assume the object remains allocated upon invocation, and to prevent memory leaks that would result if there remained no handle for a pointer component still associated with a target.

```
16   end function
17   subroutine assign(lhs,rhs)
18     class(Rmat) ,intent(inout)  :: lhs
19     type(Rmat) ,intent(in) :: rhs
20     lhs%matrix = rhs%matrix
21   end subroutine
22 end module
```

Figure 5.1(b)

```
1  program main
2    use Rmat_module
3    implicit none
4    integer ,parameter :: rows=2,cols=2
5    type(Rmat) :: a,b,c
6    allocate(a%matrix(rows,cols))
7    allocate(b%matrix(rows,cols))
8    allocate(c%matrix(rows,cols))
9    a%matrix=1.
10   b%matrix=2.
11   c = a + b ! memory leak
12   call foo(a+b,c)
13 contains
14   subroutine foo(x,y)
15     type(Rmat) ,intent(in) :: x
16     type(Rmat) ,intent(inout) :: y
17     y=x     ! memory leak
18     y=x+y   ! memory leak
19   end subroutine
20 end program
```

Figure 5.1. Memory leak example adapted from Stewart (2003).

Figure 5.2

```
1  module hermetic_module
2    implicit none
3    private
4    public :: hermetic ! Expose type and type-bound procedures
5
6    type ,abstract :: hermetic
7      private
8      integer, pointer :: temporary=>null() !Null marks non-temporary data
9    contains
10     procedure :: SetTemp   ! Mark object as temporary
11     procedure :: GuardTemp ! Increment the depth count
12     procedure :: CleanTemp ! Decrement depth count/Free memory if 1
13     procedure(final_interface) ,deferred :: force_finalization
14   end type
15
16   abstract interface
17     subroutine final_interface(this)
18       import :: hermetic
19       class(hermetic) ,intent(inout) :: this
20     end subroutine
21   end interface
```

```
22  contains
23    subroutine SetTemp(this)
24      class(hermetic) ,intent(inout) :: this
25      if (.not. associated(this%temporary)) allocate(this%temporary)
26      this%temporary = 1
27    end subroutine
28
29    subroutine GuardTemp (this)
30      class (hermetic) :: this
31      if (associated(this%temporary)) this%temporary = this%temporary + 1
32    end subroutine
33
34    subroutine CleanTemp(this)
35      class(hermetic) :: this
36      if (associated(this%temporary)) then
37        if (this%temporary > 1) this%temporary = this%temporary - 1
38        if (this%temporary == 1) then
39          call this%force_finalization()
40          deallocate(this%temporary)
41        end if
42      end if
43    end subroutine
44  end module
```

Figure 5.2. OBJECT implementation: universal memory management parent.

Figure 5.3

```
1   module vortex_module
2     use hermetic_module ,only : hermetic
3     implicit none
4     private
5     public :: vortex
6
7     integer ,parameter :: space_dim=3
8     type ,extends(hermetic) :: vortex
9       private
10      real ,dimension(space_dim) :: point
11      type(vortex) ,pointer :: next=>null()
12    contains
13      procedure :: force_finalization => manual_finalizer
14      procedure :: output
15      final :: finalize
16    end type
17
18    interface vortex
19      procedure ring
20    end interface
21
22  contains
23    function ring(radius,num_points)
24      real ,intent(in) :: radius
25      integer ,intent(in) :: num_points
26      type(vortex) ,pointer :: ring,current
27      integer :: i
```

```
28      real ,parameter :: pi = acos(-1.)
29      real :: delta,theta
30      delta = 2.*pi*radius/real(num_points) ! arclength increment
31      allocate(ring)
32      ring%point =  [radius,0.,0.]
33      current => ring
34      do i=1,num_points-1
35        allocate(current%next)
36        theta = i*delta/radius
37        current%next%point = [radius*cos(theta),radius*sin(theta),0.]
38        current => current%next
39      end do
40      current%next => ring ! point tail to head
41    end function
42
43    recursive subroutine finalize (this)
44      type(vortex), intent(inout) :: this
45      type(vortex) ,pointer :: next
46      next => this%next
47      this%next=>null() ! break the chain so recursion terminates
48      if (associated(next)) deallocate (next) ! automatic recursion
49    end subroutine
50
51    subroutine manual_finalizer(this)
52      class(vortex) ,intent(inout) :: this
53      type(vortex) ,pointer :: current,next
54      current => this%next
55      this%next=>null() ! break the chain so loop terminates
56      do while(associated(current%next))
57        next => current%next ! start after break in chain
58        deallocate(current)
59        current=>next
60      end do
61    end subroutine
62
63    subroutine output(this)
64      class(vortex) ,intent(in) ,target :: this
65      type(vortex) ,pointer :: current
66      current => this
67      do
68        print *,current%point
69        current => current%next
70        if (associated(current,this)) exit
71      end do
72    end subroutine
73  end module
```

Figure 5.3. Quantum vortex linked list with hermetic memory management.

```
1  program main
2    use vortex_module ,only : vortex
3    implicit none
4    type(vortex) ,pointer :: helium
5    helium => vortex(radius=1.,num_points=8)
```

```
6    call helium%output()
7    call helium%force_finalization()
8  end program
```

Figure 5.4. Vortex creation and destruction.

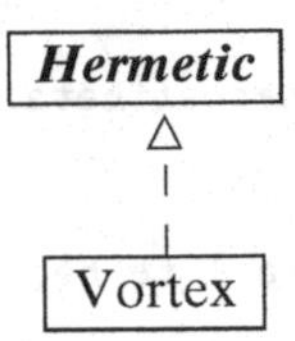

Figure 5.5. The concrete Vortex ADT implements the Hermetic interface.

5.2.2 C++ Style and Tools

This chapter marks the first substantial C++ examples in this text. We aim to put Fortran and C++ on roughly equal footing by exploiting primarily those constructs in one language that have reasonably close counterparts in the other. This strategy makes the text self-consistent and empowers readers more familiar with one language to comfortably explore the other.

Our approach requires eschewing some advanced features in both languages and restricting ourselves to current standards. Our C++ idioms might appear overly restrictive to advanced programmers who routinely link to libraries outside the core language. For example, this section's memory management approach could be greatly enhanced by the BOOST open-source C++ library[3], many parts of which are expected to enter the upcoming C++ 0x standard.

Although we seek to construct close counterparts to our Fortran implementations, we also seek to write C++ examples that appear natural to C++ programmers. This gives us freedom to use styles and naming conventions that differ from those common in Fortran. For example, in class definitions, all (private) data members' names are appended with an underscore (_), which is a common practice in C++ to indicate the names should only be referenced internally within member functions.

The next subsection demonstrates an instance in which the goals in the prior two paragraphs collide. There we construct a base class that provides tools for tracking the number of pointers to dynamically allocated memory. A common way to accomplish this in C++ involves template metaprogramming, a facility with no Fortran counterpart.

5.2.2.1 Reference Counting

Stewart's depth count scheme corresponds to the widespread reference counting approach in C++. Absent language support for garbage collection of pointer-allocated memory (although BOOST's `shared_ptr` construct provides a garbage collector similar to Fortran's allocatable variables), C++ programmers commonly

[3] http://www.boost.org

employ reference counting to obtain functionality similar to garbage collection. Reference counting refers to the strategy of keeping track of the number of references or pointers that refer to the memory associated with an object, so the memory can be freed when it is no longer referenced. In practice, reference counting helps prevent the aforementioned memory leaks in applications that rely frequently on pointers.

The significant risk of memory leaks associated with raw pointers combined with the ubiquitous need for pointers in C++ makes a reference-counted pointer (RCP) class an ideal candidate for incorporation into an OBJECT pattern. Figure 5.6 contains a RCP implementation based on C++ template classes. Figure 5.6(a) shows a basic RCP template class `pointer<>` definition. It wraps raw C++ pointers and thereby facilitates treating objects as if one were using a C++ pointer directly.

Figure 5.6(a): pointer.h

```
1  #ifndef POINTER_H_
2  #define POINTER_H_
3
4  /**
5   * A simple wrapper for raw pointer.
6   */
7
8  #include <iostream>
9  #include <unistd.h>
10
11 template <typename T>
12 class pointer
13 {
14 public:
15   pointer() : ptr_(NULL){}
16   pointer(T *other) : ptr_(other){}
17
18   void assign(T *other) { ptr_ = other; }
19
20   pointer& operator=(T * other) {
21     assign(other);
22     return *this;
23   }
24   const pointer& operator=(const T & other) {
25     assign(const_cast<T*>(&other));
26     return *this;
27   }
28   pointer& operator=(T & other) {
29     assign(&other);
30     return *this;
31   }
32
33   T* operator->() { return ptr_; }
34   const T* operator->() const { return ptr_; }
35
36   T& operator*() { return *ptr_; }
37   const T& operator*() const { return *ptr_; }
38
39   bool operator==(const T *other) const { return ptr_ == other; }
```

```
40    bool operator!=(const T *other) const { return ptr_ != other; }
41
42    T* ptr() const { return ptr_; }
43
44  private:
45    T *ptr_;
46  };
47  #endif
```

Figure 5.6(b): Ref.h

```
1   #ifndef REF_H_
2   #define REF_H_
3
4   /**
5    * A very simple invasive reference counted pointer.
6    */
7
8   #include "pointer.h"
9
10  template <typename T>
11  class Ref : virtual public pointer<T> {
12  public:
13    typedef Ref<T> Self_t;
14
15    Ref() {}
16
17    Ref(const Self_t &other) {
18      assign(other.ptr());
19      if(ptr() != NULL) ptr()->grab();
20    }
21
22    template <typename Other>
23    Ref(Ref<Other> other){
24      assign(other.ptr());
25      if(ptr() != NULL) ptr()->grab();
26    }
27
28    template <typename Other>
29    Ref(Other *other) {
30      assign(other);
31      if(ptr() != NULL) ptr()->grab();
32    }
33
34    ~Ref() {
35      if(ptr() != NULL) ptr()->release();
36    }
37
38    Self_t& operator=(const Self_t &other) {
39      if(other.ptr() != ptr()) {
40        if(ptr() != NULL) ptr()->release();
41        assign( other.ptr());
42        if(ptr() != NULL) ptr()->grab();
43      }
44      return *this;
```

```
45   }
46
47   Self_t& operator=(T *other) {
48     if(ptr() != other) {
49       if(ptr() != NULL) ptr()->release();
50       assign( other);
51       if(ptr() != NULL) ptr()->grab();
52     }
53     return *this;
54   }
55 };
56
57 template <typename Target, typename Source>
58 Ref<Target> cast(const Ref<Source> &pp) {
59   return Ref<Target>(dynamic_cast<Target*>(pp.ptr()));
60 }
61 #endif
```

Figure 5.6(c): RefBase.h

```
1  #ifndef REFBASE_H_
2  #define REFBASE_H_
3
4  /**
5   * Base for reference counted objects.
6   */
7
8  #include "Ref.h"
9
10 class RefBase {
11 public:
12   RefBase() : cnt_(0) {}
13   RefBase(const RefBase &) : cnt_(0) {}
14   virtual ~RefBase() {}
15   void grab() throw() {
16     cnt_++;
17   }
18   void release() throw() {
19     cnt_--;
20     if(! cnt_) delete this;
21   }
22 private:
23   int cnt_;
24 };
25 #endif /* ! REFBASE_H_ */
```

Figure 5.6. Reference counting in C++: (a) a template pointer base class for the reference-counted class Ref, (b) a template class for the reference-counted pointer Ref, and (c) a reference-counted object RefBase base class .

We use the RCP class from Figure 5.6(b) and the reference-counted object class of Figure 5.6(c) for the C++ examples throughout this book. The current implementation grew out of C++ sample code that first appeared in Rouson, Adalsteinsson, and Xia (2010). The presentation here provides more detailed descriptions of the structure and technical approach. In particular, we attempt to mitigate any confusion that might arise from the template syntax. Templates offer a unique and powerful technique for writing generic code that is applicable to any data types. Hence, one does not need to repeat the same algorithm for different data types. For excellent tutorials on C++ templates, see Vandevoorde and Josuttis (2003).

Highlights of the techniques employed in Figure 5.6(a) include the following:

- Constructors.
  A `pointer<T>` object can be constructed a few different ways: It can be initialized as a NULL by default or it can be initialized by a pointer. We rely on the default copy constructor generated by the compiler. Were we to define a copy constructor, it would behave identically to the compiler-generated one.

```
pointer() : ptr_(NULL){}           // default constructor
pointer(T *other) : ptr_(other){} // takes a raw pointer
```

- Assignment operators.
  The assignment operators for `pointer<>` are overloaded to have a few different forms:

```
pointer& operator=(T * other);       // RHS is a T*
const pointer& operator=(const T&); // RHS is const T&
pointer& operator=(T & other);       // RHS is T&
```

  As with the constructors, we provide no copy assignment operator and rely on the compiler to generate the correct behavior.
- Member access
  The member access operators are defined so that any `pointer<T>` object is treated just as a regular pointer. Consider the following declarations of the operators `->` and `*`. We can illustrate their usage with the following example: Given a declaration "`pointer<T> x;`", one can use pointer syntax such as `x->foo()` or `(*x).foo()`, assuming type T has a member function named `foo()`:

```
operator ->:
T* operator->();                 // member access; lvalue expr
const T* operator->() const;     // member access; rvalue expr

operator *:
T& operator*();                  // indirect; lvalue expr
const T& operator*() const;      // indirect; rvalue expr
```

- Pointer comparison.
  Since the `pointer<>` class emulates actual pointer behavior, we also need to overload the comparison operators == and != so that they are actually comparing pointer addresses. Continuing with the previous example of object `x`, the

expression x == NULL tests if x contains a NULL pointer. Similarly, x != y returns true if (the address of) the memory associated with x is not the same as that associated with y:

```
bool operator==(const T *other) const;
bool operator!=(const T *other) const;
```

- Pointer access.
  Providing access to the value of the underlying pointer is straightforward:

  ```
  T* ptr() const { return ptr_; }
  ```

Figure 5.6(b) presents the definition of a template class, Ref<>, that does the reference counting when the template parameter type is RefBase (see Figure 5.6[c]) or a subclass of RefBase. Class Ref<> is derived from pointer<> so every instance of Ref<> is also treated as a pointer. In addition, Ref<> provides techniques for managing reference counts.

- Copy constructors.
  Copy constructors increase the reference count. Whenever a new instance of Ref<T> is created, one of the copy constructors is invoked in order to copy the pointer data. Class Ref<> defines all necessary copy constructors that are used in our examples. A common theme of these copy constructors is they all call a method named grab(), which increments the reference count by one:

  ```
  // copy constructors: call grab()
  Ref(const Self_t &other);
  template <typename Other> Ref(Ref<Other> other);
  template <typename Other> Ref(Other *other);
  ```

- Destructor.
  The destructor, defined as ~Ref<>, decreases the reference count when a RCP goes out of scope. As shown in Figure 5.6(b), the destructor invokes the method release() to decrement the reference count by one. If the reference count reaches zero, then all pointers have reached the end of their lifetimes and the memory to which the pointer points is freed. The RefBase class defines the grab() and release()implementations as discussed after the next bullet:

  ```
  ~Ref();                // destructor; call release()
  ```

- Assignment operators.
  The assignment operators manage the reference counts for both the LHS and RHS. As shown below, these operators take an approach that differs slightly from the copy constructors. One can envisage the process of assigning a Ref<> pointer comprising disassociating the LHS pointer followed by the reassociating it with the RHS pointer. The implementations of the assignment operators reflect this process: release the LHS pointer first then associate (the address of) the LHS

pointer to the RHS pointer, and finally grab() the LHS pointer to update the reference count on the pointer:

```
// assignment operators:
Self_t& operator=(const Self_t&); // RHS is a Ref<T>
Self_t& operator=(T *other);      // RHS is T*
```

A helper function named cast() is also defined for Ref<>. As implied by its name, this function typecasts between two Ref<> pointers with different parameter types using C++'s dynamic_cast. The function is used in Chapter 6–8 when type cast is actually needed.

The class definition of RefBase, as shown in Figure 5.6(c), provides a base class for any object that needs reference counting. This class implements the method grab() and release(). The actual number of references to this object is stored by the attribute cnt_. A careful reader might have noticed the absence of keyword virtual in definitions of grab() and release(). This is intentional. It prevents subclasses of RefBase from overriding these methods and potentially breaking the reference counting scheme.

The combination of Ref<> and its parameter class RefBase closely corresponds to the hermetic type defined in Fortran. We also use Ref<> and RefBase in Chapters 6–8 to emulate Fortran's rank-one allocatable arrays.

5.2.2.2 Output Format for Floating Point Quantities

The I/O formatting differences between C++ and Fortran warrant a format conversion tool to facilitate direct comparison between the two languages' output. Since all of our Fortran examples use list-directed output for floating point arrays, we developed a helper class to handle the C++ float output format. Figure 5.7(a) defines this class fmt. It stores two parameters, w_ and p_, for controlling the output width and precision (the number of significant digits in the output field). Users can choose a desired width and precision when constructing a fmt object. By default, the output width is 12 characters wide and with 8 significant digits, matching roughly that in Fortran for a single float[4].

The internal member v_, which is of C++ STL's vecotr<float> type, stores the actual value of the output item. Figure 5.7(b) gives the detailed type definitions of real_t and crd_t. To simulate the list-directed output format for a Fortran real (C++ float) array, such as that produced by "print *, X", the C++ operator "<<" is overloaded, as shown from line 18 to 29 in "fmt.h". Hence, the C++ analogue to the aforementioned Fortran print statement is "std::cout << fmt(X);".

5.2.3 C++ Implementation of Vortex

Figures 5.8(a)–(b) show the C++ vortex class definition header and implementation files, respectively. Since memory management is the central theme for the vortex

[4] Contrary to Fortran's user-controlled formatting, which uses an edit descriptor such as f12.8, list-directed output formatting is controlled by compilers, which choose appropriate width and precisions depending on the data type of the output item list.

ring, the vortex class requires reference counting. This is readily seen by the fact that class vortex extends RefBase. Similar to the Fortran definition, the C++ vortex stores the vortex coordinates using an array of floats. Member variable next_, defined as type Ref<vortex>, replaces a raw C++ pointer in the linked list node structure. Since next_ does the reference counting on its own, it is not necessary to define any functionality for the destructor as is evident in the code at lines 8–9 in Figure 5.8(b)[5].

In the class definition of vortex, we include a typedef iterator to enable node traversal for insertion, deletion, update, or query. Since an iterator normally requires no reference counting when iterating through the vortex ring, we use type pointer<Ref<vortext>>. The constructor:

```
vortex(real_t radius, int num_points);
```

creates a vortex ring with the input radius and number of points, providing the same functionality as ring() in Figure 5.3.

The method destroy() initiates the vortex ring destruction. It may seem odd that vortex has an empty destructor while an additional function call is needed to initiate its destruction. Its necessity derives from the ring's circular structure: Its tail points to its head. Upon creation, there are two references to the head: one held by the external user and another that is the vortex ring's tail. Hence, the number of references to the head does not reach zero when the external user no longer needs the ring.

Figure 5.7(a): fmt.h

```
1  #ifndef _H_FMT_
2  #define _H_FMT_
3  #include "globals.h"
4  #include <iostream>
5  #include <iomanip>
6  // The fmt(...) helper class helps hide the mess that in <iomanip>
7  struct fmt {
8    explicit fmt(real_t value, int width=12, int prec=8) :
9      v_(1, value), w_(width), p_(prec)
10   {}
11   explicit fmt(crd_t value, int width=12, int prec=8) :
12     v_(value), w_(width), p_(prec)
13   {}
14
15   const crd_t v_;
16   const int w_, p_;
17 };
18 inline std::ostream& operator<<(std::ostream &os, const fmt &v) {
19   // Store format flags for the stream.
20   std::ios_base::fmtflags flags = os.flags();
21   // Force our own weird format.
22   for(crd_t::const_iterator it = v.v_.begin(); it != v.v_.end(); ++it) {
23     os << " " <<std::setw(v.w_) <<std::setprecision(v.p_) <<std::fixed
24              << *it;
```

[5] Its always a good practice to define a virtual destructor, even if it is empty, in anticipation of the class being extended later by subclasses.

```
25    }
26    // Restore original format flags.
27    os.flags(flags);
28    return os;
29  }
30  #endif //! _H_FMT_
```

Figure 5.7(b): globals.h

```
1   #ifndef _H_GLOBALS_
2   #define _H_GLOBALS_
3
4   #include "Ref.h"
5   #include "RefBase.h"
6   #include <vector>
7
8   typedef float real_t;
9   typedef std::vector<real_t> crd_t;
10
11  #endif //!_H_GLOBALS
```

Figure 5.7. Output tools: (a) C++ emulation of Fortran's list-directed output and (b) global header file defining real_t and crd_t.

Figure 5.8(a): vortex.h

```
1   #ifndef _VORTEX_H_
2   #define _VORTEX_H_
3
4   #include "RefBase.h"
5   #include "globals.h"
6   #include <vector>
7
8   class vortex : virtual public RefBase
9   {
10      public:
11          typedef Ref<vortex> ptr_t;
12          typedef pointer<ptr_t> iterator;
13          vortex();
14          vortex(real_t radius, int num_points);
15
16          virtual ~vortex();
17
18          void destroy();
19          virtual void output() const;
20
21      private:
22          crd_t   point_;
23          ptr_t   next_;
24  };
25
26  #endif
```

Figure 5.8(b): vortex.cpp

```
1  #include <cmath>
2  #include "vortex.h"
3  #include "fmt.h"
4
5  vortex::vortex()
6  {}
7
8  vortex::~vortex()
9  {}
10
11 void vortex::destroy()
12 {
13     // the following line start destruction of the vortex ring
14     // breaks the ring and triggers the chain action
15     this->next_ = NULL;
16 }
17
18 vortex::vortex(real_t radius, int num_points)
19 {
20     const real_t pi = 3.1415926f;
21     real_t delta;
22     iterator current;
23
24     delta = 2.0f*pi*radius/real_t(num_points);
25
26     this->point_.push_back(radius);
27     this->point_.push_back(0.0f);
28     this->point_.push_back(0.0f);
29
30     this->next_ = new vortex();
31
32     current = this->next_;
33
34     for (int i = 0; i < num_points-1; ++i)
35     {
36         crd_t temp_point;
37         real_t theta;
38
39         theta = real_t(i+1) * delta/radius;
40
41         temp_point.push_back(radius*cos(theta));
42         temp_point.push_back(radius*sin(theta));
43         temp_point.push_back(0.0);
44
45         (*current)->point_ = temp_point;
46
47         (*current)->next_ = new vortex();
48
49         current = (*current)->next_;
50     }
51
52     *current = this; // this will cause the last memory allocation
53                      // returned by the new operation in the last
54                      // iteration to be automatically deleted
```

```
55  }
56
57  void vortex::output() const
58  {
59      iterator current;
60      crd_t tmp;
61
62      tmp = this->point_;
63
64      std::cout << fmt(tmp) << std::endl;
65
66      current = this->next_;
67
68      do
69      {
70          tmp = (*current)->point_;
71          std::cout << fmt(tmp) << std::endl;
72
73          current = (*current)->next_;
74      } while (*current != this);
75  }
```

Figure 5.8(c): main.cpp

```
1   #include <iostream>
2   #include "vortex.h"
3
4   int main ()
5   {
6       typedef vortex::ptr_t ptr_t;
7
8       ptr_t helium = ptr_t(new vortex(1.0, 8));
9
10      helium->output();
11      helium->destroy();
12
13      return 0;
14  }
15
```

Figure 5.8. Quantum vortex class definition.

In the aforementioned scenario, the reference-counting scheme would only lead to ring destruction for a ring that is no longer circular, that is, if something breaks the ring. The function `destroy()` does just that, as shown at line 15 in Figure 5.8(b). By breaking the ring at the first link, it effectively sets the reference count for the second vortex point to zero. This initiates the destruction of the second point, which in turns initiates the destruction of the third point, and so forth. The destruction process thus cascades throughout the entire ring until it reaches the head, at which point only the external user holds a reference to the head. Figure 5.8(c) provides a simple program demonstrating a vortex ring's creation and destruction.

5.3 The Consequences

Any design pattern involves trade-offs between its benefits and its drawbacks. The OBJECT pattern benefits a software package by ensuring the availability of important functionality. The project leadership can ensure access to that functionality by mandating that all classes in the project extend the OBJECT hierarchy.

Although not exploited in the current chapter, the second benefit of the OBJECT pattern derives from the availability of a universal way to reference objects in the project. Since all classes extend the OBJECT class hierarchy, any object in the project could be referenced by an entity declared as an instance of the object class at the base of the hierarchy. Chapter 11 demonstrates the utility of such a capability.

The aforementioned benefits can be double-edged swords. The first benefit might provide a false sense of security that breaks down when developers make no use of the functionality provided by the base hierarchy. The second benefit might lead to a lot of type guarding in Fortran or type-switching in C++ if liberal passing of the base type confronts frequent use of the additional functionality provided in a child type.

Finally, universal mandates might also be viewed as universal handcuffs. Judicious use of OBJECT implies minimalistic construction. A base hierarchy might have the best chance of feeling nonconstricting when the functionality it provides is not domain-specific and simply meets general programming needs. Needs addressed by a well-designed OBJECT are likely to be low level in the sense that they could be candidates for inclusion in the language itself. This certainly is the case with garbage collection, which Fortran includes for `allocatable` objects and C++ provides for some STL `vector` objects.

5.4 Related Patterns

Since the intention is for all classes in a package to extend OBJECT, it is in some sense the "largest" of the patterns discussed in this book and therefore relates to every other. In the context of solving differential equations, the OBJECT hierarchy will likely be the immediate ancestor of an `integrand` object, an ABSTRACT CALCULUS pattern implementation that facilitates integration of differential equations. Chapter 6 explores this scenario.

In some settings (and possibly in less polite company), we facetiously refer to the abstract class at the base of an OBJECT hierarchy as a GOD pattern instance, the notion being that a GOD is the ultimate progenitor of all parents. Each remaining abstract class in an OBJECT would then be DEMIGOD. The advantage of this terminology accrues from any clarity added by differentiating between the classes and roles because OBJECT often represents a multiabstraction pattern.

Finally, the linked list data structure chosen to implement this chapter's quantum vortex codes naturally inspires thoughts of the GoF ITERATOR pattern. An ITERATOR implementation facilitates traversing an ordered collection of objects. A typical implementation invokes methods an object provides for returning the beginning, next, and ending elements while remaining agnostic with respect to the underlying infrastructure, which might equally well be a linked list, an array, or any ordered collection of objects with a defined beginning and ending. Therein lies the rub: A circular linked list has no unambiguous beginning or end. Although one might create artificial beginning and ending points by breaking a link as Section 5.2 demonstrates in the context of object destruction, doing so merely to traverse the list would be overkill at best and catastrophic at worst, given the possibility of starting the destruction cascade. A second approach might involve storing references to arbitrarily chosen beginning and ending objects. We opted to avoid these approaches in favor of simplicity.

EXERCISES

1. Draw a UML sequence diagram for a three-point ring undergoing the object destruction cascade described in Section 5.2.
2. Write a `remeshable_vortex` implementation that extends the `vortex` ADT, adding an `insert` point insertion procedure and a `remove` point removal procedure.
3. Add the requisite operators in the `vortex` type to facilitate time advancement via ADT calculus based on the local induction approximation.

6 The Abstract Calculus Pattern

"All professions are conspiracies against the laity."
George Bernard Shaw

6.1 The Problem

The context of ABSTRACT CALCULUS is the construction of numerical software that approximates various differential and integral forms. Two pairs of conflicting forces arise in this context. In the first pair, the low-level nature of the mathematical constructs provided by mainstream programming languages constrains the design of most scientific programs. A desire for syntax and semantics that naturally represents the much richer mathematical language of scientists and engineers opposes this constraint.

The C++ language contains native scalar and one-dimensional array variables. The C++ STL extends these in vectors with nice properties such as automatic memory management, including sizing and resizing upon assignment. Fortran 2003 provides similar capabilities with its multidimensional allocatable array construct. It also provides numerous useful intrinsic procedures for determining array properties including size, shape, maximum element, and minimum element, as well as intrinsic procedures and operators for combining arrays into sums, matrix vector products, and other derived information. It is common in scientific and engineering work to build up from these native constructs a set of array classes with a variety of additionally useful methods (Barton and Nackman 1994; Heroux et al. 2005). Nonetheless, the resulting objects model very low-level mathematical entities in the sense that one typically arrives at these entities after fairly involved derivations from, and approximations to, much higher-level constructs. In physics-based simulations, these higher-level constructs include scalar, vector, and tensor fields along with the various operations these admit: gradients, divergences, curls, surface and volume integrals, and others. The path from these high-level abstractions to simple arrays is littered with assumptions and approximations that make the code less general.

At an even higher level of abstraction, one uses the available mathematical constructs to model physical constructs. These include fluid and solid continuua, particles, electrical and thermal currents, and a host of other materials and phenomena.

One might imagine generalizing the various mathematical operators to operate on a whole fluid, particle, or conductor such that the time derivative of the physical entity implies the time derivative of all of the mathematical entities that describe its state.

The second pair of opposing forces arises in the context of code reuse. With the proliferation of scientific simulations as research-and-design tools and the concomitant proliferation of codes, one rarely writes custom applications from scratch. More commonly, one leverages existing capabilities by coupling one discipline's software applications to those in related disciplines or by building atop highly optimized libraries that provide specialized services such as nonlinear iteration, sparse matrix inversion, and stiff system integration. As discussed in Sections 1.2–1.3, linking to conventional, procedural libraries increases the data dependencies that bedevil developers during debugging (cf. Press et al. 2002). This imposes an opposing desire to preserve data privacy and, if possible, to avoid passing data altogether.

6.2 The Solution

ABSTRACT CALCULUS resolves the conflicts from the previous section by writing abstract specifications for software constructs that model high-level mathematical constructs. These specifications comprise stateless, abstract classes that support a set of arithmetic, differential, and integral operators. As such, they simply delineate the allowable operations and operands of a calculus, leaving implementors to construct concrete, descendant types and define the algorithms that carry out the operations.

Clients can use the abstract class to write abstract expressions that perform sophisticated mathematical calculations without exposing the concrete type involved in the calculation. At runtime, such a procedure can accept any concrete descendant and invoke the requisite methods on the received object.

Consider the integration of the Lorenz equation system (4.5) over some time interval. Suppose we encapsulate the equation system in a Lorenz ADT. Rather than passing a Lorenz object's state vector into a conventional time integration procedure, we might consider defining an abstract Integrand class that specifies the operations required to integrate objects via ADT calculus. The integration algorithm could then be written to operate on an Integrand and serve this same purpose for any classes that extend Integrand. In this context, ABSTRACT CALCULUS also goes by the name SEMI-DISCRETE pattern in reference to its use for solving the systems of ODEs that result from semidiscrete numerical approximations to PDEs (Rouson, Adalsteinsson, and Xia 2010).

Since we aim to write an abstract integration procedure without knowledge of the classes to be integrated, the procedure lives logically outside any class definition. It might be packaged with, though not a part of, the Integrand abstract class. UML uses the *package* concept to represent collections of elements, including classes, interfaces, collaborations, use cases, and diagrams. UML packages may be composed of other UML elements, including other packages. Packaging also forms a namespace: a list of unique identifiers for the elements contained in the package.

Figure 6.1 describes a Lorenz system integrator design. Figure 6.1(a) employs the UML package graphical element. A tab showing the package name Integrator adorns the package. The Integrator contents, designated as public, include the Integrand

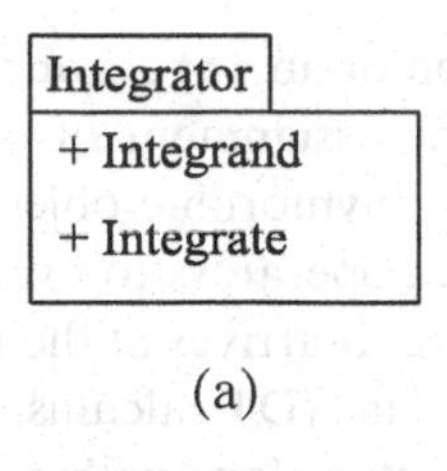

(a)

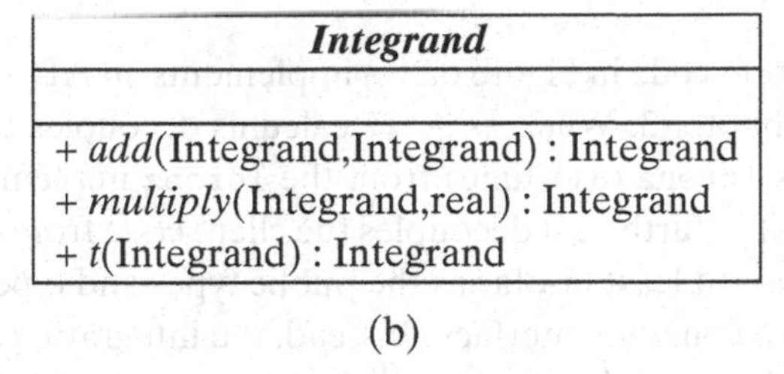

(b)

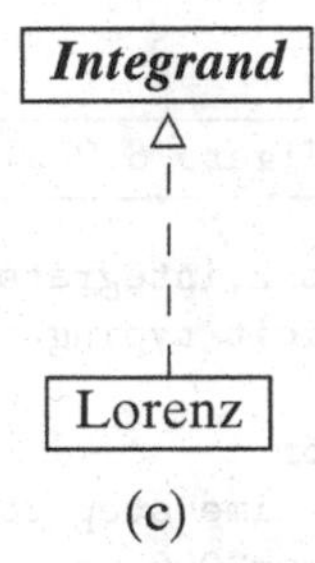

(c)

Figure 6.1. Lorenz system integration: (a) integration package (Integrator), (b) abstract Integrand interface, and (c) class diagram.

class and the Integration procedure. Figure 6.1(b) depicts the Integrand class interface. It contains the addition, multiplication, and time differentiation operations required to implement Runge-Kutta marching schemes, in which context authors often refer to the time differentiation operation as "function evaluation" because it involves evaluating the RHS function in an ODE system, namely $\vec{R}$ in equation (4.4). UML depicts abstract operation names such as *add* and *multiply* in italics and abstract class names such as ***Integrand*** in bold italics.

6.2.1 Fortran Implementation

In Fortran, the `module` construct provides the namespace encapsulation mechanism. Abstract classes correspond to derived types with the `abstract` attribute. An abstract derived type can have deferred type-bound procedure bindings that conform to an `abstract interface`. Figure 6.2 exploits these language features to implement the design described in Figure 6.1.

Figure 6.2(a) declares a `lorenz` object, instantiates it by invoking its like-named constructor, and steps it forward in time by looping over calls to `integrate`. Figure 6.2(b) encapsulates an `integrand` abstract type and a polymorphic `integrate` procedure. The `integrand` definition includes deferred bindings for addition of two integrand objects, multiplication of an integrand by a real scalar, time differentiation of an integrand, and the assignment of one integrand to another. The integrate procedure accepts a polymorphic object along with a real, scalar time step and uses the aforementioned operators to overwrite the object with a version of itself one time step in the future. It arrives at the future time step instance via the explicit Euler algorithm expressed in ADT calculus. Figure 6.2(c) defines the lorenz type, which implements a constructor along with each of the deferred bindings and an output method.

The Lorenz system code in Figure 6.2(c) implements an ADT calculus much like that presented in Chapter 3. Whereas ADT calculus decouples the client code (the code that integrates lorenz over time) from the lorenz implemenation, ABSTRACT CALCULUS goes one step further: It decouples the client code from depending even on the lorenz interface – at least insofar as the public types and type-bound procedure interfaces comprise a concrete interface. Instead, the integration code depends only on an abstract interface specification to which the concrete interface conforms.

Figure 6.2(a)

```
 1 program main
 2   use lorenz_module ,only : lorenz,integrate
 3   implicit none   ! Prevent implicit typing
 4
 5   type(lorenz)        :: attractor
 6   integer             :: step  ! time step counter
 7   integer ,parameter :: num_steps=2000, &
 8                          space_dimension=3 ! phase space dimension
 9   real    ,parameter :: sigma=10.,rho=28.,beta=8./3.,&
10                          dt=0.01 ! Lorenz parameters and time step size
11   real    ,parameter ,dimension(space_dimension) &
12                       :: initial_condition=(/1.,1.,1./)
13
14   attractor = lorenz(initial_condition,sigma,rho,beta)
15   print *,attractor%output()
16   do step=1,num_steps
17     call integrate(attractor,dt)
18     print *,attractor%output()
19   end do
20 end program
```

Figure 6.2(b)

```
 1 module integrand_module
 2   implicit none       ! Prevent implicit typing
 3   private             ! Hide everything by default
 4   public :: integrate ! expose time integration procedure
 5
 6   type ,abstract ,public :: integrand
 7   contains
 8     procedure(time_derivative      ) ,deferred :: t
 9     procedure( symmetric_operator  ) ,deferred :: add
10     procedure( symmetric_assignment) ,deferred :: assign
11     procedure(asymmetric_operator  ) ,deferred :: multiply
12
```

```
13      generic :: operator(+) => add ! Map operators to proceures
14      generic :: operator(*) => multiply
15      generic :: assignment(=) => assign
16    end type
17
18    abstract interface
19      function time_derivative(this) result(dState_dt)
20        import :: integrand
21        class(integrand) ,intent(in)  :: this
22        class(integrand) ,allocatable :: dState_dt
23      end function time_derivative
24      function symmetric_operator(lhs,rhs) result(operator_result)
25        import :: integrand
26        class(integrand) ,intent(in)  :: lhs,rhs
27        class(integrand) ,allocatable :: operator_result
28      end function symmetric_operator
29      function asymmetric_operator(lhs,rhs) result(operator_result)
30        import :: integrand
31        class(integrand) ,intent(in)  :: lhs
32        class(integrand) ,allocatable :: operator_result
33        real                ,intent(in) :: rhs
34      end function asymmetric_operator
35      subroutine symmetric_assignment(lhs,rhs)
36        import :: integrand
37        class(integrand) ,intent(in)    :: rhs
38        class(integrand) ,intent(inout) :: lhs
39      end subroutine symmetric_assignment
40    end interface
41
42  contains
43    subroutine integrate(model,dt)
44      class(integrand) :: model
45      real ,intent(in) :: dt         ! time step size
46      model = model + model%t()*dt ! Explicit Euler formula
47    end subroutine
48  end module integrand_module
```

Figure 6.2(c)

```
1   module lorenz_module
2     use integrand_module ,only : integrand,integrate
3     implicit none
4
5     private              ! Hide everything by default
6     public :: integrate ! Expose time integration procedure
7     public :: lorenz
8
9     type ,extends(integrand) :: lorenz
10      private
11      real ,dimension(:) ,allocatable :: state    ! solution vector
12      real :: sigma ,rho ,beta                     ! Lorenz parameters
13    contains
14       procedure ,public :: t        => dLorenz_dt
15       procedure ,public :: add      => add_lorenz
16       procedure ,public :: multiply => multiply_lorenz
17       procedure ,public :: assign   => assign_lorenz
```

```
18       procedure ,public :: output
19     end type
20
21     interface lorenz
22       procedure constructor
23     end interface
24
25   contains
26
27     type(lorenz) function constructor(initial_state,s,r,b)
28       real ,dimension(:) ,intent(in)  :: initial_state
29       real               ,intent(in)  :: s ,r ,b  ! passed values for
30                                                   ! sigma, rho and beta
31       constructor%state=initial_state
32       constructor%sigma=s
33       constructor%rho=r
34       constructor%beta=b
35     end function
36
37     function output(this) result(coordinates)
38       class(lorenz)      ,intent(in)  :: this
39       real ,dimension(:) ,allocatable :: coordinates
40       coordinates = this%state
41     end function output
42
43     ! time derivative: encapsulates Lorenz equations
44     function dLorenz_dt(this) result(dState_dt)
45       class(lorenz)           ,intent(in)  :: this
46       class(integrand) ,allocatable :: dState_dt
47       type(lorenz)            ,allocatable :: delta
48
49       allocate(delta)
50       allocate(delta%state(size(this%state)))
51       ! 1st lorenz equation
52       delta%state(1)=this%sigma*( this%state(2) -this%state(1))
53       ! 2nd lorenz equation
54       delta%state(2)=this%state(1)*(this%rho-this%state(3))-this%state(2)
55       ! 3rd lorenz equation
56       delta%state(3)=this%state(1)*this%state(2)-this%beta*this%state(3)
57       ! hold Lorenz parameters constant over time
58       delta%sigma=0.
59       delta%rho=0.
60       delta%beta=0.
61       call move_alloc (delta, dState_dt)
62     end function
63
64     function add_Lorenz(lhs,rhs) result(sum) ! add two Lorenz objects
65       class(lorenz)           ,intent(in)  :: lhs
66       class(integrand) ,intent(in)  :: rhs
67       class(integrand) ,allocatable :: sum
68       type(lorenz)            ,allocatable :: local_sum
69
70       allocate (lorenz :: local_sum)
71       select type(rhs)
72         class is (lorenz)
```

```
73            local_sum%state = lhs%state + rhs%state
74            local_sum%sigma = lhs%sigma + rhs%sigma
75            local_sum%rho   = lhs%rho   + rhs%rho
76            local_sum%beta  = lhs%beta  + rhs%beta
77          class default
78            stop 'add_Lorenz: rhs argument type not supported'
79        end select
80        call move_alloc(local_sum, sum)
81      end function
82
83      ! multiply a Lorenz object by a real scalar
84      function multiply_Lorenz(lhs,rhs) result(product)
85        class(lorenz) ,intent(in)   :: lhs
86        real          ,intent(in)   :: rhs
87        class(integrand) ,allocatable :: product
88        type(lorenz)                ,allocatable :: local_product
89
90        allocate (local_product)
91        local_product%state = lhs%state*rhs
92        local_product%sigma = lhs%sigma*rhs
93        local_product%rho   = lhs%rho  *rhs
94        local_product%beta  = lhs%beta *rhs
95        call move_alloc(local_product, product)
96      end function
97
98      ! assign one lorenz object to another
99      subroutine assign_lorenz(lhs,rhs)
100       class(lorenz)              ,intent(inout) :: lhs
101       class(integrand)  ,intent(in)        :: rhs
102
103         select type(rhs)
104           class is (lorenz)
105             lhs%state = rhs%state
106             lhs%sigma = rhs%sigma
107             lhs%rho   = rhs%rho
108             lhs%beta  = rhs%beta
109           class default
110             stop 'assign_lorenz: rhs argument type not supported'
111         end select
112     end subroutine
113   end module lorenz_module
```

Figure 6.2. Fortran implementation of ABSTRACT CALCULUS applied to a Lorenz system: (a) Integration test for the Lorenz system; (b) Integrand class; (c) Lorenz class.

Figure 6.2(c) represents the first occurrence of the `select type` construct and `move_alloc` intrinsic procedure in this text. A confluence of circumstances likely to occur in most abstract calculi necessitates these language facilities. First, the deferred binding `add` in the abstract type `integrand` in Figure 6.2(b) specifies an abstract interface `symmetric_operator` that forces procedures implementing the deferred binding to accept a dummy argument declared as `class(integrand)`. This restriction cannot apply to the passed-object dummy argument `lhs`, which must be declared as `class(lorenz)` in order for the compiler to differentiate this implementation of the deferred binding from other derived types' implementations. The restriction applies to the other argument `rhs`.

Inside the `select type` construct, the code accesses `lorenz` components not included in the `integrand` abstract type specification. Since this code would fail for any extensions of `integrand` that do not contain these components, the language semantics require guarding against this possibility. Although the type determination happens at runtime, the `class is(lorenz)` type guard statement ensures that the guarded code only executes when `rhs` is a `lorenz` object or an instance of a type that extends `lorenz` and therefore contains the required components.

Were `add_Lorenz` to compute its result (`sum`) directly, an additional (nested) type guard statement would be required to ensure that the result (delcared as `class(integrand)`) has the requisite components. The desired result type is known, so this level of type guarding feels somewhat superfluous even though the deferred binding requires it. We therefore eliminate the need for it by declaring and computing a `local_sum` of type `lorenz` and transferring its allocation to the actual result `sum` via the `move_alloc` intrinsic procedure. After the `move_alloc` invocation at the end of `add_Lorenz`, `sum` will be associated with the memory originally associated with `local_sum`, whereas the latter will be unallocated. This `move_alloc` invocation ensures that the chosen method for avoiding extra type guarding does not impose additional dynamic memory utilization. For this reason, the `move_alloc` intrinsic will also play a significant role in memory recycling in Chapter 10.

In Fortran, the `select type` construct provides the only means for declared runtime type conversions because the language contains no explicit type casting facility. Similar to a `select case` construct, a `select type` allows a programmer to specify a list of possible blocks of code to execute based on the possible dynamic types of the object at runtime. Each constituent block is bound with a statement, which can be `type is` or `class is`, specifying a unique data type. A special type guard statement, `class default`, also provides an error-handling facility in case all specified types fail to match the dynamic type of the object[1]. The Fortran 2003 standard also guarantees that, at most, one of the constituent blocks executes. If none of the blocks is selected and no `class default` statement is provided, the program simply exits the construct quietly and continues to subsequent lines. Although the `select type` construct guards against crashes, it remains the programmer's responsibility to insert a `class default` case to detect and handle the scenario in which no block is selected.

In C++, the corresponding technique involves attempting a `dynamic_cast` to some useful type. If the ultimate type proves incompatible with the type being cast, however, `dynamic_cast` either returns `NULL` for casts involving a pointer or throws an exception for casts involving a reference. Handling such exceptions necessitates a conditional test (e.g., an `if` statement) or a exception-handling code (i.e., a `try-catch` block). This design approach suffers from fragility: It relies heavily on an assumption that all programmers will use safe practices. Neglecting the conditional or the exception handling can lead to catastrophic results wherein the code crashes or an incorrect result goes undetected until some later (and potentially distant) point in the code.

[1] Although C++ programmers commonly decry the lack of exception handling in Fortran, this scenario exemplifies the general rule that exception handling is less necessary in Fortran than in C++.

Even though both language constructs rely on runtime type information (RTTI) and both have pitfalls, a `select type`, in general, is safer than a `dynamic_cast` in that a `select type` only performs a downcast along the class hierarchy tree based upon the actual dynamic type (the runtime type) of the object or pointer. Downcasts are the most common situations arising from need for explicit type changes (casts) so that the programmer can access additional data members and functions of an extended type. In C++, `dynamic_cast` operators can be used for downcasts as well as upcasts. A common casting error occurs when one casts to a type on a different branch of an inheritance hierarchy[2]. The distinction between the two constructs is also apparent from the language design point of view: C++'s `dynamic_cast` relies solely on programmers' knowledge of the actual runtime type to which an object is cast, whereas Fortran's `select type` asks programmers to supply a list of possible dynamic types with corresponding execution blocks to be selected. Therefore, `dynamic_cast` is more flexible to use at the cost of placing a greater number of responsibilities on programmers.

6.2.2 C++ Implementation

Figure 6.3 provides a C++ implementation of ABSTRACT CALCULUS, following the same structure as the Fortran code in Figure 6.2. One significant difference is the inclusion of exception handling in the form of `try-catch` blocks in the `main.cpp` in Figure 6.3(a). With exception handling, runtime errors in `integrate(attractor, dt)` cause the program control to transfer to error-handling code blocks. The code blocks that catch and handle the error can be local to the routine in which the error occurred or somewhere further up the call tree in code that invoked the routine containing the `try`. Fortran lacks a corresponding capability.

Figure 6.3(a): main.cpp

```
1  #include "lorenz.h"
2  #include "fmt.h"
3  #include <iostream>
4
5  int main () {
6    using namespace std;
7    typedef lorenz::ptr_t ptr_t;
8
9    const int   num_steps=2000, space_dimension=3;
10   const float sigma=10, rho=28, beta=8.0/3.0, dt=0.01;
11   const crd_t initial_condition(space_dimension, 1.0);
12
13   ptr_t       attractor = ptr_t(new lorenz(initial_condition,
14                                   sigma,rho,beta));
15   const crd_t &output   = attractor->output();
16
17   try {
18     std::cout << fmt(output, 12, 10) << "\n";
```

[2] An upcast followed by a downcast can end up with cross-branch casting.

```
19
20      for (int step = 1; step <= num_steps; ++step) {
21        integrate (attractor, dt);
22        std::cout << fmt(output, 12, 10) << "\n";
23      }
24    } catch(std::exception &e) {
25      std::cerr << "Error exit following exception of type " << e.what()
26                << "\n";
27      return EXIT_FAILURE;
28    } catch(...) {
29      std::cerr << "Error exit following an unknown exception type\n";
30      return EXIT_FAILURE;
31    }
32    return EXIT_SUCCESS;
33  }
```

Figure 6.3(b): integrand.h

```
1   #ifndef __H_integrand__
2   #define __H_integrand__ 1
3
4   #include "globals.h"
5
6   class integrand : virtual public RefBase {
7   public:
8     typedef Ref<integrand> ptr_t;
9
10    virtual ~integrand();
11
12    virtual ptr_t d_dt() const = 0;
13    virtual void  operator+=(ptr_t) = 0;
14    virtual ptr_t operator*(real_t) const = 0;
15
16  protected:
17    integrand(const integrand&);
18    integrand();
19  };
20
21  void integrate (integrand::ptr_t, real_t);
22
23  #endif
```

Figure 6.3(c): integrand.cpp

```
1   #include "integrand.h"
2
3   typedef integrand::ptr_t ptr_t;
4
5   integrand::integrand() : RefBase() {
6   }
7
8   integrand::integrand(const integrand&) : RefBase(){
9   }
10
11  integrand::~integrand() {
12  }
13
```

```
14 void integrate (ptr_t model, real_t dt) {
15   *model += *(model->d_dt()) * dt;
16 }
```

Figure 6.3(d): lorenz.h

```
1  #ifndef __H_LORENZ__
2  #define __H_LORENZ__ 1
3
4  #include "integrand.h"
5
6  class lorenz : public integrand {
7  public:
8    typedef Ref<lorenz> ptr_t;
9    lorenz ();
10   lorenz (const crd_t, real_t sigma, real_t rho, real_t beta);
11   // Default copy and assignment operators are just fine for this type.
12
13 public:
14   integrand::ptr_t   d_dt() const;
15   void               operator+=(integrand::ptr_t other);
16   integrand::ptr_t   operator*(float val) const;
17
18   const crd_t&              output() const;
19
20   virtual ~lorenz();
21
22 private:
23   crd_t state_;     // solution vector.
24   float sigma_, rho_, beta_;
25 };
26
27 #endif
```

Figure 6.3(e): lorenz.cpp

```
1  #include <iostream>
2  #include <exception>
3
4  #include "lorenz.h"
5
6  using namespace std;
7
8  struct LorenzError : public std::exception {
9    virtual ~LorenzError() throw() {}
10 };
11
12 // default constructor
13 lorenz::lorenz ()
14 {}
15
16 // constructor using each element
17 lorenz::lorenz (const crd_t initial_state, real_t s, real_t r, real_t b)
18        :  state_(initial_state), sigma_(s), rho_(r), beta_(b)
19 {}
20
21 const crd_t& lorenz::output() const {
```

```
22    return state_;
23  }
24
25  integrand::ptr_t lorenz:: d_dt() const
26  {
27    ptr_t result = ptr_t(new lorenz);
28    result->state_.resize(3);
29    result->state_.at(0) = sigma_*(state_.at(1) - state_.at(0));
30    result->state_.at(1) = state_.at(0)*(rho_ - state_.at(2))
31                         - state_.at(1);
32    result->state_.at(2) = state_.at(0)*state_.at(1) - beta_*state_.at(2);
33    return result;
34  }
35
36
37  void lorenz::operator+=(integrand::ptr_t rhs) {
38    ptr_t other = cast<lorenz>(rhs);
39    if(other == NULL) {
40      std::cerr << "lorenz::operator+=:  Failed dynamic cast\n";
41      throw LorenzError();
42    }
43    if(other->state_.size() != this->state_.size()) {
44      std::cerr << "lorenz::operator+=:  Non-identical dimensions.\n";
45      throw LorenzError();
46    }
47
48    for(size_t i = 0; i < state_.size(); ++i) {
49      state_.at(i) += other->state_.at(i);
50    }
51  }
52
53  integrand::ptr_t lorenz::operator*(real_t rhs) const
54  {
55    ptr_t result = ptr_t(new lorenz(*this));
56    for(size_t i = 0; i < result->state_.size(); ++i) {
57      result->state_.at(i) *= rhs;
58    }
59    return result;
60  }
61
62  lorenz::~lorenz()
63  {}
```

Figure 6.3. C++ implementation of an abstract calculus for the Lorenz system: (a) Integration test, (b) Integrand declaration, (c) Integrand implementation, (d) Lorenz declaration, and (e) Lorenz implementation.

6.3 The Consequences

In ABSTRACT CALCULUS, obvious benefits accrue from the simplicity, clarity, and generality of the interface. Only a few operators must be defined to construct any new dynamical system. Their purpose very closely mirrors the corresponding operators in the mathematical statement of the system and the time marching algorithm. Often the internals of each operator relies on sufficiently simple calculations that they can delegate their work to highly optimized versions of linear algebra libraries such as BLAS (Blackford et al. 2002) and LAPACK (Barker et al. 2001) or scalable parallel libraries such as ScaLAPACK (Blackford et al. 1997), PETSc (Balay et al. 2007), or Trilinos (Heroux et al. 2005). We detail the process of building atop Trilinos in Chapter 11.

A common thread running through many design patterns is their fostering of loose couplings between abstractions. Unlike procedural time-integration libraries and even many object-oriented ones, `integrate()` never sees the data it integrates. It therefore relies on no assumptions about the data's layout in memory, its type, or its precision. The programmer retains the freedom to restructure the data or change its type or precision with absolutely no impact on the time-integration code.

Sometimes, one developer's benefit is another's drawback. For example, if implemented naively, one potential drawback of ABSTRACT CALCULUS lies in making it more challenging for the compiler to optimize. Section 3.3 delineated several of the corresponding challenges for ADT calculus and potential workarounds. Naturally, these same potential pitfalls and solutions apply to ABSTRACT CALCULUS, because this pattern simply creates a unified interface to an ADT calculus supported by any classes that implement the interface.

The benefits of the ABSTRACT CALCULUS over type-specific ADT calculus stem from the uniformity the pattern imposes across the elements of a package and the resulting ability to vary the dynamic type of a mathematical entity at execution time. The cost paid for this is the runtime overhead involved in discovering that type on the fly and, more importantly, the seeming impossibility of performing the static analysis of the code required for the more advanced optimization strategies – particularly interprocedural optimizations. This returns again to the theme of coarse-grained data structure design that facilitates large numbers of computations inside each operator, tied to a sound understanding of theoretical operation-count predictions and a disciplined approach to empirical profiling.

Ultimately the highest virtue to which ABSTRACT CALCULUS aspires harkens back to *Timeless Way* (Section 4.2.1), which attempts to breathe life into designs by releasing one's innate understanding of the application domain. For those trained in scientific simulation, the underlying mathematics eventually feels innate. ABSTRACT CALCULUS takes the conversations that historically occurred at the blackboard to the screen. This is a dialogue spoken in pattern languages. Although each person possesses a pattern language unique to their perspective and life experiences (some prefer vector notation, others tensor notation), providing an elementary set of common operations enables users to extend the classes so developed in ways that reflect their predilections. In doing so, it draws back the curtain and attracts software users into a dialogue previously limited to expert developers.

6.4 Related Patterns

ABSTRACT CALCULUS and PUPPETEER, the two domain-specific patterns in this text, have close counterparts in the GoF patterns. Most design patterns discussions list such pattern relationships under the heading "Also Known As." Outside of the scientific simulation domain, a pattern with the same structural form and purpose as ABSTRACT CALCULUS would be known as the GoF FACADE pattern. Much as the facade described in Section 4.2.3 presents a simplified and unified interface to two complicated subsystems, so does ABSTRACT CALCULUS greatly simplify and unify the top-level interface to a multiphysics solver. As formulated, this pattern supports writing expressions on a single class of objects at a time without reference to, or a mechanism for, combining classes.

Chapter 8 outlines how to aggregate multiple classes into one entity, a PUPPETEER, for writing ABSTRACT CALCULUS expressions. From one viewpoint, a PUPPETEER serves merely as the concrete side of the double-faceted FACADE structure described in section 4.2.3. It could therefore be treated as part of an ABSTRACT CALCULUS. However, it plays an additional role that warrants treating it as a separate pattern. Chapter 8 explains further.

The ordering of our patterns chapters draws inspiration from A Pattern Language (Alexander et al. 1975). Just as they started broadest patterns for towns and regions and proceeded to smaller patterns that complete the larger ones, ABSTRACT CALCULUS sets functionality requirements that propagate downward throughout the remainder of the design. In one sense, the remainder of Part II concerns laying out patterns necessary to complete an ABSTRACT CALCULUS in a complex, multiphysics setting.

Were we presenting a complete pattern language in the style of Alexander et al. (1977), we would express the relationships between patterns in a graph with nodes representing patterns and edges drawn between the nodes to indicate which patterns complete which other patterns. Instead, we focus Part II on a core set of patterns that would likely all be involved in completing a solver.

EXERCISES

1. Write an abstract type or pure virtual class that defines the operators and methods required to time advance the heat conduction equation used throughout Part I of this text.
2. Write a concrete ADT that implements the interface defined in Exercise 1 and solves the heat conduction equations. Explore varying approaches to code reuse: (a) aggregate an instance of one of the heat conduction ADT's from Part I and delegate calculations to that object, and (b) modify one of the earlier heat conduction ADTs so that it extends the interface defined in Exercise 1 and directly associates any deferred type-bound procedures or pure virtual functions with those in the modified ADT. Discuss the benefits and drawbacks of each approach in terms of code construction costs and maintainability.

7 The Strategy and Surrogate Patterns

> "However beautiful the strategy, you should occasionally look at the results."
>
> Winston Churchill

7.1 The Problem

This chapter introduces the GoF STRATEGY pattern along with a Fortran-specific, enabling pattern: SURROGATE. In scientific programming, one finds context for a STRATEGY when the choice of numerical algorithms must evolve dynamically. Multiphysics modeling, for example, typically implies multi-numerics modeling. As the physics changes, so must the numerical methods.

A problem arises when the software does not separate its expression of the physics from its expression of the discrete algorithms. In the Lorenz system, for example, equation (4.5) expresses the physics, whereas equation (4.6) expresses the discretization. Consider a concrete ADT that solves the same equations as the abstractions presented in Section 6.2 and extends the `integrand` ADT but requires a new time integration algorithm. The extended type must overload the `integrate` name. This would be a simple task when changing one algorithm in one particular ADT. However, when numerous algorithms exist, one faces the dilemma of either putting all possible algorithms in the parent type or leaving to extended types the task of each implementing their own algorithms. Section 7.3 explains the adverse impact the first option has on code maintainability. The second option could lead to redundant (and possibly inconsistent) implementations.

The Lorenz equation solver described in Chapter 6 uses explicit Euler time advancement. That solver updates the solution vector at each time step without explicitly storing a corresponding time coordinate. Such an approach might be appropriate when one requires only a representative set of points in the problem phase space to calculate geometrical features of the solution such as the fractal dimension of its strange attractor. By contrast, if one desires temporal details of the solution trajectory, then it might be useful to create an extended type that stores a time stamp and uses a marching algorithm with higher-order accuracy in time such as 2nd-order

Runge-Kutta (RK2). RK2 advances the solution in time according to:

$$\vec{V}' = \vec{V}^n + \vec{R}\left(\vec{V}^n\right)\frac{\Delta t}{2} \quad (7.1)$$

$$\vec{V}^{n+1} = \vec{V}^n + \vec{R}\left(\vec{V}'\right)\Delta t \quad (7.2)$$

which has been written in a predictor/corrector form involving two substeps but can also easily be combined into a single step.

Encapsulating the state vector $\vec{V}$ in an object and time advancing that object via ABSTRACT CALCULUS makes sense if the overloaded and defined operations in that calculus have reasonable mathematical interpretations when applied to each state vector component. More specifically, the overloaded forms of equations (7.1)–(7.2) have meaningful interpretations if each component of the state vector satisfies a differential equation. Time is a state vector component in the current context, so we augment the governing equations (4.5) with a fourth equation:

$$\frac{d\tau}{dt} = 1 \quad (7.3)$$

where τ is the object's time stamp. The challenge is to extend the `lorenz` type by adding the new component τ, adding its governing equation (7.3), and adding the ability to select an integration strategy dynamically at runtime.

Obviously the driving force in this context is the need for flexibility. An opposing force is a desire for simplicity. Placing the responsibility for implementing new numerical quadrature algorithm in the hands of each `integrand` descendant requires little effort on the part of the `integrand` designer and is thus the simplest approach from a design standpoint. Conversely, requiring the `integrand` designer to implement each new algorithm requires greater effort on the part of the `integrand` designer, but still retains the simplicity of having all the algorithms in one abstraction and under the control of that abstraction's developers.

The next section discusses how STRATEGY resolves the conflicting forces in the current design context. The Fortran implementation of STRATEGY provides the context for the SURROGATE pattern. We therefore discuss the problem a SURROGATE solves in the Fortran implementation section 7.2.1.

7.2 The Solution

The GoF resolved this problem with the STRATEGY pattern. They stated the essence of this pattern as follows:

> Define a family of algorithms, encapsulate each one, and make them interchangeable. Strategy lets the algorithm vary independently from clients that use it.

A STRATEGY severs the link between algorithms and data. Data objects delegate operations to Strategy classes that apply appropriate algorithms to the data.

Our STRATEGY implementation defines a `timed_lorenz` type that extends the `lorenz` type. As before, the `lorenz` concrete type extends the `integrand` abstract type. In addition to the deferred bindings that `integrand` stipulates must be implemented by its extended types, it now contains a reference to an abstract derived type, `strategy`, to which `integrand` delegates the responsibility to provide a type-bound `integrate()` procedure. The `strategy` defines only the interface (via deferred binding) for the time-integration method, leaving its own extended types to provide actual quadrature schemes. In applying the strategy pattern, one passes a reference to a `lorenz` dynamical system as an actual argument to the `integrate()` method of the `strategy` object. Again, the program syntax mirrors the mathematical syntax:

```
class(surrogate) ,intent(inout) :: this
real ,intent(in) :: dt
...
this_half = this + this%t()*(0.5*dt)
this = this + this_half%t()*dt
```

We explain the role of the new `surrogate` type next.

Figure 7.1 depicts a UML class model of our Fortran STRATEGY implementation, including an empty abstract `surrogate` type, instances of which substitute for all references to `integrand` objects and their extended types when referenced inside the `strategy` hierarchy. Declaring all such instances with the Fortran `class` keyword defers to runtime the resolution of their ultimate dynamic type, which can be any descendant of the `surrogate` type. The role played by the `surrogate` provides an example of a Fortran-specific design pattern. A SURROGATE circumvents Fortran's prohibition against circular references, wherein one module's reference to a second module via a `use` statement precludes referencing the first module in the second via a reciprocal `use`. In C++, one avoids such circular references by using *forward references*. Forward references allow one ADT to declare another and then to manipulate references to the declared ADT without knowing its definition.

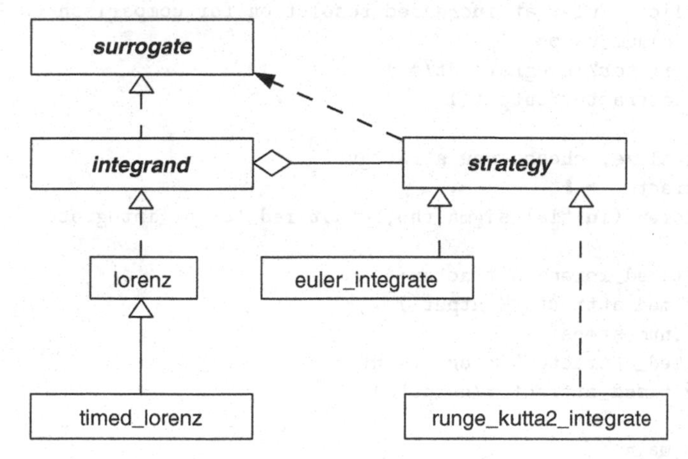

Figure 7.1. Fortran class model for applying the STRATEGY and SURROGATE patterns to extend the Lorenz model.

7.2.1 Fortran Implementation

Figure 7.2 provides STRATEGY and SURROGATE pattern demonstrations in Fortran. The figure lists the codes top-down, starting with `main` in Figure 7.2(a) and progressing through the UML class diagram of Figure 7.1 from the `surrogate` in Figure 7.2(b), down through the `integrand` in Figure 7.2(c), and the `strategy` in Figure 7.2(d). From there, the listing proceeds down the left-hand branch of the class diagram through the `lorenz` implementation of Figure 7.2(e) and the `timed_lorenz` implementation of Figure 7.2(f). Figures 7.2(g) and 7.2(h) traverse the right branch, comprising the `explicit_euler` and `runge_kutta_2nd` implementations.

Figure 7.2(a): main.f03

```
1  program main
2    use lorenz_module ,only : lorenz
3    use timed_lorenz_module ,only : timed_lorenz
4    use explicit_euler_module ,only : explicit_euler
5    use runge_kutta_2nd_module ,only : runge_kutta_2nd
6    implicit none ! Prevent implicit typing
7
8    type(explicit_euler)  :: lorenz_integrator       !Integration strategy
9    type(runge_kutta_2nd) :: timed_lorenz_integrator !Integration strategy
10   type(lorenz)          :: attractor ! Lorenz equation/state abstraction
11   type(timed_lorenz)    :: timed_attractor ! Time-stamped abstraction
12   integer               :: step            ! Time step counter
13   integer ,parameter    :: num_steps=2000,space_dimension=3
14
15   ! Lorenz parameters and step size
16   real    ,parameter    :: sigma=10.,rho=28.,beta=8./3.,dt=0.01
17   real    ,parameter ,dimension(space_dimension) &
18                         :: initial=(/1.,1.,1./)
19
20   ! Initialize and choose strategy
21   attractor = &
22     lorenz(initial,sigma,rho,beta,lorenz_integrator)
23   print *,'lorenz attractor:'
24   print *,attractor%output()
25
26   ! Run explicit Euler at increased resolution for comparison to RK2
27   do step=1,4*num_steps
28     call attractor%integrate(dt/4.)
29     print *,attractor%output()
30   end do
31   ! Re-initialize, choose new strategy
32   timed_attractor = &
33     timed_lorenz(initial,sigma,rho,beta,timed_lorenz_integrator)
34   print *,''
35   print *,'timed_lorenz attractor:'
36   print *,timed_attractor%output()
37   do step=1,num_steps
38     call timed_attractor%integrate(dt)
39     print *,timed_attractor%output()
40   end do
41 end program main
```

Figure 7.2(b): surrogate.f03

```
1  module surrogate_module
2    implicit none                    ! Prevent implicit typing
3    private                          ! Hide everything by default
4    public :: surrogate
5
6    ! This stateless type serves only for purposes of extension by other
7    ! types.  In such a role, it can serve as a substitute for the child
8    ! type when that type is inaccessible because of Fortran's
9    ! prohibition against circular references.
10
11   type ,abstract :: surrogate
12   end type
13 end module
```

Figure 7.2(c): integrand.f03

```
1  module integrand_module
2    use surrogate_module ,only : surrogate ! Integrand parent
3    use strategy_module ,only : strategy   ! Integration strategy parent
4    implicit none ! Prevent implicit typing
5    private       ! Hide everything by default
6
7    type ,abstract ,public, extends(surrogate) :: integrand
8      private
9      class(strategy), allocatable :: quadrature
10   contains
11     procedure, non_overridable :: integrate   ! Time integrator
12     procedure, non_overridable :: set_quadrature
13     procedure, non_overridable :: get_quadrature
14     procedure(time_derivative) ,deferred :: t ! Time derivative that
15                                               ! evaluates evolution
16                                               ! equations
17
18     procedure(symmetric_operator) ,deferred :: add
19     procedure(asymmetric_operator),deferred :: multiply
20     procedure(symmetric_assignment) ,deferred :: assign
21
22     ! Map operators to corresponding procedures
23     generic                                  :: operator(+) => add
24     generic                                  :: operator(*) => multiply
25     generic                                  :: assignment(=) => assign
26   end type
27
28   abstract interface
29     function time_derivative(this) result(dState_dt)
30       import :: integrand
31       class(integrand) ,intent(in)  :: this
32       class(integrand) ,allocatable :: dState_dt
33     end function time_derivative
34     function symmetric_operator(lhs,rhs) result(operator_result)
35       import :: integrand
36       class(integrand) ,intent(in)  :: lhs,rhs
37       class(integrand) ,allocatable :: operator_result
38     end function symmetric_operator
39     function asymmetric_operator(lhs,rhs) result(operator_result)
```

```
40        import :: integrand
41        class(integrand) ,intent(in)  :: lhs
42        class(integrand) ,allocatable :: operator_result
43        real             ,intent(in) :: rhs
44      end function asymmetric_operator
45      subroutine symmetric_assignment(lhs,rhs)
46        import :: integrand
47        class(integrand) ,intent(in)    :: rhs
48        class(integrand) ,intent(inout) :: lhs
49      end subroutine symmetric_assignment
50    end interface
51
52  contains
53
54    subroutine set_quadrature (this, s)
55      class(integrand), intent(inout) :: this
56      class(strategy), intent(in) :: s
57
58      if (allocated(this%quadrature)) deallocate (this%quadrature)
59
60      allocate (this%quadrature, source=s)
61    end subroutine
62
63    function get_quadrature (this) result (this_strategy)
64      class(integrand), intent(in) :: this
65      class(strategy), allocatable :: this_strategy
66
67      allocate (this_strategy, source=this%quadrature)
68    end function
69
70    subroutine integrate(model,dt)
71      class(integrand) :: model        ! integrand
72      real ,intent(in)        :: dt    ! time step size
73      if (allocated(model%quadrature)) then
74        call model%quadrature%integrate(model,dt)
75      else
76        stop 'integrate: no integration procedure available.'
77      end if
78    end subroutine
79  end module integrand_module
```

Figure 7.2(d): strategy.f03

```
1   module strategy_module
2     ! Substitute for integrand (avoiding circular references)
3     use surrogate_module ,only : surrogate
4
5     implicit none               ! Prevent implicit typing
6     private                     ! Hide everything by default
7
8     ! Abstract time integration strategy
9     type, abstract ,public :: strategy
10    contains
11      ! Abstract integration procedure interface
12      procedure(integrator_interface), nopass, deferred :: integrate
```

```
13   end type strategy
14
15   abstract interface
16     subroutine integrator_interface(this,dt)
17       import :: surrogate
18       class(surrogate) ,intent(inout) :: this ! integrand
19       real             ,intent(in)    :: dt   ! time step size
20     end subroutine
21   end interface
22 end module
```

Figure 7.2(e): lorenz.f03

```
1  module lorenz_module
2    use strategy_module ,only : strategy   ! time integration strategy
3    use integrand_module ,only : integrand ! Abstract integrand
4    implicit none ! Prevent implicit typing
5    private       ! Hide everything by default
6
7    public :: integrand ! Expose integrand type
8    public :: lorenz    ! Expose lorenz type and constructor
9
10   type ,extends(integrand) ,public :: lorenz
11     private
12     real ,dimension(:) ,allocatable :: state   ! solution vector
13     real :: sigma ,rho ,beta                   ! Lorenz parameters
14   contains
15      procedure ,public :: t  => dlorenz_dt ! Time deriv. (eval. RHS)
16      procedure ,public :: add        => add_lorenz
17      procedure ,public :: multiply   => multiply_lorenz
18      procedure ,public :: assign     => assign_lorenz
19      procedure ,public :: output             ! Accessor: return state
20   end type
21
22   interface lorenz
23     procedure constructor
24   end interface
25
26 contains
27
28   ! Constructor: allocate and initialize
29   type(lorenz) function constructor(initial_state,s,r,b,this_strategy)
30     real ,dimension(:) ,intent(in)  :: initial_state
31     real               ,intent(in)  :: s ,r ,b
32     class(strategy)    ,intent(in)  :: this_strategy
33     ! constructor%state is automatically allocated by the assignment
34     constructor%state=initial_state
35     constructor%sigma=s
36     constructor%rho=r
37     constructor%beta=b
38     call constructor%set_quadrature(this_strategy)
39   end function
40
41   ! Time derivative (specifies evolution equations)
42   function dLorenz_dt(this) result(dState_dt)
```

```
43      class(lorenz)    ,intent(in)  :: this
44      class(integrand) ,allocatable :: dState_dt
45      type(lorenz)     ,allocatable :: local_dState_dt
46      allocate(local_dState_dt)
47      call local_dState_dt%set_quadrature(this%get_quadrature())
48      allocate(local_dState_dt%state(size(this%state)))
49      ! 1st Lorenz equation
50      local_dState_dt%state(1) = this%sigma*(this%state(2)-this%state(1))
51      ! 2nd Lorenz equation
52      local_dState_dt%state(2) = this%state(1)*(this%rho-this%state(3)) &
53                                 -this%state(2)
54      ! 3rd Lorenz equation
55      local_dState_dt%state(3) = this%state(1)*this%state(2)            &
56                                 -this%beta*this%state(3)
57      local_dState_dt%sigma    = 0.
58      local_dState_dt%rho      = 0.
59      local_dState_dt%beta     = 0.
60      ! transfer the allocation from local_dState_dt to dState_dt
61      call move_alloc(local_dState_dt,dState_dt)
62    end function
63
64    ! Add two instances of type lorenz
65    function add_lorenz(lhs,rhs) result(sum)
66      class(lorenz)    ,intent(in)  :: lhs
67      class(integrand) ,intent(in)  :: rhs
68      class(integrand) ,allocatable :: sum
69      type(lorenz)     ,allocatable :: local_sum
70      select type(rhs)
71        class is (lorenz)
72          allocate(local_sum)
73          call local_sum%set_quadrature(lhs%get_quadrature())
74          ! local_sum%state is automatically allocated by assignment
75          local_sum%state = lhs%state + rhs%state
76          local_sum%sigma = lhs%sigma + rhs%sigma
77          local_sum%rho   = lhs%rho   + rhs%rho
78          local_sum%beta  = lhs%beta  + rhs%beta
79        class default
80          stop 'assig_lorenz: unsupported class'
81      end select
82      ! no additional allocation needed by using move_alloc
83      call move_alloc(local_sum,sum)
84    end function
85
86    ! Multiply an instance of lorenz by a real scalar
87    function multiply_lorenz(lhs,rhs) result(product)
88      class(lorenz)    ,intent(in)  :: lhs
89      real             ,intent(in)  :: rhs
90      class(integrand) ,allocatable :: product
91      type(lorenz)     ,allocatable :: local_product
92      allocate(local_product)
93      call local_product%set_quadrature(lhs%get_quadrature())
94      ! local_product%state is automatically allocated by assignment
95      local_product%state = lhs%state* rhs
96      local_product%sigma = lhs%sigma* rhs
97      local_product%rho   = lhs%rho   * rhs
```

```
98      local_product%beta  = lhs%beta * rhs
99      ! avoid unnecessary memory allocation by using move_alloc
100     call move_alloc(local_product,product)
101   end function
102
103   ! Assign one instance to another
104   subroutine assign_lorenz(lhs,rhs)
105     class(lorenz)    ,intent(inout) :: lhs
106     class(integrand) ,intent(in)    :: rhs
107     select type(rhs)
108       class is (lorenz)
109         ! let assignment automatically allocate lhs%state
110         lhs%state = rhs%state
111         lhs%sigma = rhs%sigma
112         lhs%rho   = rhs%rho
113         lhs%beta  = rhs%beta
114         call lhs%set_quadrature(rhs%get_quadrature())
115       class default
116         stop 'assign_lorenz: unsupported class'
117     end select
118   end subroutine
119
120   ! Accessor: return state
121   function output(this) result(coordinates)
122     class(lorenz)       ,intent(in)  :: this
123     real ,dimension(:) ,allocatable :: coordinates
124     ! assignment allocates coordinates automatically
125     coordinates = this%state
126   end function output
127 end module lorenz_module
```

Figure 7.2(f): timed_lorenz.f03

```
1  module timed_lorenz_module
2    use lorenz_module  ,only :lorenz,integrand ! Parent and grandparent
3    use strategy_module,only :strategy          ! Time integration strategy
4    implicit none              ! Prevent implicit typing
5    private                    ! Hide everything by default
6    public :: integrand        ! Expose abstract integrand and integrator
7    public :: timed_lorenz
8
9    type ,extends(lorenz) :: timed_lorenz
10     private
11     real :: time                             ! time stamp
12   contains
13     procedure ,public :: t => dTimed_lorenz_dt ! time deriv. (eval. RHS)
14     procedure ,public :: add      => add_timed_lorenz
15     procedure ,public :: multiply  => multiply_timed_lorenz
16     procedure ,public :: assign    => assign_timed_lorenz
17     procedure ,public :: output                ! accessor: return state
18   end type timed_lorenz
19
20   interface timed_lorenz
21     procedure constructor
22   end interface
```

```
23
24  contains
25
26    ! Constructor: allocate and initialize state
27    type(timed_lorenz) function constructor(initial,s,r,b,this_strategy)
28      real ,dimension(:)  ,intent(in)  :: initial
29      real                ,intent(in)  :: s ,r ,b ! Lorenz parameters:
30                                                  ! sigma, rho and beta
31      class(strategy),intent(in) :: this_strategy ! marching algorithm
32      constructor%lorenz = lorenz &
33        (initial_state=initial,s=s,r=r,b=b,this_strategy=this_strategy)
34      constructor%time    = 0.
35    end function
36
37    ! time derivative (expresses evolution equations)
38    function dTimed_lorenz_dt(this) result(dState_dt)
39      class(timed_lorenz),intent(in)  :: this
40      class(integrand)   ,allocatable :: dState_dt
41      type(timed_lorenz) ,allocatable :: local_dState_dt
42
43      allocate(local_dState_dt)
44      local_dState_dt%time   = 1.                     ! dt/dt = 1.
45      local_dState_dt%lorenz = this%lorenz%t()        ! delegate to parent
46      ! avoid unnecessary memory allocation
47      call move_alloc(local_dState_dt,dState_dt)
48    end function
49
50    ! add two instances of timed_lorenz
51    function add_timed_lorenz(lhs,rhs) result(sum)
52      class(timed_lorenz)      ,intent(in)  :: lhs
53      class(integrand)         ,intent(in)  :: rhs
54      class(integrand)         ,allocatable :: sum
55      type(timed_lorenz)       ,allocatable :: local_sum
56
57      select type(rhs)
58        class is (timed_lorenz)
59          allocate(local_sum)
60          local_sum%time   = lhs%time   + rhs%time
61          local_sum%lorenz = lhs%lorenz + rhs%lorenz
62        class default
63          stop 'add_timed_lorenz: type not supported'
64      end select
65
66      ! avoid unnecessary memory allocation
67      call move_alloc(local_sum,sum)
68    end function
69
70    ! multiply one instance of timed_lorenz by a real scalar
71    function multiply_timed_lorenz(lhs,rhs) result(product)
72      class(timed_lorenz)      ,intent(in)  :: lhs
73      real                     ,intent(in)  :: rhs
74      class(integrand)         ,allocatable :: product
75      type(timed_lorenz)       ,allocatable :: local_product
76
77      allocate(local_product)
```

```
78      local_product%time   = lhs%time   * rhs
79      local_product%lorenz = lhs%lorenz* rhs
80
81      ! transfer allocation from local_product to result product
82      call move_alloc(local_product,product)
83    end function
84
85    ! assign one instance of timed_lorenz to another
86    subroutine assign_timed_lorenz(lhs,rhs)
87      class(timed_lorenz)     ,intent(inout) :: lhs
88      class(integrand)        ,intent(in)    :: rhs
89      select type(rhs)
90        class is (timed_lorenz)
91          lhs%time   = rhs%time
92          lhs%lorenz = rhs%lorenz
93        class default
94          stop 'assign_timed_lorenz: type not supported'
95      end select
96    end subroutine
97
98    ! return state
99    function output(this) result(coordinates)
100     class(timed_lorenz) ,intent(in)  :: this
101     real ,dimension(:)  ,allocatable :: coordinates
102
103     ! assignment automatically allocates coordinates
104     coordinates = [ this%time, this%lorenz%output() ]
105   end function
106 end module timed_lorenz_module
```

Figure 7.2(g): explicit_euler.f03

```
1   module explicit_euler_module
2     use surrogate_module ,only : surrogate ! time integration strategy
3     use strategy_module ,only : strategy   ! time integration strategy
4     use integrand_module ,only : integrand ! abstract integrand
5
6     implicit none ! Prevent implicit typing
7     private       ! Hide everything by default
8
9     ! 1st-order explicit time integrator
10    type, extends(strategy) ,public :: explicit_euler
11    contains
12      procedure, nopass :: integrate
13    end type
14
15  contains
16
17    subroutine integrate(this,dt) ! Time integrator
18      class(surrogate) ,intent(inout) :: this ! integrand
19      real             ,intent(in)    :: dt   ! time step size
20      select type (this)
21        class is (integrand)
22          this = this + this%t()*dt ! Explicit Euler formula
23        class default
```

```
24         stop 'integrate: unsupported class.'
25       end select
26     end subroutine
27   end module
```

Figure 7.2(h): runge_kutta_2nd.f03

```
 1 module runge_kutta_2nd_module
 2   use surrogate_module,only : surrogate
 3   use strategy_module ,only : strategy   ! time integration strategy
 4   use integrand_module,only : integrand  ! abstract integrand
 5
 6   implicit none                          ! Prevent implicit typing
 7   private                                ! Hide everything by default
 8
 9   ! 2nd-order Runge-Kutta time integration
10   type, extends(strategy) ,public :: runge_kutta_2nd
11   contains
12     procedure, nopass :: integrate       ! integration procedure
13   end type
14
15 contains
16
17   ! Time integrator
18   subroutine integrate(this,dt)
19     class(surrogate) ,intent(inout) :: this      ! integrand
20     real             ,intent(in)    :: dt        ! time step size
21     class(integrand) ,allocatable   :: this_half ! function evaluation
22                                                  ! at interval t+dt/2.
23
24     select type (this)
25       class is (integrand)
26         allocate(this_half,source=this)
27         this_half = this + this%t()*(0.5*dt)     ! predictor step
28         this      = this + this_half%t()*dt      ! corrector step
29       class default
30         stop 'integrate: unsupported class'
31     end select
32   end subroutine
33 end module
```

Figure 7.2. Fortran strategy pattern implementation for Lorenz and timed Lorenz systems: (a) integration test of Lorenz and time-stamped Lorenz systems, (b) abstract surrogate, (d) abstract strategy, (c) abstract integrand, (e) Lorenz ADT, (f) time-stamped Lorenz ADT, (g) explicit Euler strategy, and (h) RK2 strategy.

One construct appearing for the first time in this text is the `non_overridable` attribute applied to the `integrate` procedure binding at line 11 in Figure 7.2(c). This attribute ensures that types that extend `integrand` retain its implementation of that binding. This ensures consistency in the sense that extended types will not be able to overload the `integrate` name. Each will instead have access to the same collection of integration methods implemented in a separate inheritance hierarchy.

Another construct making its first appearance is the `nopass` attribute appearing on line 12 in Figures 7.2(d), (g), and (h). It instructs the compiler to not pass as an argument the object on which the type-bound procedure is invoked. Instead, the user must pass the argument (i.e., the integrand) explicitly. This simply forces the procedure invocation syntax, say, integrate(x,dt), to mirror the corresponding mathematical syntax $\int x dt$. Although one must prepend x% to the procedure invocation, this could be avoided by defining a module procedure that in turn invokes the type-bound procedure.

The previously used intrinsic procedure allocate appears in a new form: sourced allocation. This form appears first at line 60 in Figure 7.2(c):

```
allocate (this%quadrature, source=s)
```

where the new source argument ensures that this%quadrature takes on the value and dynamic type of s, which must be of a concrete strategy type. Another sourced allocation occurs at line 26 in Figure 7.2(h):

```
allocate(this_half,source=this)
```

after which the immediately subsequent line overwrites this_half and thereby negates the utility of copying the value of this. Fortran 2008 provides a new molded allocation of the form:

```
allocate(this_half,mold=this)
```

which copies only the dynamic type of this, not its value.

Beginning with this chapter, the software problems being solved exhibit sufficient complexity that the code length exceeds the length of the related text. This increases the expectation that the codes will be intelligible in a self-contained way. We therefore increase the frequency with which we document our intent in comments. We also continue to attempt to write code that comments itself in the manner described in Chapter 1. Also, the complexity of the examples in this chapter begin to show the real power and benefit of UML, as Figure 7.1 succinctly summarizes the structural relationships between the different parts of the code in Figure 7.2.

Our strategies for dealing with code length evolve as the trend toward longer codes continues unabated in subsequent chapters. In Chapter 8, we shift the code listings to the ends of the relevant sections to preserve continuity in the text. In Chapters 11 and 12, where we discuss multipackage, open-source projects with code listings in the tens of thousands of lines, we shift to an even greater reliance on UML supplemented by illustrative code snippets.

7.2.2 C++ Implementation

Figure 7.3 provides C++ STRATEGY code corresponding to the Fortran code listed in Figure 7.2. As shown at line 7 in Figure 7.3(b), the strategy class forward references the integrand class in its class definition, thus obviating the need for a SURROGATE C++ class. In addition, because strategy extends class RefBase, it becomes eligible as a parameter type for the reference counted pointer, Ref<>. Class integrand, as shown at line 28 in Figure 7.3(c), uses a such reference counted strategy for its data member quadrature_. This marks the first use of a reference counted pointer in our C++ examples that imitates the behaviors of a polymorphic allocatable component in a derived type definition in Fortran. The corresponding Fortran declaration is found at line 9 of Figure 7.2(c).

Figure 7.3(a): main.cpp

```
1  #include "timed_lorenz.h"
2  #include "explicit_euler.h"
3  #include "runge_kutta_2nd.h"
4  #include "fmt.h"
5  #include <iostream>
6
7  int main ()
8  {
9    typedef lorenz::ptr_t ptr_t;
10
11   static const int num_steps=2000;
12   const real_t  sigma=10., rho=28., beta=8./3., dt=0.01;
13   crd_t initial_condition(3, 1.0);
14
15   Ref<lorenz> attractor = new lorenz(initial_condition, sigma, rho, beta,
16                                      new explicit_euler);
17   std::cout << " lorenz attractor:\n"
18        << fmt(attractor->coordinate(), 12, 9) << std::endl;
19   for (int step = 0; step < 4*num_steps; ++step)
20   {
21     attractor->integrate(0.25*dt);
22     std::cout << fmt(attractor->coordinate(), 12, 9) << std::endl;
23   }
24
25   Ref<timed_lorenz> timed_attractor
26     = new timed_lorenz(initial_condition, sigma, rho, beta,
27                                         new runge_kutta_2nd);
28   std::cout << "\n timed_lorenz attractor:\n"
29             << fmt(timed_attractor->get_time(), 12, 9) << " "
30             << fmt(timed_attractor->coordinate(), 12, 9) << std::endl;
31   for (int i = 0; i < num_steps; ++i)
32   {
33     timed_attractor->integrate(dt);
34     std::cout << fmt(timed_attractor->get_time(), 12, 9) << " "
35          << fmt(timed_attractor->coordinate(), 12, 9) << std::endl;
36   }
37
38   return 0;
39 }
```

Figure 7.3(b): strategy.h

```
1  #ifndef _H_STRATEGY_
2  #define _H_STRATEGY_
3
4  #include "globals.h"
5  #include "RefBase.h"
6
7  class integrand;
8  class strategy : public RefBase {
9  public:
```

```
10    typedef Ref<strategy> ptr_t;
11    typedef Ref<integrand> model_t;
12
13    virtual ~strategy() {}
14    virtual void integrate (model_t this_obj, real_t dt) const = 0;
15  };
16
17  #endif //!_H_STRATEGY_
```

Figure 7.3(c): integrand.h

```
1   #ifndef _H_INTEGRAND_
2   #define _H_INTEGRAND_
3
4   #include "strategy.h"
5   #include "RefBase.h"
6   #include "globals.h"
7
8   class integrand : virtual public RefBase
9   {
10  public:
11    typedef Ref<integrand> ptr_t;
12    typedef strategy::ptr_t strategy_t;
13
14    integrand(strategy_t);
15    integrand(const integrand&);
16    virtual ~integrand();
17
18    void set_strategy (strategy_t);
19    strategy_t get_strategy () const;
20    void integrate (real_t);
21
22    virtual ptr_t clone() const = 0;
23    virtual ptr_t d_dt() const = 0;
24    virtual ptr_t operator+=(ptr_t) = 0;
25    virtual ptr_t operator*=(real_t) = 0;
26
27  private:
28    strategy_t quadrature_;
29  };
30
31  #include "model_ops.h"
32
33  #endif //!_H_integrand_
```

Figure 7.3(d): model_ops.h

```
1   #ifndef _H_MODEL_OPS_
2   #define _H_MODEL_OPS_
3
4   inline integrand::ptr_t
5   operator+(integrand::ptr_t a, integrand::ptr_t b)
6   {
7     integrand::ptr_t tmp = a->clone();
8     *tmp += b;
9     return tmp;
10  }
```

```
11
12  inline integrand::ptr_t operator*(integrand::ptr_t a,
13                                    real_t b)
14  {
15    integrand::ptr_t tmp = a->clone();
16    *tmp *= b;
17    return tmp;
18  }
19
20  #endif //! _H_MODEL_OPS_
```

Figure 7.3(e): integrand.cpp

```
1   #include "integrand.h"
2   #include <exception>
3
4   struct integrand_error : public std::exception
5   {
6     virtual ~integrand_error() throw() {}
7   };
8
9   typedef integrand::ptr_t ptr_t;
10  typedef integrand::strategy_t strategy_t;
11
12  integrand::integrand(strategy_t quad) :
13    RefBase(), quadrature_(quad)
14  {}
15
16  integrand::integrand(const integrand& other) :
17    RefBase(), quadrature_(other.quadrature_)
18  {}
19
20  integrand::~integrand()
21  {}
22
23  void integrand::set_strategy (strategy_t quad)
24  {
25    quadrature_ = quad;
26  }
27
28  strategy_t integrand::get_strategy () const
29  {
30    return quadrature_;
31  }
32
33  void integrand::integrate (real_t dt)
34  {
35    quadrature_->integrate(this, dt);
36  }
```

Figure 7.3(f): lorenz.h

```
1   #ifndef _H_LORENZ_
2   #define _H_LORENZ_
3
4   #include "integrand.h"
5
```

```
 6  class lorenz : public integrand
 7  {
 8  public:
 9    typedef integrand::ptr_t ptr_t;
10    typedef integrand::strategy_t strategy_t;
11
12    lorenz(const crd_t&, real_t sigma, real_t rho, real_t beta,
13           strategy_t);
14    virtual ~lorenz();
15
16    virtual ptr_t clone() const;
17    virtual ptr_t d_dt() const;
18    virtual ptr_t operator+=(ptr_t);
19    virtual ptr_t operator*=(real_t);
20
21    void set_coordinate(const crd_t&);
22    const crd_t& coordinate() const;
23    real_t sigma() const;
24    real_t rho() const;
25    real_t beta() const;
26
27  private:
28    crd_t state_;
29    real_t sigma_, rho_, beta_;
30  };
31
32  #endif // !_H_LORENZ_
```

Figure 7.3(g): lorenz.cpp

```
 1  #include "lorenz.h"
 2  #include "fmt.h"
 3  #include <exception>
 4
 5  struct lorenz_error : public std::exception
 6  {
 7    virtual ~lorenz_error() throw() {}
 8  };
 9
10  typedef lorenz::ptr_t ptr_t;
11
12  typedef lorenz::strategy_t strategy_t;
13
14  lorenz::lorenz(const crd_t& ste, real_t s, real_t r, real_t b,
15                 strategy_t str) :
16    integrand(str), state_(ste), sigma_(s), rho_(r), beta_(b)
17  {}
18
19  lorenz::~lorenz()
20  {}
21
22  ptr_t lorenz::clone() const
23  {
24    return ptr_t(new lorenz(*this));
25  }
26
```

```
27 ptr_t lorenz::d_dt() const
28 {
29   crd_t new_state(3);
30   new_state.at(0) = sigma_ * (state_.at(1) - state_.at(0));
31   new_state.at(1) = state_.at(0) * (rho_ - state_.at(2)) - state_.at(1);
32   new_state.at(2) = state_.at(0) * state_.at(1) - beta_ * state_.at(2);
33   return ptr_t(new lorenz(new_state, sigma_, rho_, beta_,
34                           get_strategy()));
35 }
36
37 ptr_t lorenz::operator+=(ptr_t inval)
38 {
39   Ref<lorenz> other = cast<lorenz>(inval);
40   if((other == NULL) || (state_.size() != other->state_.size())) {
41     std::cerr << "lorenz::operator+=:  Invalid input argument\n";
42     throw lorenz_error();
43   }
44   size_t size = state_.size();
45   for(size_t i = 0; i < size; ++i) {
46     state_.at(i) += other->state_.at(i);
47   }
48   return ptr_t(this);
49 }
50
51 ptr_t lorenz::operator*=(real_t val)
52 {
53   size_t size = state_.size();
54   for(size_t i = 0; i < size; ++i) {
55     state_.at(i) *= val;
56   }
57   return ptr_t(this);
58 }
59
60 void lorenz::set_coordinate(const crd_t& state)
61 {
62   state_ = state;
63 }
64
65 const crd_t& lorenz::coordinate() const
66 {
67   return state_;
68 }
69
70 real_t lorenz::sigma() const
71 {
72   return sigma_;
73 }
74
75 real_t lorenz::rho() const
76 {
77   return rho_;
78 }
79
80 real_t lorenz::beta() const
81 {
```

```
82    return beta_;
83  }
```

Figure 7.3(h): timed_lorenz.h

```
1   #ifndef _H_TIMED_LORENZ_
2   #define _H_TIMED_LORENZ_
3
4   #include "strategy.h"
5   #include "lorenz.h"
6
7
8   class timed_lorenz : public lorenz {
9   public:
10    typedef lorenz::ptr_t ptr_t;
11    typedef lorenz::strategy_t strategy_t;
12
13    timed_lorenz(const crd_t&, real_t sigma, real_t rho, real_t beta,
14                 strategy_t, double t_init = 0);
15    virtual ~timed_lorenz();
16    virtual ptr_t clone() const;
17    virtual ptr_t d_dt() const;
18    virtual ptr_t operator+=(ptr_t);
19    virtual ptr_t operator*=(real_t);
20
21    void   set_time (real_t);
22    real_t get_time() const;
23  private:
24    real_t time_;
25  };
26
27  #endif //!_H_TIMED_LORENZ_
```

Figure 7.3(i): timed_lorenz.cpp

```
1   #include "timed_lorenz.h"
2   #include <exception>
3
4   struct timed_lorenz_error : public std::exception
5   {
6     virtual ~timed_lorenz_error() throw() {}
7   };
8
9   typedef timed_lorenz::ptr_t ptr_t;
10  typedef timed_lorenz::strategy_t strategy_t;
11
12  timed_lorenz::timed_lorenz(const crd_t& ste, real_t s, real_t r,
13                             real_t b, strategy_t strat, double t_init) :
14    lorenz(ste, s, r, b, strat), time_(t_init)
15  {}
16
17  timed_lorenz::~timed_lorenz()
18  {}
19
20  ptr_t timed_lorenz::clone() const
21  {
22    return ptr_t(new timed_lorenz(*this));
```

```
23 }
24
25 ptr_t timed_lorenz::d_dt() const
26 {
27   Ref<lorenz> parent = cast<lorenz>(lorenz::d_dt());
28   return ptr_t(new timed_lorenz(parent->coordinate(), parent->sigma(),
29                parent->rho(), parent->beta(),
30                parent->get_strategy(), 1.0));
31 }
32
33 ptr_t timed_lorenz::operator+=(ptr_t inval)
34 {
35   Ref<timed_lorenz> other = cast<timed_lorenz>(inval);
36   if(other == NULL) {
37     std::cerr << "timed_lorenz::operator+=:  Invalid input type\n";
38     throw timed_lorenz_error();
39   }
40   lorenz::operator+=(other);
41   time_ += other->time_;
42   return ptr_t(this);
43 }
44
45 ptr_t timed_lorenz::operator*=(real_t val)
46 {
47   lorenz::operator*=(val);
48   time_ *= val;
49   return ptr_t(this);
50 }
51
52 void timed_lorenz::set_time (real_t t)
53 {
54   time_ = t;
55 }
56
57 real_t timed_lorenz::get_time() const
58 {
59   return time_;
60 }
```

Figure 7.3(j): explicit_euler.h

```
1  #ifndef _H_EXPLICIT_EULER_
2  #define _H_EXPLICIT_EULER_
3
4  #include "strategy.h"
5  #include "integrand.h"
6
7  class explicit_euler : public strategy
8  {
9  public:
10   virtual ~explicit_euler();
11   virtual void integrate (model_t this_obj, real_t dt) const;
12 };
13
14 #endif //!_H_EXPLICIT_EULER_
```

Figure 7.3(k): explicit_euler.cpp

```
#include "explicit_euler.h"
#include "integrand.h"
#include <exception>

explicit_euler::~explicit_euler()
{}

void explicit_euler::integrate (model_t this_obj, real_t dt) const {
  *this_obj += this_obj->d_dt() * dt;
}
```

Figure 7.3(l): runge_kutta_2nd.h

```
#ifndef _H_RUNGE_KUTTA_2ND_
#define _H_RUNGE_KUTTA_2ND_

#include "strategy.h"
#include "integrand.h"

class runge_kutta_2nd : public strategy {
public:
  virtual ~runge_kutta_2nd();
  virtual void integrate(model_t this_obj, real_t dt) const;
};

#endif
```

Figure 7.3(m): runge_kutta_2nd.cpp

```
#include <iostream>
#include <exception>

#include "integrand.h"
#include "runge_kutta_2nd.h"

using namespace std;

runge_kutta_2nd::~runge_kutta_2nd()
{}

void runge_kutta_2nd::integrate (model_t this_obj, real_t dt) const
{
  model_t this_half = this_obj + this_obj->d_dt() * (0.5*dt);//predictor
  *this_obj += this_half->d_dt() * dt; // corrector
}
```

Figure 7.3. C++ strategy implementation for Lorenz and timed Lorenz systems: (a) Lorenz and time-stamped Lorenz integration tests, (b) abstract strategy class definition using forward references to an abstract integrand class, (c) operator definitions (+ and *) for integrands, (d) abstract integrand definition, (e) abstract integrand implementation, (f) Lorenz definition, (g) Lorenz implementation, (h) time-stamped Lorenz definition, (i) time-stamped Lorenz implementation, (j) explicit Euler strategy definition, (k) explicit Euler strategy implementation, (l) RK2 strategy definition, and (m) RK2 strategy integration.

7.3 The Consequences

Because STRATEGY uses aggregation instead of inheritance, it allows better decoupling of the classes that focus on data (the context) from those that focus on algorithms (behaviors). One obvious advantage is that strategies can provide varying implementations (or algorithms) for the same behavior, thus giving users more freedom to choose at runtime based on the problem at hand. Compared to patterns that employ inheritance as the means for maintaining algorithms (e.g., the INTERPRETER pattern of the GoF), a STRATEGY is much easier to understand and extend. Since each concrete strategy class implements one particular algorithm for the context, the use of this pattern also encourages programmers to avoid lumping many different algorithms into one class that often lead to unmanageable code.

This pattern is commonly used in applications where a family of related algorithms or behaviors exists for the context (data). Examples given by the GoF include the register allocation schemes and instruction scheduling policies used in the compiler optimization code in the Register Transfer Language (RTL) systems. The application of strategy patterns yielded great flexibility for the optimizer in targeting different machine architectures.

As with every design pattern, STRATEGY pattern also has drawbacks. One potential disadvantage is that in applying the pattern, a user must be aware of the differences among algorithms in order to select the appropriate one. This sometimes becomes a burden as programmers must acquire knowledge on various algorithms. Another potential shortcoming of strategies lies in the possible communication overhead between data objects and algorithm objects. Although this overhead can normally be minimized by careful design of the strategy interfaces, a naïve implementer of the strategy pattern may nevertheless attempt to design the interfaces with many unnecessary parameters to pass between data and algorithms. Thus achieving a balance between data-algorithm coupling and communication overhead should always be one of the goals for designing a good STRATEGY.

7.4 Related Patterns

In the context of writing differential equation solvers, the STRATEGY pattern relates closely to ABSTRACT CALCULUS. From differentiation and integration to addition and subtraction, virtually every operation performed in an ABSTRACT CALCULUS serves as an umbrella for a collection of strategies. In numerical analysis, approximations to derivatives and integrals come in many varieties with the varying orders of accuracy, stencil widths, phase errors, stability criterion, and symmetries. Any situation in which the ability to vary these properties provides fertile ground for a STRATEGY to take root. In this sense, a STRATEGY helps complete an ABSTRACT CALCULUS.

In Fortran, SURROGATE and STRATEGY are inextricably bound. The former completes the latter, the relationship being sufficiently close that one can consider a SURROGATE to be part of a STRATEGY. The uses for a SURROGATE, however, extend to all situations in which the class relationships in a design suggest the need for circular references. As evidenced by the frequency with which the question of how to facilitate such references appears in the comp.lang.fortran Usenet newsgroup.[1]

[1] http://groups.google.com/group/comp.lang.fortran

In a multiphysics simulation context, one needs the next chapter's PUPPETEER pattern to complete a STRATEGY. The PUPPETEER hides the single-physics submodels by aggregating their ADTs. It provides a single ADT on which the STRATEGY can operate without knowledge of the individual sub-models.

EXERCISES

1. Runge-Kutta-Fehlberg (RKF) algorithms provide one context for dynamically varying a time integration method. By alternating between two schemes with different orders of accuracy, RKF methods facilitate time step size control based on error estimation. For example, a 4th-order-accurate Runge-Kutta-Fehlberg method (RKF45) uses a 5th-order-accurate step for error estimation. In advancing a differential equation $dy/dt = f(t,y)$ from a time t_i over a time step h_i to a time $t_{i+1} \equiv t_i + h_i$, RKF45 constructs both approximations from one set of function evaluations (Morris 2008):

$$K_1 = hf(t_i, y_i)$$
$$K_2 = hf\left(t_i + \frac{1}{4}h, y_i + \frac{1}{4}K_1\right)$$
$$K_3 = hf\left(t_i + \frac{3}{8}h, y_i + \frac{3}{32}K_1 + \frac{9}{32}K_2\right)$$
$$K_4 = hf\left(t_i + \frac{12}{13}h, y_i + \frac{1932}{2197}K_1 - \frac{7200}{2197}K_2 + \frac{7296}{2197}K_3\right)$$
$$K_5 = hf\left(t_i + h, y_i + \frac{439}{216}K_1 - 8K_2 + \frac{3680}{513}K_3 - \frac{845}{4104}K_4\right)$$
$$K_6 = hf\left(t_i + \frac{1}{2}h, y_i - \frac{8}{27}K_1 + 2K_2 - \frac{3544}{2565}K_3 - \frac{1849}{4104}K_4 - \frac{11}{40}K_5\right)$$

which yield the 4th- and 5th-order-accurate schemes

$$y_{i+1}^{RK4} = y_i + \frac{25}{216}K_1 + \frac{1408}{2565}K_3 + \frac{2197}{4101}K_4 - \frac{1}{5}K_5$$
$$y_{i+1}^{RK5} = y_i + \frac{16}{135}K_1 + \frac{6656}{12825}K_3 + \frac{28561}{56430}K_4 - \frac{9}{50}K_5 + \frac{2}{55}K_5$$

Estimating the error as $e \approx y_{i+1}^{RK4} - y_{i+1}^{RK5}$, one adjusts the time step size just before each step according to:

$$h_i = \begin{cases} Sh_{i-1}\left|\frac{e_{\text{tol}}}{e}\right|^{0.20} & \text{if } e \leq e_{\text{tol}} \\ Sh_{i-1}\left|\frac{e_{\text{tol}}}{e}\right|^{0.25} & \text{if } e > e_{\text{tol}} \end{cases}$$

where S is a safety factor just below unity and e_{tol} is the error tolerance. Design a STRATEGY for solving the Lorenz system (4.5) using the RKF45 algorithm. Express your design in a UML class diagram. Given $S =$ and $e_{\text{tol}} =$, write an implementation of your design.

2. In many situations, finite difference approximations perform best when the stencil employed to approximate derivatives at a given point is skewed toward the direction from which information is propagating. Labeling a given direction in one dimension "forward" and the opposite direction "backward," it is desirable to be able to switch between the forward and backward differencing schemes described in Section A.5.1. Design a STRATEGY for switching between these two schemes at runtime. Express your design in a UML class diagram. Write an implementation of your design.

8 The Puppeteer Pattern

"Never be afraid to try something new. Remember, amateurs built the ark. Professionals built the Titanic."

Miss Piggy

8.1 The Problem

While the ABSTRACT CALCULUS and STRATEGY patterns apply to the integration of a single physics abstraction, our chief concern lies in linking multiple abstractions. This poses at least two significant software design problems. The first involves how to facilitate interabstraction communication. The GoF addressed interabstraction communication with the MEDIATOR pattern. When N objects interact, a software architect can reduce the $N(N-1)+$ associations between the objects to $2N$ associations by employing a MEDIATOR.

The MEDIATOR association count stems from the requirements that the Mediator know each communicating party and those parties know the Mediator. For example, in a MEDIATOR implementation presented by Gamma et al. (1995), the sender passes a reference to itself to the MEDIATOR. The sender must be aware of the MEDIATOR in order to know where to send the message. Likewise, the MEDIATOR must be aware of the sender in order to invoke methods on the sender via the passed reference. Figure 8.1 illustrates the associations in an atmospheric boundary layer model, wherein the air, ground, and cloud ADTs might solve equation sets for the airflow, ground transpiration, and discrete droplet motion, respectively.

A second and conceptually more challenging problem concerns how one assembles quantities of an inherently global nature – that is, information that can only be determined with simultaneous knowledge of the implementation details of each of the single-physics abstractions. This cuts to the heart of the question of whether ABSTRACT CALCULUS can be considered a general software design philosophy.

We illustrate this second dilemma with an important and common mathematical problem: implicit time advancement of nonlinear equation systems. Of particular concern is the desire to calculate cross-coupling terms without violating abstractions by exposing their data. Consider marching the ordinary differential equation

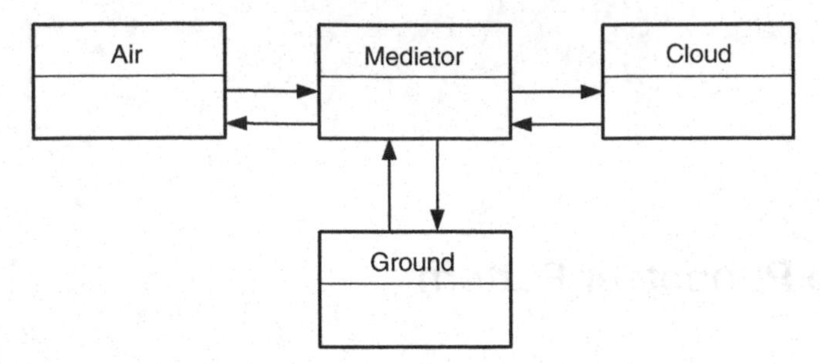

Figure 8.1. Associations in the Mediator pattern.

system (4.4) forward in time according to the trapezoidal rule:

$$\vec{V}^{n+1} = \vec{V}^n + \frac{\Delta t}{2}\left[\vec{R}\left(\vec{V}^n\right) + \vec{R}\left(\vec{V}^{n+1}\right)\right] \tag{8.1}$$

where the presence of $\vec{V}^{n+1}$ on the RHS makes iteration necessary when $\vec{R}$ contains nonlinearities. One generally poses the problem in terms of finding the roots of a residual vector $\vec{f}$ formed by gathering the terms in equation (8.1):

$$\vec{f}(\vec{V}^{n+1}) \equiv \vec{V}^{n+1} - \left\{\vec{V}^n + \frac{\Delta t}{2}\left[\vec{R}\left(\vec{V}^n\right) + \vec{R}\left(\vec{V}^{n+1}\right)\right]\right\} \tag{8.2}$$

where $\vec{V}^n$ is known from a previous time step.

The difficulty arises in finding the roots of $\vec{f}$ using Jacobian-based iteration methods. Consider Newton's method. Defining $\vec{y}^m$ as the mth iterative approximation to $\vec{V}^{n+1}$, Newton's method can be expressed as:

$$J\,\Delta\vec{y}^m \equiv -\vec{f}(\vec{y}^m) \tag{8.3}$$

$$\vec{y}^{m+1} \equiv \vec{y}^m + \Delta\vec{y}^m \tag{8.4}$$

$$J_{ij} \equiv \left.\frac{\partial f_i}{\partial y_j}\right|_{\vec{y}=\vec{y}^m} = \left.\frac{\partial f_i}{\partial V_j^{n+1}}\right|_{\vec{V}^{n+1}=\vec{y}^m} = \delta_{ij} - \frac{\Delta t}{2}\left[\frac{\partial R_i(\vec{V}^{n+1})}{\partial V_j^{n+1}}\right]_{\vec{V}^{n+1}=\vec{y}^m} \tag{8.5}$$

where $\mathbf{J}$ is the Jacobian, R_i is the RHS of the ith governing equation, and δ_{ij} is the Kronecker delta. Equation (8.3) represents a linear algebraic system. Equation (8.4) represents vector addition.

Now imagine that $\vec{f}$, $\vec{y}$, and $\vec{R}$ are partitioned such that different elements are hidden behind interfaces to different abstractions. In particular, let us partition $\vec{R}$ such that $\vec{R} \equiv \{\vec{a} \quad \vec{c} \quad \vec{g}\}^T$, where the partitions separate the elements of $\vec{R}$ corresponding to the air, cloud, and ground ADTs, respectively. For present purposes, each partition can be thought of as a 1D vector containing the RHS of one component equation from the Lorenz system (4.5). With this notation, one can rewrite the Jacobian in equation (8.3) as:[1]

$$\mathbf{J} \equiv \mathbf{I} - \frac{\Delta t}{2}\frac{\partial(\vec{a},\vec{c},\vec{g})}{\partial(\vec{\alpha},\vec{\chi},\vec{\gamma})} \tag{8.6}$$

[1] In equation (8.4) and elsewhere, we abbreviate a common notation for Jacobians in which the list of $\vec{f}$ components appears in the numerator and the list of $\vec{y}$ components appears in the denominator. We do not mean for the arrows to connote vectors that transform as first-order tensors.

where $\vec{\alpha}$, $\vec{\chi}$, and $\vec{\gamma}$ are the air, cloud, and ground state vectors, respectively, so $\vec{y} \equiv \{\vec{\alpha} \quad \vec{\chi} \quad \vec{\gamma}\}^T$, and where **I** is the identity matrix.

The second dilemma presents itself in the need to calculate cross terms such as $\partial\vec{a}/\partial\vec{\chi}$. The question is, "How does the air ADT differentiate its components of $\vec{R}$ with respect to the state vector partition $\vec{\chi}$ when that partition lies hidden inside the cloud abstraction?" Even if one solves this puzzle, a more perplexing one remains: Where does **J** live? Since the precise form of equations (8.5)–(8.6) varies with the choice of time integration algorithms, the natural place to construct **J** is inside the time integration procedure. In an ABSTRACT CALCULUS implementation, however, that procedure maintains a blissful ignorance of the governing equations inside the dynamical system it integrates as exemplified by the `integrate` procedures in Chapter 6.

It appears the conflicting forces driving the software problems in this book climax in the current chapter's context, potentially forcing the violation of abstractions. The desire to maintain a modular architecture with local, private data confronts the need to handle intermodule dependencies. These dependencies force rigidly specified communication and complicate the calculation of inherently global entities.

A Showdown at the OOP Corral

The ABSTRACT CALCULUS edifice threatens to collapse under the weight of the problems posed here. Modularity and data privacy present significant impediments to coupling independently developed abstractions without exposing their implementation details. Of prime concern are how to facilitate interabstraction communication and how to assemble inherently global quantities.

8.2 The Solution

A PUPPETEER encapsulates references to each time-integrable system involved in the simulation. When one of the `integrate` procedures of Chapter 6, for example, receives a PUPPETEER argument, that PUPPETEER controls the behavior of all other integrable systems in the simulation. It does so by delegating operations and serving as an intermediary for communications.

The PUPPETEER designer also ensures consistency in the communication format. She might do so by reducing, or requiring the puppets to reduce, each datum communicated to a type intrinsic to the language. In doing so, the PUPPETEER would obviate the need to expose information about the data structures each ADT employs. When an application calls for more nuanced data structures than the language supports – for example, compressed storage of sparse matrices – the PUPPETEER designer might instead enforce that all puppets encapsulate message information in a data structure tailored to the given application domain. In distributed computing applications, for example, the Trilinos Epetra objects discussed in Chapters 5 and 11 is one likely candidate for encapsulating data that all ADTs would be required to accept.

A PUPPETEER exploits the regularity and predictability of interabstraction communications defined formally in the terms that couple the governing equations. The developer of each ADT requests the requisite coupling terms by placing them in the argument lists of the procedures that define each dynamical system. The PUPPETEER

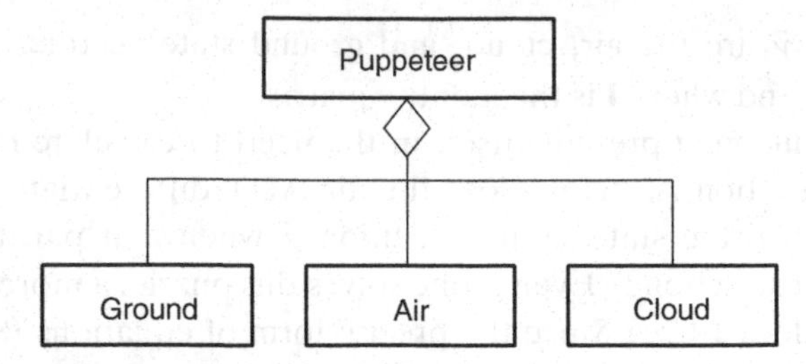

Figure 8.2. Aggregation associations in the Puppeteer pattern.

developer fulfills these requests by calling accessors made public by the other dynamical systems in the simulation. As depicted in Figure 8.2, the PUPPETEER uses object aggregation to reduce the aforementioned number of interabstraction associations from $2N = 6$ to $N = 3$: the PUPPETEER knows a datum's sender and recipient, but they need not know the PUPPETEER.

The dialogue the PUPPETEER mediates in the atmospheric boundary layer model can be paraphrased as follows:

Puppeteer to Air: Please pass me a vector containing the partial derivatives of each of your governing equations' RHS with respect to each element in your state vector.

Air to Puppeteer: Here is the requested vector ($\partial\vec{a}/\partial\vec{\alpha}$). You can tell the dimension of my state by the size of this vector.

Puppeteer to Air: I also know that your governing equations depend on the Cloud state vector because your interface requests information that I retrieved from a Cloud. Please pass me a vector analogous to the previous one but differentiated with respect to the Cloud state information I passed to you.

Puppeteer note to self: Since I did not pass any information to the Air object from the Ground object, I will set the cross-terms corresponding to $\partial\vec{a}/\partial\vec{g}$ to zero. I'll determine the number of zero elements by multiplying the size of $\partial\vec{a}/\partial\vec{\alpha}$ by the size of $\partial\vec{g}/\partial\vec{\gamma}$ after I receive the latter from my Ground puppet.

The PUPPETEER then holds analogous dialogues with its Cloud and Ground puppets, after which the PUPPETEER passes an array containing $\partial\vec{R}/\partial\vec{V}$ to `integrate`. The latter procedure uses the passed array to form **J**, which it then passes to a PUPPETEER for use in inverting the matrix system (equation 8.3). Most importantly, `integrate` does so without violating the PUPPETEER's data privacy, and the PUPPETEER responds without violating the privacy of its puppets. The PUPPETEER's construction is based solely on information from the public interfaces of each puppet. Figure 8.3 details the construction of $\partial\vec{R}/\partial\vec{V}$ in a UML sequence diagram. Note the `coordinate()` method calls return solution variables. The next three calls return diagonal blocks of $\partial\vec{R}/\partial\vec{V}$. The subsequent six calls return off-diagonal blocks. The final two steps allocate and fill $\partial\vec{R}/\partial\vec{V}$.

8.2.1 Fortran Implementation

Figure 8.4 provides the definition of the `atmosphere` type, which acts as a PUPPETEER in our prototype atmospheric boundary-layer simulation involving the Lorenz

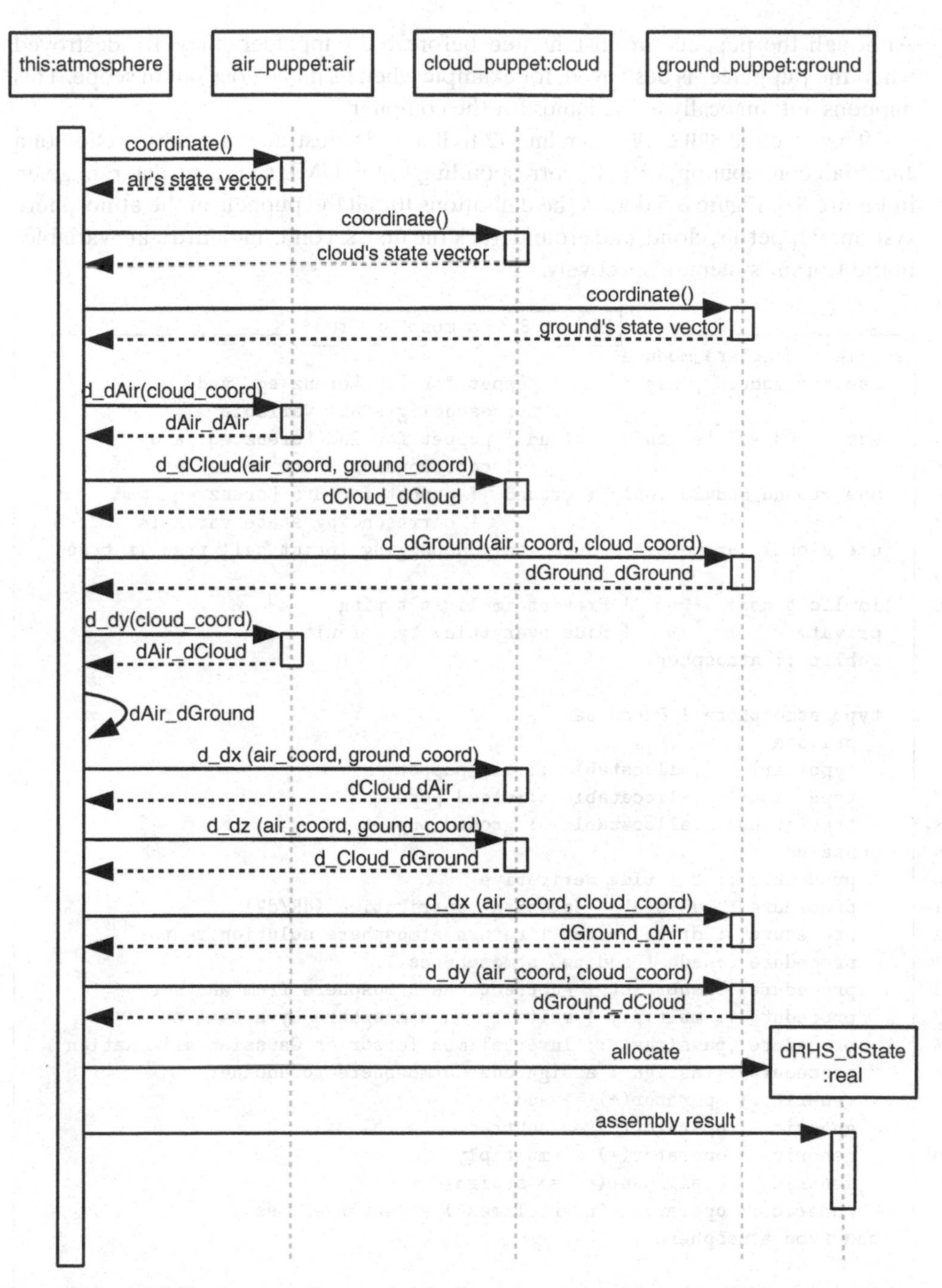

Figure 8.3. A UML sequence diagram for the Puppeteer's Jacobian contribution.

equations partitioned in the manner described in the previous section. Here the `atmosphere` contains references to all its puppets, which are of type `air`, `cloud`, and `ground`. The puppets' data can be constructed independently after the `atmosphere` object is created.

Lines 16–18 in Figure 8.4 show the composition of the `atmosphere` type. The `atmosphere` object relays all information using Fortran allocatable scalar data.

Although the puppets are instantiated before the puppeteer, they are destroyed when the puppeteer is destroyed, for example when its name goes out of scope. This happens automatically at the hands for the compiler.

The function dRHS_dV (from line 72 to line 153) illustrates the construction of a Jacobian contribution, $\partial\vec{R}/\partial\vec{V}$, corresponding to the UML sequence diagram given in Figure 8.3. Figure 8.5 details the definitions for all the puppets in the atmosphere system. Puppet air, cloud, and ground track the first, second, and third state variables in the Lorenz system, respectively.

Figure 8.4: atmosphere.f03

```
1  module atmosphere_module
2    use air_module ,only : air ! puppet for 1st Lorenz eq. and
3                               ! corresonding state variable
4    use cloud_module ,only : cloud ! puppet for 2nd Lorenz eq. and
5                                   ! corresonding state variable
6    use ground_module ,only : ground  ! puppet for 3rd Lorenz eq. and
7                                      ! corresonding state variable
8    use global_parameters_module ,only:debugging !print call tree if true
9
10  !implicit none         ! Prevent implicit typing
11   private               ! Hide everything by default
12   public :: atmosphere
13
14   type atmosphere ! Puppeteer
15     private
16     type(air)    ,allocatable :: air_puppet
17     type(cloud)  ,allocatable :: cloud_puppet
18     type(ground) ,allocatable :: ground_puppet
19   contains
20     procedure :: t ! time derivative
21     procedure :: dRHS_dV ! Jacobian contribution (dR/dV)
22     procedure :: state_vector ! return atmosphere solution vector
23     procedure :: add ! add two atmospheres
24     procedure :: subtract ! subtract one atmosphere from another
25     procedure :: multiply ! multiply an atmosphere by a real scalar
26     procedure ,pass(rhs) :: inverseTimes !abstract Gaussian elimination
27     procedure :: assign ! assign one  atmosphere to another
28     generic :: operator(+) => add
29     generic :: operator(-) => subtract
30     generic :: operator(*) => multiply
31     generic :: assignment(=) => assign
32     generic :: operator(.inverseTimes.) => inverseTimes
33   end type atmosphere
34
35   interface atmosphere
36     procedure constructor
37   end interface
38
39 contains
40
41   ! constructor for an atmosphere object
42   type(atmosphere) function constructor &
43     (air_target,cloud_target,ground_target)
44     type(air)     ,allocatable ,intent(inout) :: air_target
```

```
45    type(cloud)  ,allocatable ,intent(inout) :: cloud_target
46    type(ground) ,allocatable ,intent(inout) :: ground_target
47    if (debugging) print *,'    atmosphere%construct(): start'
48    ! transfer allocations from puppets to Puppeteer
49    call move_alloc(air_target, constructor%air_puppet)
50    call move_alloc(ground_target, constructor%ground_puppet)
51    call move_alloc(cloud_target, constructor%cloud_puppet)
52    if (debugging) print *,'    atmosphere%construct(): end'
53  end function
54
55  ! time derivative (evolution equations)
56  function t(this) result(dState_dt)
57    class(atmosphere) ,intent(in)  :: this
58    class(atmosphere) ,allocatable :: dState_dt
59    type(atmosphere)  ,allocatable :: delta
60    if (debugging) print *,'    atmosphere%t(): start'
61    allocate(delta)
62    delta%air_puppet= this%air_puppet%t(this%cloud_puppet%coordinate())
63    delta%cloud_puppet = this%cloud_puppet%t( &
64      this%air_puppet%coordinate(),this%ground_puppet%coordinate())
65    delta%ground_puppet = this%ground_puppet%t( &
66      this%air_puppet%coordinate(),this%cloud_puppet%coordinate())
67    call move_alloc(delta, dState_dt)
68    if (debugging) print *,'    atmosphere%t(): end'
69  end function
70
71  ! atmosphere contribution to Jacobian
72  function dRHS_dV(this) result(dRHS_dState)
73    class(atmosphere)    ,intent(in)  :: this
74
75    ! Sub-blocks of dR/dV array
76    real ,dimension(:,:) ,allocatable ::                          &
77                         dAir_dAir ,dAir_dCloud ,dAir_dGround,     &
78                         dCloud_dAir ,dCloud_dCloud ,dCloud_dGround,&
79                         dGround_dAir ,dGround_dCloud ,dGround_dGround
80
81    real ,dimension(:,:) ,allocatable :: dRHS_dState
82
83    real ,dimension(:)   ,allocatable ::                          &
84                  air_coordinate, cloud_coordinate, ground_coordinate
85
86    integer :: air_eqs ,air_vars,cloud_eqs ,cloud_vars , &
87               ground_eqs ,ground_vars ,i ,j ,rows ,cols
88
89    if (debugging) print *,'atmosphere%dRHS_dV(): start'
90
91      ! Calculate matrices holding partial derivative of puppet
92      ! evolution equation right-hand sides with respect to the
93      ! dependent variables of each puppet.
94
95        air_coordinate  = this%air_puppet%coordinate()
96        cloud_coordinate= this%cloud_puppet%coordinate()
97        ground_coordinate=this%ground_puppet%coordinate()
98
99        ! compute diagonal block submatrix
```

```
100          dAir_dAir       = this%air_puppet%d_dAir(cloud_coordinate)
101          dCloud_dCloud   = this%cloud_puppet%d_dCloud(air_coordinate,&
102                                                        ground_coordinate)
103          dGround_dGround = this%ground_puppet%d_dGround(air_coordinate,&
104                                                          cloud_coordinate)
105          air_eqs     = size(dAir_dAir,1)          ! submatrix rows
106          air_vars    = size(dAir_dAir,2)          ! submatrix columns
107          cloud_eqs   = size(dCloud_dCloud,1)      ! submatrix rows
108          cloud_vars  = size(dCloud_dCloud,2)      ! submatrix columns
109          ground_eqs  = size(dGround_dGround,1)    ! submatrix rows
110          ground_vars = size(dGround_dGround,2)    ! submatrix columns
111
112          ! compute off-diagonal values
113          dAir_dCloud    = this%air_puppet%d_dy(cloud_coordinate)
114          dAir_dGround = reshape(source=(/(0.,i=1,air_eqs*ground_vars)/)&
115                                 ,shape=(/air_eqs,ground_vars/))
116          dCloud_dAir    = this%cloud_puppet%d_dx(air_coordinate,    &
117                                                  ground_coordinate)
118          dCloud_dGround = this%cloud_puppet%d_dz(air_coordinate,    &
119                                                  ground_coordinate)
120          dGround_dAir   = this%ground_puppet%d_dx(air_coordinate,   &
121                                                   cloud_coordinate)
122          dGround_dCloud = this%ground_puppet%d_dy(air_coordinate,   &
123                                                   cloud_coordinate)
124          rows=air_eqs+cloud_eqs+ground_eqs
125          cols=air_vars+cloud_vars+ground_vars
126          allocate(dRHS_dState(rows,cols))
127
128          ! Begin result assembly
129          dRHS_dState(1:air_eqs, 1:air_vars)                         =    &
130                       dAir_dAir
131          dRHS_dState(1:air_eqs, air_vars+1:air_vars+cloud_vars) =    &
132                       dAir_dCloud
133          dRHS_dState(1:air_eqs, air_vars+cloud_vars+1:cols)     =    &
134                       dAir_dGround
135          dRHS_dState(air_eqs+1:air_eqs+cloud_eqs, 1:air_vars)   =    &
136                       dCloud_dAir
137          dRHS_dState(air_eqs+1:air_eqs+cloud_eqs,                    &
138                       air_vars+1:air_vars+cloud_vars)           =    &
139                       dCloud_dCloud
140          dRHS_dState(air_eqs+1:air_eqs+cloud_eqs,                    &
141                       air_vars+cloud_vars+1:cols)               =    &
142                       dCloud_dGround
143          dRHS_dState(air_eqs+cloud_eqs+1:rows, 1:air_vars)      =    &
144                       dGround_dAir
145          dRHS_dState(air_eqs+cloud_eqs+1:rows,                       &
146                       air_vars+1:air_vars+cloud_vars)           =    &
147                       dGround_dCloud
148          dRHS_dState(air_eqs+cloud_eqs+1:rows,                       &
149                       air_vars+cloud_vars+1:cols)               =    &
150                       dGround_dGround
151        ! result assembly finishes
152      if (debugging) print *,'    atmosphere%dRHS_dV(): end'
153    end function dRHS_dV
154
```

```
155   ! assemble and return solution vector
156   function state_vector(this) result(phase_space)
157     class(atmosphere)  ,intent(in)  :: this
158     real ,dimension(:) ,allocatable :: state
159     real ,dimension(:) ,allocatable :: x,y,z,phase_space
160     integer                         :: x_start,y_start,z_start
161     integer                         :: x_end  ,y_end  ,z_end
162     x = this%air_puppet%coordinate()
163     x_start=1
164     x_end=x_start+size(x)-1
165     y = this%cloud_puppet%coordinate()
166     y_start=x_end+1
167     y_end=y_start+size(y)-1
168     z = this%ground_puppet%coordinate()
169     z_start=y_end+1
170     z_end=z_start+size(z)-1
171     allocate(phase_space(size(x)+size(y)+size(z)))
172     phase_space(x_start:x_end) = x
173     phase_space(y_start:y_end) = y
174     phase_space(z_start:z_end) = z
175   end function
176
177   function add(lhs,rhs) result(sum)
178     class(atmosphere) ,intent(in)  :: lhs
179     class(atmosphere) ,intent(in)  :: rhs
180     class(atmosphere) ,allocatable :: sum
181     if (debugging) print *,'  atmosphere%add(): start'
182     allocate(sum)
183     sum%air_puppet    = lhs%air_puppet    + rhs%air_puppet
184     sum%cloud_puppet  = lhs%cloud_puppet  + rhs%cloud_puppet
185     sum%ground_puppet = lhs%ground_puppet + rhs%ground_puppet
186     if (debugging) print *,'   atmosphere%add(): end'
187   end function
188
189   function subtract(lhs,rhs) result(difference)
190     class(atmosphere)        ,intent(in)  :: lhs
191     class(atmosphere) ,intent(in)  :: rhs
192     class(atmosphere) ,allocatable :: difference
193     if (debugging) print *,' atmosphere%subtract(): start'
194     allocate(difference)
195     difference%air_puppet = lhs%air_puppet - rhs%air_puppet
196     difference%cloud_puppet = lhs%cloud_puppet - rhs%cloud_puppet
197     difference%ground_puppet = lhs%ground_puppet - rhs%ground_puppet
198     if (debugging) print *,'    atmosphere%subtract(): end'
199   end function
200
201   ! Solve linear system Ax=b by Gaussian elimination
202   function inverseTimes(lhs,rhs) result(product)
203     class(atmosphere) ,intent(in) :: rhs
204     class(atmosphere) ,allocatable :: product
205     real ,dimension(:,:) ,allocatable ,intent(in) :: lhs
206     real ,dimension(:)   ,allocatable :: x,b
207     real ,dimension(:,:) ,allocatable :: A
208     real ,parameter   :: pivot_tolerance=1.0E-02
209     integer :: row,col,n,p ! p=pivot row/col
```

```
210      real :: factor
211
212      n=size(lhs,1)
213      b = rhs%state_vector()
214      if (n /= size(lhs,2) .or. n /= size(b)) &
215        stop 'atmosphere.f03: ill-posed matrix problem in inverseTimes()'
216      allocate(x(n))
217      A = lhs
218      do p=1,n-1          ! Forward elimination
219        if (abs(A(p,p))<pivot_tolerance) &
220          stop 'invert: use an algorithm with pivoting'
221        do row=p+1,n
222          factor=A(row,p)/A(p,p)
223          forall(col=p:n)
224            A(row,col) = A(row,col) - A(p,col)*factor
225          end forall
226          b(row) = b(row) - b(p)*factor
227        end do
228      end do
229      x(n) = b(n)/A(n,n) ! Back substitution
230      do row=n-1,1,-1
231        x(row) = (b(row) - sum(A(row,row+1:n)*x(row+1:n)))/A(row,row)
232      end do
233      allocate(product,source=rhs)
234      product%air_puppet = air(x(1),x(2))
235      call product%cloud_puppet%set_coordinate(x(3))
236      call product%ground_puppet%set_coordinate(x(4))
237    end function
238
239    ! multiply atmosphere object by a real scalar
240    function multiply(lhs,rhs) result(product)
241      class(atmosphere) ,intent(in)  :: lhs
242      real              ,intent(in)  :: rhs
243      class(atmosphere) ,allocatable :: product
244      if (debugging) print *,'    atmosphere%multiply(): start'
245      allocate(product)
246      product%air_puppet    = lhs%air_puppet    * rhs
247      product%cloud_puppet  = lhs%cloud_puppet  * rhs
248      product%ground_puppet = lhs%ground_puppet * rhs
249      if (debugging) print *,'    atmosphere%multiply(): end'
250    end function
251
252    ! assign one atmosphere object to another
253    subroutine assign(lhs,rhs)
254      class(atmosphere) ,intent(inout) :: lhs
255      class(atmosphere) ,intent(in)    :: rhs
256      if (debugging) print *,'    atmosphere%assign(): start'
257      lhs%air_puppet    = rhs%air_puppet    ! automatic allocation
258      lhs%cloud_puppet  = rhs%cloud_puppet  ! automatic allocation
259      lhs%ground_puppet = rhs%ground_puppet ! automatic allocation
260      if (debugging) print *,'    atmosphere%assign(): end'
261    end subroutine
262  end module atmosphere_module
```

Figure 8.4. Fortran implementation of an atmosphere puppeteer.

Figure 8.5(a): air.f03

```
1  module air_module
2    use global_parameters_module ,only : debugging !print call treeif true
3    implicit none ! Prevent implicit typing
4    private       ! Hide everything by default
5    public :: air
6
7    ! Number of evolution equations/variables exposed to the outside world
8    ! (via Jacobian sub-block shapes):
9    integer ,parameter :: num_eqs=2,num_vars=2
10
11   ! This type tracks the evolution of the first state variable in the
12   ! Lorenz system according to the first Lorenz equation.  It also
13   ! tracks the corresponding paramater (sigma) according to the
14   ! differential equation d(sigma)/dt=0. For illustrative purposes, this
15   ! implementation exposes the number of state variables (2) to the
16   ! puppeteer without providing direct access to them or exposing
17   ! anything about their layout, storage location or identifiers (x and
18   ! sigma).  Their existence is apparent in the rank (2) of the matrix
19   ! d_dAir() returns as its diagonal Jacobian element contribution.
20
21   type air
22     private
23     real :: x,sigma !1st Lorenz equation solution variable and parameter
24   contains
25     procedure :: coordinate! accessor: return phase-space coordinates
26     procedure :: t ! time derivative ( evaluate RHS of Lorenz equation)
27     procedure :: d_dAir ! contribution to diagonal Jacobian element
28     procedure :: d_dy   ! contribution to off-diagonal Jacobian element
29     procedure ,private :: add ! add two instances
30     procedure ,private :: subtract ! subtract one instance from another
31     procedure ,private :: multiply ! multiply instance by a real scalar
32     ! map defined operators to corresponding procedures
33     generic   :: operator(+) => add
34     generic   :: operator(-) => subtract
35     generic   :: operator(*) => multiply
36   end type
37
38   interface air
39     procedure constructor
40   end interface
41
42 contains
43   ! constructor: allocate and initialize components
44   type(air) function constructor(x_initial,s)
45     real            ,intent(in)  :: x_initial
46     real            ,intent(in)  :: s
47     if (debugging) print *,'      air%construct: start'
48     constructor%x = x_initial
49     constructor%sigma  = s
50     if (debugging) print *,'      air%construct: end'
51   end function
52
53   ! accessor: returns phase-space coordinates
54   function coordinate(this) result(return_x)
```

```
55      class(air)            ,intent(in)  :: this
56      real  ,dimension(:) ,allocatable :: return_x
57      return_x = [ this%x ,this%sigma ]
58    end function
59
60    ! time derivative (evolution equations)
61    function t(this,y) result(dx_dt)
62      class(air)          ,intent(in) :: this
63      real ,dimension(:) ,intent(in) :: y
64      type(air)                       :: dx_dt
65      if (debugging) print *,'        air%t: start'
66      dx_dt%x=this%sigma*(y(1)-this%x)
67      dx_dt%sigma=0.
68      if (debugging) print *,'        air%t: end'
69    end function
70
71    ! contribution to diagonal Jacobian element
72    function d_dAir(this,y) result(dRHS_dx)
73      class(air)    ,intent(in)                      :: this
74      real          ,dimension(:,:) ,allocatable :: dRHS_dx
75      real              ,dimension(:) ,allocatable :: y
76      if (debugging) print *,'          air%d_dAir: start'
77     !allocate(dRHS_dx(num_eqs,num_vars))
78     !dRHS_dx = [ d{sigma*(y-x)}/dx   d{sigma*(y-x)}/dsigma ]
79     !          [ d{0}/dx             d{0}/dsigma           ]
80      if (size(y) /= 1) stop 'd_dAir: invalid y size'
81      dRHS_dx = reshape(source=(/-this%sigma,0.,y(1)-this%x,0./),      &
82                shape=(/num_eqs,num_vars/))
83      if (debugging) print *,'          air%d_dAir: end'
84    end function
85
86    ! contribution to off-diagonal Jacobian element
87    function d_dy(this,y) result(dRHS_dy)
88      class(air)    ,intent(in)                      :: this
89      real          ,dimension(:,:) ,allocatable :: dRHS_dy
90      real          ,dimension(:)   ,allocatable :: y
91
92      if (debugging) print *,'          air%d_dy: start'
93      allocate(dRHS_dy(num_eqs,size(y)))
94     !dRHS_dy = [ d{sigma*(y(1)-x(1))}/dy(1)  0  ... 0]
95     !          [ d{0}/dy(1)                  0  ... 0]
96      dRHS_dy = 0.
97      dRHS_dy(1,1) = this%sigma
98      if (debugging) print *,'          air%d_dy: end'
99    end function
100
101   function add(lhs,rhs) result(sum) ! add two instances
102     class(air) ,intent(in) :: lhs,rhs
103     type(air)              :: sum
104
105     if (debugging) print *,'          air%add: start'
106     sum%x     = lhs%x     + rhs%x
107     sum%sigma = lhs%sigma + rhs%sigma
108     if (debugging) print *,'          air%add: end'
109   end function
```

```
110
111     ! subtract one instance from another
112     function subtract(lhs,rhs) result(difference)
113       class(air) ,intent(in) :: lhs,rhs
114       type(air)              :: difference
115       if (debugging) print *,'    air%subtract: start'
116       difference%x     = lhs%x     - rhs%x
117       difference%sigma = lhs%sigma - rhs%sigma
118       if (debugging) print *,'    air%subtract: end'
119     end function
120
121     ! multiply an instance by a real scalar
122     function multiply(lhs,rhs) result(product)
123       class(air) ,intent(in) :: lhs
124       real       ,intent(in) :: rhs
125       type(air)              :: product
126       if (debugging) print *,'    air%multiply: start'
127       product%x     = lhs%x     *rhs
128       product%sigma = lhs%sigma*rhs
129       if (debugging) print *,'    air%multiply: end'
130     end function
131   end module air_module
```

Figure 8.5(b): cloud.f03

```
1     module cloud_module
2       use global_parameters_module ,only :debugging !print call tree if true
3
4       implicit none   ! Prevent implicit typing
5       private         ! Hide everything by default
6       public :: cloud
7
8       ! This type tracks the evolution of the second state variable in the
9       ! Lorenz system according to the second Lorenz equation.  It also
10      ! tracks the corresponding paramater (rho) according to the
11      ! differential equation d(rho)/dt=0. For illustrative purposes, this
12      ! implementation does not expose the number of state variables (2) to
13      ! the puppeteer because no iteration is required and the need for
14      ! arithmetic operations on rho is therefore an internal concern.  The
15      ! rank of the matrix d_dCloud() returns is thus 1 to reflect the only
16      ! variable on which the puppeteer needs to iterate when handling
17      ! nonlinear couplings in implicit solvers.
18
19      ! Number of evolution equations/variables exposed to the outside
20      ! world (via Jacobian sub-block shapes):
21      integer ,parameter :: num_eqs=1,num_vars=1
22
23      type cloud
24        private
25        real :: y,rho ! 2nd Lorenz equation solution variable and parameter
26      contains
27        procedure :: set_coordinate ! accessor: set phase-space coordinate
28        procedure :: coordinate! accessor: return phase-space coordinate
29        procedure :: t          ! time derivative
30        procedure :: d_dCloud  ! contribution to diagonal Jacobian element
31        procedure :: d_dx  ! contribution to off-diagonal Jacobian element
```

```
32      procedure :: d_dz  ! contribution to off-diagonal Jacobian element
33      procedure ,private :: add      ! add two instances
34      procedure ,private :: subtract ! subtract one instance from another
35      procedure ,private :: multiply ! multiply instance by a real scalar
36      ! map defined operators to corresponding procedures
37      generic   ,public  :: operator(+) => add
38      generic   ,public  :: operator(-) => subtract
39      generic   ,public  :: operator(*) => multiply
40    end type cloud
41
42    interface cloud
43      procedure constructor
44    end interface
45
46  contains
47    ! constructor: allocate and initialize components
48    type(cloud) function constructor(y_initial,r)
49      real ,intent(in)  :: y_initial
50      real ,intent(in)  :: r
51      if (debugging) print *,'     cloud: start'
52      constructor%y = y_initial
53      constructor%rho = r
54      if (debugging) print *,'     cloud: end'
55    end function
56
57    ! accessor: set component
58    subroutine set_coordinate(this,y_update)
59      class(cloud) ,intent(inout) :: this
60      real ,intent(in) :: y_update
61      this%y = y_update
62    end subroutine
63
64    ! accessor (returns phase-space coordinate)
65    function coordinate(this) result(return_y)
66      class(cloud)          ,intent(in)  :: this
67      real ,dimension(:) ,allocatable :: return_y
68      return_y = [this%y]
69    end function
70
71    ! contribution to diagonal Jacobian element
72    function d_dCloud(this,x_ignored,z_ignored) result(dRHS_dy)
73      class(cloud)                            ,intent(in) :: this
74      real ,dimension(:)   ,allocatable ,intent(in) :: x_ignored,z_ignored
75      real ,dimension(:,:) ,allocatable             :: dRHS_dy
76      if (debugging) print *,'     cloud%d_dCloud: start'
77      allocate(dRHS_dy(num_eqs,num_vars))
78     !dRHS_dy(1) = [ d{x(1)*(rho-z(1))-y(1)}/dy(1) ]
79      dRHS_dy(1,1) = -1.
80      if (debugging) print *,'     cloud%d_dCloud: end'
81    end function
82
83    ! contribution to off-diagonal Jacobian element
84    function d_dx(this,x,z) result(dRHS_dx)
85      class(cloud)                           ,intent(in) :: this
86      real ,dimension(:)   ,allocatable ,intent(in) :: x,z
```

```
87      real ,dimension(:,:) ,allocatable              :: dRHS_dx
88      if (debugging) print *,'      cloud%d_dx: start'
89      allocate(dRHS_dx(num_eqs,size(x)))
90     !dRHS_dx = [ d{x(1)*(rho-z(1))-y}/dx(1)  0  ... 0]
91      dRHS_dx = 0.
92      dRHS_dx(1,1) = this%rho-z(1)
93      if (debugging) print *,'      cloud%d_dx: end'
94    end function
95
96    ! contribution to off-diagonal Jacobian element
97    function d_dz(this,x,z) result(dRHS_dz)
98      class(cloud)                    ,intent(in) :: this
99      real ,dimension(:)    ,allocatable ,intent(in) :: x,z
100     real ,dimension(:,:) ,allocatable              :: dRHS_dz
101     if (debugging) print *,'      cloud%d_dz: start'
102     allocate(dRHS_dz(num_eqs,size(z)))
103    !dRHS_dz = [ d{x(1)*(rho-z(1))-y(1)}/dz(1)  0  ... 0]
104     dRHS_dz = 0.
105     dRHS_dz(1,1) = -x(1)
106     if (debugging) print *,'      cloud%d_dz: end'
107   end function
108
109   ! time derivative (evolution equations)
110   function t(this,x,z) result(dy_dt)
111     class(cloud)          ,intent(in) :: this
112     real ,dimension(:) ,intent(in) :: x,z
113     type(cloud)                       :: dy_dt
114     if (debugging) print *,'      cloud%t: start'
115     dy_dt%y = x(1)*(this%rho-z(1))-this%y
116     dy_dt%rho = 0.
117     if (debugging) print *,'      cloud%t: end'
118   end function
119
120   function add(lhs,rhs) result(sum) ! add two instances
121     class(cloud) ,intent(in) :: lhs,rhs
122     type(cloud)               :: sum
123     if (debugging) print *,'      cloud%add: start'
124     sum%y   = lhs%y   + rhs%y
125     sum%rho = lhs%rho + rhs%rho
126     if (debugging) print *,'      cloud%add: end'
127   end function
128
129   ! subtract one instance from another
130   function subtract(lhs,rhs) result(difference)
131     class(cloud) ,intent(in) :: lhs,rhs
132     type(cloud)               :: difference
133     if (debugging) print *,'      cloud%subtract: start'
134     difference%y   = lhs%y   - rhs%y
135     difference%rho = lhs%rho - rhs%rho
136     if (debugging) print *,'      cloud%subtract: end'
137   end function
138
139   ! multiply an instance by a real scalar
140   function multiply(lhs,rhs) result(product)
141     class(cloud) ,intent(in) :: lhs
```

```
142      real           ,intent(in) :: rhs
143      type(cloud)               :: product
144      if (debugging) print *,'      cloud%multiply: start'
145      product%y   = lhs%y* rhs
146      product%rho = lhs%rho* rhs
147      if (debugging) print *,'      cloud%multiply: end'
148    end function
149  end module cloud_module
```

Figure 8.5(c): ground.f03

```
1  module ground_module
2    use global_parameters_module ,only :debugging !print call tree if true
3
4    implicit none ! Prevent implicit typing
5    private       ! Hide everything by default
6    public :: ground
7
8    ! This type tracks the evolution of the third state variable in the
9    ! Lorenz system according to the third Lorenz equation.  It also
10   ! tracks the corresponding paramater (beta) according to the
11   ! differential equation d(beta)/dt=0. For illustrative purposes,
12   ! this implementation does not expose the number of state variables(2)
13   ! to the puppeteer because no iteration is required and the need for
14   ! arithmetic operations on beta is therefore an internal concern.  The
15   ! rank of the matrix d_dGround() returns is thus 1 to reflect the
16   ! only variable on which the puppeteer needs to iterate when handling
17   ! nonlinear couplings in implicit solvers.
18
19   ! Number of evolution equations/variables exposed to the outside
20   ! world (via Jacobian sub-block shapes):
21   integer ,parameter :: num_eqs=1,num_vars=1
22
23   type ground
24     private
25     real :: z,beta ! 3rd Lorenz equation solution variable and parameter
26   contains
27     procedure :: set_coordinate ! accessor set phase-space coordinate
28     procedure :: coordinate ! accessor: return phase-space coordinate
29     procedure :: t          ! time derivative
30     procedure :: d_dGround  ! contribution to diagonal Jacobian element
31     procedure :: d_dx   ! contribution to off-diagonal Jacobian element
32     procedure :: d_dy   ! contribution to off-diagonal Jacobian element
33     procedure ,private :: add ! add two instances
34     procedure ,private :: subtract ! subtract one instance from another
35     procedure ,private :: multiply ! multiply instance by a real scalar
36     ! map defined operators to corresponding procedures
37     generic ,public  :: operator(+) => add
38     generic ,public  :: operator(-) => subtract
39     generic ,public  :: operator(*) => multiply
40   end type
41
42   interface ground
43     procedure constructor
44   end interface
45
```

```
46  contains
47
48    ! constructor: allocate and initialize components
49    type(ground) function constructor(z_initial,b)
50      real ,intent(in) :: z_initial
51      real ,intent(in) :: b
52      if (debugging) print *,'      ground%construct: start'
53      constructor%z = z_initial
54      constructor%beta = b
55      if (debugging) print *,'      ground%construct: end'
56    end function
57
58    ! accessor (returns phase-space coordinate)
59    function coordinate(this) result(return_z)
60      class(ground)         ,intent(in) :: this
61      real ,dimension(:) ,allocatable :: return_z
62      return_z = [ this%z ]
63    end function
64
65    ! accessor (returns phase-space coordinate)
66    subroutine set_coordinate(this,z_update)
67      class(ground) ,intent(inout)  :: this
68      real ,intent(in) :: z_update
69      this%z = z_update
70    end subroutine
71
72    ! time derivative (evolution equations)
73    function t(this,x,y) result(dz_dt)
74      class(ground)       ,intent(in) :: this
75      real ,dimension(:) ,intent(in) :: x,y
76      type(ground)                   :: dz_dt
77      if (debugging) print *,'      ground%t: start'
78      dz_dt%z = x(1)*y(1) - this%beta*this%z
79      dz_dt%beta = 0.
80      if (debugging) print *,'      ground%t: end'
81    end function
82
83    ! contribution to diagonal Jacobian element
84    function d_dGround(this,x_ignored,y_ignored) result(dRHS_dz)
85      class(ground)                           ,intent(in) :: this
86      real ,dimension(:)   ,allocatable ,intent(in) :: x_ignored,y_ignored
87      real ,dimension(:,:) ,allocatable                :: dRHS_dz
88      if (debugging) print *,'      ground%d_dGround: start'
89     !dRHS_dz = [ d{x(1)*y(1) - beta*z}/dz(1) ]
90      allocate(dRHS_dz(num_eqs,num_vars))
91      dRHS_dz(1,1) = -this%beta
92      if (debugging) print *,'      ground%d_dGround: end'
93    end function
94
95    ! contribution to off-diagonal Jacobian element
96    function d_dx(this,x,y) result(dRHS_dx)
97      class(ground)                           ,intent(in) :: this
98      real ,dimension(:)   ,allocatable ,intent(in) :: x,y
99      real ,dimension(:,:) ,allocatable                :: dRHS_dx
100     if (debugging) print *,'      ground%d_dx: start'
```

```
101      allocate(dRHS_dx(num_eqs,size(x)))
102     !dRHS_dz = [ d{x(1)*y(1) - beta*z(1)}/dx(1)   0  ...  0 ]
103      dRHS_dx=0.
104      dRHS_dx(1,1) = y(1)
105      if (debugging) print *,'      ground%d_dx: end'
106    end function
107
108    ! contribution to off-diagonal Jacobian element
109    function d_dy(this,x,y) result(dRHS_dy)
110      class(ground)                       ,intent(in) :: this
111      real ,dimension(:)    ,allocatable ,intent(in) :: x,y
112      real ,dimension(:,:) ,allocatable              :: dRHS_dy
113      if (debugging) print *,'      ground%d_dy: start'
114      allocate(dRHS_dy(num_eqs,size(y)))
115     !dRHS_dz = [ d{x(1)*y(1) - beta*z(1)}/dy(1)   0  ...  0 ]
116      dRHS_dy = 0.
117      dRHS_dy(1,1) = x(1)
118      if (debugging) print *,'      ground%d_dy: end'
119    end function
120
121    function add(lhs,rhs) result(sum) ! add two instances
122      class(ground) ,intent(in) :: lhs,rhs
123      type(ground)              :: sum
124      if (debugging) print *,'      ground%add: start'
125      sum%z    = lhs%z    + rhs%z
126      sum%beta = lhs%beta + rhs%beta
127      if (debugging) print *,'      ground%add: end'
128    end function
129
130    ! subtract one instance from another
131    function subtract(lhs,rhs) result(difference)
132      class(ground) ,intent(in) :: lhs,rhs
133      type(ground)              :: difference
134      if (debugging) print *,'      ground%subtract: start'
135      difference%z    = lhs%z    - rhs%z
136      difference%beta = lhs%beta - rhs%beta
137      if (debugging) print *,'      ground%subtract: end'
138    end function
139
140    ! multiply an instance by a real scalar
141    function multiply(lhs,rhs) result(product)
142      class(ground) ,intent(in) :: lhs
143      real        ,intent(in)  :: rhs
144      type(ground)              :: product
145      if (debugging) print *,'      ground%multiply: start'
146      product%z    = lhs%z    * rhs
147      product%beta = lhs%beta* rhs
148      if (debugging) print *,'      ground%multiply: end'
149    end function
150  end module ground_module
```

Figure 8.5. Fortran 2003 implementation of puppets for an atmosphere system. (a) The air puppet class. It tracks the first-state variable in the Lorenz system. (b) The cloud puppet class. It tracks the second-state variable in the Lorenz system. (c) The ground puppet class. It tracks the third-state variable in the Lorenz system.

Figure 8.6(a) illustrates the process of a puppeteer (the atmosphere) working with its puppets (the air, cloud, and ground) for an atmosphere system. Figure 8.6 (b) provides a global debug flag that can be used to turn on addition debugging information when running the application.

8.2.2 A C++ Tool: 2D Allocatable Arrays

The use of Fortran two-dimensional allocatable real array is simulated using a `mat_t` class in C++. The definition of this class is shown in Figure 8.7. Internally the actual data, denoted by data member `data_`, is stored using an STL's `vector` of type `float`. Two additional attributes, `r_` and `c_`, are used to represent the shape (row and column) of a two dimensional array. Some member functions emulate the behaviors of an allocatable array in Fortran. For instance, the auto-allocation and auto-reallocation of a 2D allocatable arrays in an assignment are achieved by `resize()` and `clear_resize()`, respectively.[2]

One syntactic distinction we note here is the difference in denoting a two-dimensional array and its element ordering by subscripts between Fortran and C++. Both languages represent a 2D array as a sequence of scalars arranged in a rectangular pattern; for example, it can be viewed as a matrix composed of rows and columns. However, the same subscript values normally result in completely different element indices in the array by the two languages. One well-understood difference is the index base: C++ is zero-based indexing (subscript starts from zero), whereas by default, Fortran uses one-based indexing (subscript value starts from one) although Fortran allows one to use arbitrary value as lower bound.

Figure 8.6(a): main.f03

```
1   program main
2     use air_module ,only : air
3     use cloud_module  ,only :cloud
4     use ground_module ,only : ground
5     use atmosphere_module ,only : atmosphere
6     use global_parameters_module ,only :debugging !print call tree if true
7
8     implicit none ! Prevent implicit typing
9
10    ! This code integrates the Lorenz equations over time using separate
11    ! abstractions for equation and hiding the coupling of those
12    ! abstractions inside an abstraction that follows the Puppeteer design
13    ! pattern of Rouson, Adalsteinsson and Xia (ACM TOMS 37:1, 2010).
14
15    type(air)     ,allocatable :: sky   !puppet for 1st Lorenz equation
16    type(cloud)   ,allocatable :: puff  !puppet for 2nd Lorenz equation
17    type(ground)  ,allocatable :: earth !puppet for 3rd Lorenz equation
18    type(atmosphere)           :: boundary_layer  ! Puppeteer
19    integer                    :: step            ! time step
```

[2] In class definition for mat_t, we only provide a minimum set of definitions of functions and operators that are actually used throughout the example. It is by no means to be used as a realistic class definition for two-dimensional allocatable array. Readers are encouraged to extend this class so it becomes more complete.

```
20   integer ,parameter          :: num_steps=1000  ! total time steps
21   real    ,parameter          :: x=1.,y=1.,z=1.  ! initial conditions
22   real                        :: t               ! time coordinate
23                                  ! Lorenz parameters
24   real    ,parameter          :: sigma=10.,rho=28.,beta=8./3.,dt=.02
25
26   if (debugging) print *,'main: start'
27   allocate (sky, puff, earth)
28   sky = air(x,sigma)
29   puff = cloud(y,rho)
30   earth = ground(z,beta)
31
32   ! transfer allocations into puppeteer
33   boundary_layer = atmosphere(sky,puff,earth)
34   ! all puppets have now been deallocated
35
36   t=0.
37   write(*,'(f10.4)',advance='no') t
38   print *,boundary_layer%state_vector()
39   do step=1,num_steps
40     call integrate(boundary_layer,dt)
41     t = t + dt
42     write(*,'(f10.4)',advance='no') t
43     print *,boundary_layer%state_vector()
44   end do
45   if (debugging) print *,'main: end'
46
47 contains
48   ! abstract Trapezoidal rule integration
49   subroutine integrate(integrand,dt)
50     type(atmosphere) ,intent(inout) :: integrand
51     real             ,intent(in)    :: dt
52     type(atmosphere) ,allocatable   :: integrand_estimate,residual
53     integer ,parameter :: num_iterations=5
54     integer :: newton_iteration,num_equations,num_statevars,i,j
55     real ,dimension(:,:) ,allocatable :: dRHS_dState,jacobian,identity
56
57     if (debugging) print *,'  integrate: start'
58     allocate(integrand_estimate, source=integrand)
59     allocate(residual)
60     do newton_iteration=1,num_iterations
61       dRHS_dState   = integrand_estimate%dRHS_dV()
62       num_equations = size(dRHS_dState,1)
63       num_statevars = size(dRHS_dState,2)
64       if (num_equations /= num_statevars) &
65         stop 'integrate: ill-posed problem.'
66       identity = reshape( &
67         source=(/((0.,i=1,num_equations),j=1,num_statevars)/), &
68         shape=(/num_equations,num_statevars/) )
69       forall(i=1:num_equations) identity(i,i)=1.
70       jacobian = identity - 0.5*dt*dRHS_dState
71       residual = integrand_estimate - &
72         ( integrand + (integrand%t() + integrand_estimate%t())*(0.5*dt))
73       integrand_estimate = integrand_estimate - &
74         (jacobian .inverseTimes. residual)
```

```
75        end do
76        integrand = integrand_estimate
77        if (debugging) print *,'  integrate: end'
78      end subroutine
79    end program main
```

Figure 8.6(b): global_parameters.f03

```
1    module global_parameters_module
2      ! parameter that is used to control the debugging.  The call tree
3      ! information is printed if value of debugging is true
4      logical ,parameter :: debugging=.false.
5    end module
```

Figure 8.6. Fortran 2003 implementation of puppeteer pattern applied to an atmosphere system. (a) Time-integration test of Lorenz system using puppeteer pattern. (b) A global debugging flag.

Another difference is the element ordering with regard to dimensions. Fortran's element sequence ordering (traversal rule) is column-based, whereas C++ is row-based. This second difference can sometimes be problematic in applications mixing the two languages. Consider the following declaration of a 3×3 matrix x,

```
Fortran:
    real    x(3,3)    ! Fortran declaration
C++:
    float   x[3][3]; // C++ declaration
```

In Fortran, x(3,1) refers to the third element of x, and x(1,3) refers to seventh element of x. The corresponding C++ expressions are x[0][2] and x[2][0], respectively.

In designing `mat_` class, we choose zero-based notation because it is the most natural way to express array subscripts to C++ programmers. We also adhere to Fortran's convention in element ordering. The computation of element order is shown in the definitions of `operator()` in figure 8.7(b) as follows:

```
mat_t::value_type mat_t::operator()(int r, int c) const {
  // error checking for invalid subscript range; code omitted
  return data_.at(c*r_ + r);
}
```

Thus for the aforementioned 3×3 matrix, mat_(2, 0) corresponds to Fortran's x(3, 1), and mat_(0, 2) corresponds to x(1, 3).

Figure 8.7(a): mat.h

```
1    #ifndef _H_MAT_
2    #define _H_MAT_
3
4    #include "globals.h"
5    #include <iostream>
6    #include <iomanip>
```

```
7
8   class mat_t {
9   public:
10    typedef crd_t::value_type value_type;
11    typedef crd_t::reference  reference;
12
13    mat_t();
14    mat_t(int rows, int cols);
15    void clear();
16    void resize(int rows, int cols);
17    void clear_resize(int rows, int cols, value_type value = 0);
18    void identity(int rows);
19    int rows() const;
20    int cols() const;
21    value_type operator()(int r, int c) const;
22    reference operator()(int r, int c);
23    void set_submat(int r, int c, const mat_t &other);
24    mat_t& operator-=(const mat_t&);
25    mat_t& operator*=(real_t);
26
27  private:
28    int r_, c_;
29    crd_t data_;
30  };
31
32  inline mat_t operator-(const mat_t &a, const mat_t &b) {
33    mat_t retval(a);
34    retval -= b;
35    return retval;
36  }
37
38  inline mat_t operator*(real_t value, const mat_t &matrix) {
39    mat_t retval(matrix);
40    retval *= value;
41    return retval;
42  }
43
44  struct dim_t {
45    const int eqs;
46    const int vars;
47
48    dim_t(int eqcnt, int varcnt) :
49      eqs(eqcnt), vars(varcnt)
50    {}
51  };
52
53  inline std::ostream& operator<<(std::ostream &os, const mat_t &mat) {
54    std::ios_base::fmtflags flags = os.flags();
55    for(int r = 0; r < mat.rows(); ++r) {
56      os << "[";
57      for(int c = 0; c < mat.cols(); ++c) {
58        os << " " <<std::setw(12) <<std::setprecision(8)
59               <<std::fixed <<mat(r,c);
60      }
61      os << "]\n";
```

```
62     }
63     os.flags(flags);
64     return os;
65   }
66 
67   #endif // !_H_MAT_
```

Figure 8.7(b): mat.cpp

```
1    #include "mat.h"
2    #include <exception>
3    #include <iostream>
4 
5    struct matrix_error : public std::exception {
6      virtual ~matrix_error() throw() {}
7    };
8 
9    mat_t::mat_t() :
10     r_(0), c_(0)
11   {}
12 
13   mat_t::mat_t(int rows, int cols) {
14     this->resize(rows, cols);
15   }
16 
17   void mat_t::clear() {
18     r_ = c_ = 0;
19     data_.clear();
20   }
21 
22   void mat_t::resize(int rows, int cols) {
23     if(rows < 0 || cols < 0) {
24       std::cerr << "mat_t::resize:  Rows and columns must be >= 0.\n";
25       throw matrix_error();
26     }
27     if(! data_.empty()) {
28       // Copy data over.
29     }
30     else {
31       // common case.
32       data_.resize(rows*cols);
33       r_ = rows;
34       c_ = cols;
35     }
36   }
37 
38   void mat_t::clear_resize(int rows, int cols, value_type value) {
39     if(rows < 0 || cols < 0) {
40       std::cerr << "mat_t::clear_resize:  Rows and columns must be >= 0\n";
41       throw matrix_error();
42     }
43     data_.resize(rows*cols);
44     r_ = rows;
45     c_ = cols;
46     std::fill(data_.begin(), data_.end(), value);
47   }
```

```
48
49  void mat_t::identity(int size) {
50    this->clear_resize(size, size, 0);
51    for(int i = 0; i < size; ++i) {
52      this->operator()(i,i) = 1;
53    }
54  }
55
56  int mat_t::rows() const {
57    return r_;
58  }
59
60  int mat_t::cols() const {
61    return c_;
62  }
63
64  mat_t::value_type mat_t::operator()(int r, int c) const {
65    if(r < 0 || r >= r_ || c < 0 || c >= c_) {
66      std::cerr << "mat_t::operator(): Invalid index (" << r << ", " << c
67                << "). Bounds are (" << r_ << ", " << c << ")\n";
68      throw matrix_error();
69    }
70    return data_.at(c*r_ + r);
71  }
72
73  mat_t::reference mat_t::operator()(int r, int c) {
74    if(r < 0 || r >= r_ || c < 0 || c >= c_) {
75      std::cerr << "mat_t::operator(): Invalid index (" << r << ", " << c
76                << "). Bounds are (" << r_ << ", " << c << ")\n";
77      throw matrix_error();
78    }
79    return data_.at(c*r_ + r);
80  }
81
82  void mat_t::set_submat(int startrow, int startcol, const mat_t &other) {
83    for(int r = 0; r < other.rows(); ++r) {
84      for(int c = 0; c < other.cols(); ++c) {
85        this->operator()(r+startrow, c+startcol) = other(r,c);
86      }
87    }
88  }
89
90  mat_t& mat_t::operator-=(const mat_t &other) {
91    if(this->rows() != other.rows() || this->cols() != other.cols()) {
92      std::cerr <<
93       "mat_t::operator-=: Matrices must be of identical size.\n";
94      throw matrix_error();
95    }
96    const size_t size = data_.size();
97    for(size_t i = 0; i < size; ++i)
98      data_[i] -= other.data_[i];
99    return *this;
100 }
101
102 mat_t& mat_t::operator*=(real_t value) {
```

```
103    for(crd_t::iterator it = data_.begin(); it != data_.end(); ++it)
104      *it *= value;
105    return *this;
106  }
```

Figure 8.7. A mat_t class in C++ that simulates the functionality of 2D allocatable real arrays in Fortran. (a) Class definition for mat_t, which simulates Fortran's two-dimensional allocatable array of type real. (b) Implementation of class mat_t.

8.2.3 C++ Implementation

Figure 8.8 shows the C++ definition for the atmosphere class, the puppeteer in a weather system. In Figure 8.9, we give C++ implementations for air, cloud, and ground classes. These are the puppets of the atmosphere system. Figure 8.10 shows the C++ puppeteer corresponding to Fortran puppeteer given in Figure 8.6. Again the similarity to the Fortran counterpart is obvious.

Figure 8.8(a): atmosphere.h

```
1   #ifndef _H_ATMOSPHERE_
2   #define _H_ATMOSPHERE_
3
4   #include "integrand.h"
5   #include "air.h"
6   #include "cloud.h"
7   #include "ground.h"
8
9   class atmosphere : public integrand {
10  public:
11    typedef integrand::ptr_t ptr_t;
12    typedef Ref<atmosphere> self_t;
13
14    atmosphere(const air&, const cloud&, const ground&);
15    virtual ~atmosphere();
16
17    // The following methods do dynamic allocation.
18    virtual ptr_t d_dt() const;
19    virtual void  dRHS_dV(mat_t&) const;
20    virtual ptr_t clone() const;
21    virtual ptr_t  inverse_times(const mat_t&) const;
22    virtual crd_t state_vector() const;
23
24    // The following methods are destructive updates.
25    virtual ptr_t operator+=(ptr_t);
26    virtual ptr_t operator-=(ptr_t);
27    virtual ptr_t operator*=(real_t);
28
29  private:
30    air air_;
31    cloud cloud_;
32    ground ground_;
33  };
34
35  #endif //!_H_ATMOSPHERE_
```

Figure 8.8(b): atmosphere.cpp

```
1  #include "atmosphere.h"
2  #include <exception>
3  #include <cmath>
4
5  struct atmosphere_error : public std::exception {
6    virtual ~atmosphere_error() throw() {}
7  };
8
9  typedef atmosphere::ptr_t ptr_t;
10
11 atmosphere::atmosphere(const air &a, const cloud &c, const ground &g) :
12   air_(a), cloud_(c), ground_(g)
13 {}
14
15 atmosphere::~atmosphere() {
16 }
17
18 ptr_t atmosphere::d_dt() const {
19   return
20     ptr_t(new atmosphere(air_.d_dt(cloud_.coordinate()),
21          cloud_.d_dt(air_.coordinate(),ground_.coordinate()),
22          ground_.d_dt(air_.coordinate(),cloud_.coordinate())));
23 }
24
25 void atmosphere::dRHS_dV(mat_t &result) const {
26   // Figure out the required dimensions.
27   dim_t adim = air_.dimensions();
28   dim_t cdim = cloud_.dimensions();
29   dim_t gdim = ground_.dimensions();
30   // Resize the result matrix.
31   result.clear_resize(adim.eqs  + cdim.eqs  + gdim.eqs,
32          adim.vars + cdim.vars + gdim.vars);
33   if(result.rows() != result.cols()) {
34     std::cerr << "atmosphere::dRHS_dV:  Ill-formed problem:  total of "
35         << result.rows() << " equations and " << result.cols()
36         << " variables\n";
37     throw atmosphere_error();
38   }
39   // dAir/dAir
40   result.set_submat(0, 0, air_.d_dAir(cloud_.coordinate()));
41   // dAir/dCloud
42   result.set_submat(0, adim.vars, air_.d_dy(cloud_.coordinate()));
43   // dAir/dGround is all zero -- skipping that one.
44   // dCloud/dAir
45   result.set_submat(adim.eqs, 0,
46         cloud_.d_dx(air_.coordinate(),ground_.coordinate()));
47   // dCloud/dCloud
48   result.set_submat(adim.eqs, adim.vars,
49         cloud_.d_dCloud(air_.coordinate(),ground_.coordinate()));
50   // dCloud/dGround
51   result.set_submat(adim.eqs, adim.vars+cdim.vars,
52         cloud_.d_dz(air_.coordinate(),ground_.coordinate()));
53   // dGround/dAir
54   result.set_submat(adim.eqs+cdim.eqs, 0,
```

```
55            ground_.d_dx(air_.coordinate(),cloud_.coordinate()));
56      // dGround/dCloud
57      result.set_submat(adim.eqs+cdim.eqs, adim.vars,
58            ground_.d_dy(air_.coordinate(),cloud_.coordinate()));
59      // dGround/dGround
60      result.set_submat(adim.eqs+cdim.eqs, adim.vars+cdim.vars,
61            ground_.d_dGround(air_.coordinate(),cloud_.coordinate()));
62    }
63
64    ptr_t atmosphere::clone() const {
65      return ptr_t(new atmosphere(*this));
66    }
67
68    crd_t atmosphere::state_vector() const {
69      const crd_t &cc = cloud_.coordinate(), &gc = ground_.coordinate();
70      crd_t state_space = air_.coordinate();
71      state_space.insert(state_space.end(), cc.begin(), cc.end());
72      state_space.insert(state_space.end(), gc.begin(), gc.end());
73      return state_space;
74    }
75
76
77    ptr_t atmosphere::inverse_times(const mat_t &lhs) const {
78      static const real_t pivot_tolerance = 1e-2;
79
80      const int n = lhs.rows();
81      crd_t b = this->state_vector();
82      if((n != lhs.cols()) || (size_t(n) != b.size())) {
83        std::cerr <<"integrand::inverse_times:  ill-posed matrix problem\n";
84        throw atmosphere_error();
85      }
86      crd_t x(n);
87      mat_t A(lhs);
88      for(int p = 0; p < n-1; ++p) { // forward elimination
89        if(fabs(A(p,p)) < pivot_tolerance) {
90          std::cerr << "integrand::inverse_times:  "
91        << "use an algorithm with pivoting\n";
92          throw atmosphere_error();
93        }
94        for(int row = p+1; row < n; ++row) {
95          real_t factor = A(row,p) / A(p,p);
96          for(int col = p; col < n; ++col) {
97      A(row,col) = A(row,col) - A(p,col)*factor;
98          }
99          b.at(row) = b.at(row)- b.at(p)*factor;
100       }
101     }
102     x.at(n-1) = b.at(n-1) / A(n-1,n-1); // back substitution
103     for(int row = n-1; row >= 0; --row) {
104       real_t the_sum = 0;
105       for(int col = row+1; col < n; ++col) {
106         the_sum += A(row,col) * x.at(col);
107       }
108       x.at(row) = (b.at(row) - the_sum) / A(row,row);
109     }
```

```
110   return ptr_t(new atmosphere(air(x.at(0), x.at(1)),
111            cloud(x.at(2), cloud_.rho()),
112            ground(x.at(3), ground_.beta())));
113 }
114
115 ptr_t atmosphere::operator+=(ptr_t other) {
116   self_t added = cast<atmosphere>(other);
117   if(other == NULL) {
118     std::cerr << "atmosphere::operator+=:  Invalid input type\n";
119     throw atmosphere_error();
120   }
121   air_    += added->air_;
122   cloud_  += added->cloud_;
123   ground_ += added->ground_;
124   return ptr_t(this);
125 }
126
127 ptr_t atmosphere::operator-=(ptr_t other) {
128   self_t subbed = cast<atmosphere>(other);
129   if(other == NULL) {
130     std::cerr << "atmosphere::operator-=:  Invalid input type\n";
131     throw atmosphere_error();
132   }
133   air_    -= subbed->air_;
134   cloud_  -= subbed->cloud_;
135   ground_ -= subbed->ground_;
136   return ptr_t(this);
137 }
138
139 ptr_t atmosphere::operator*=(real_t value) {
140   air_    *= value;
141   cloud_  *= value;
142   ground_ *= value;
143   return ptr_t(this);
144 }
```

Figure 8.8. Atmosphere class definitions in C++. (a) Class definition for atmosphere. (b) Implementation of class atmosphere. (c) Definition for time-integration operation.

Figure 8.9(a): air.h

```
1  #ifndef _H_AIR_
2  #define _H_AIR_
3
4  #include "mat.h"
5
6  class air {
7  public:
8    air(real_t x, real_t sigma);
9
10   const crd_t& coordinate() const;
11   air d_dt(const crd_t&) const;
12   mat_t d_dAir(const crd_t&) const;
13   mat_t d_dy(const crd_t&) const;
14   air& operator+=(const air&);
```

```
15    air& operator-=(const air&);
16    air& operator*=(real_t);
17
18    inline dim_t dimensions() const { return dim_t(dim_, dim_); }
19
20  private:
21    static const int dim_;
22    crd_t x_;  // sigma is stored at x_[1]
23  };
24
25  #endif //!_H_AIR_
```

Figure 8.9(b): air.cpp

```
1   #include "air.h"
2
3   const int air::dim_ = 2;
4
5   air::air(real_t x, real_t sigma) {
6     x_.push_back(x);
7     x_.push_back(sigma);
8   }
9
10  const crd_t& air::coordinate() const {
11    return x_;
12  }
13
14  air air::d_dt(const crd_t &y) const {
15    return air((x_.at(1) * (y.at(0) - x_.at(0))), 0);
16  }
17
18  mat_t air::d_dAir(const crd_t& y) const {
19    mat_t result(dim_, dim_);
20    result(0, 0) = -x_.at(1);
21    result(0, 1) = y.at(0) - x_.at(0);
22    return result;
23  }
24
25  mat_t air::d_dy(const crd_t &y) const {
26    mat_t result(dim_, y.size());
27    result(0, 0) = x_.at(1);
28    return result;
29  }
30
31  air& air::operator+=(const air &other) {
32    x_.at(0) += other.x_.at(0);
33    x_.at(1)   += other.x_.at(1);
34    return *this;
35  }
36
37  air& air::operator-=(const air &other) {
38    x_.at(0) -= other.x_.at(0);
39    x_.at(1)   -= other.x_.at(1);
40    return *this;
41  }
42
```

```
43  air& air::operator*=(real_t value) {
44    x_.at(0) *= value;
45    x_.at(1) *= value;
46    return *this;
47  }
```

Figure 8.9(c): cloud.h

```
1   #ifndef _H_CLOUD_
2   #define _H_CLOUD_
3
4   #include "mat.h"
5
6   class cloud {
7   public:
8     cloud(real_t y, real_t rho);
9
10    const crd_t& coordinate() const;
11    real_t rho() const;
12    cloud d_dt(const crd_t&, const crd_t&) const;
13    mat_t d_dCloud(const crd_t&, const crd_t&) const;
14    mat_t d_dx(const crd_t&, const crd_t&) const;
15    mat_t d_dz(const crd_t&, const crd_t&) const;
16    cloud& operator+=(const cloud&);
17    cloud& operator-=(const cloud&);
18    cloud& operator*=(real_t);
19
20    inline dim_t dimensions() const { return dim_t(dim_, dim_); }
21
22  private:
23    static const int dim_;
24    crd_t y_;
25    real_t rho_;
26  };
27
28  #endif //!_H_CLOUD_
```

Figure 8.9(d): cloud.cpp

```
1   #include "cloud.h"
2
3   const int cloud::dim_ = 1;
4
5   cloud::cloud(real_t y, real_t rho) :
6     y_(1, y), rho_(rho)
7   {}
8
9   const crd_t& cloud::coordinate() const {
10    return y_;
11  }
12
13  real_t cloud::rho() const {
14    return rho_;
15  }
16
17  cloud cloud::d_dt(const crd_t &x, const crd_t &z) const {
18    return cloud((x.at(0) * (rho_ - z.at(0)) - y_.at(0)), 0);
```

```
19 }
20
21 mat_t cloud::d_dCloud(const crd_t&, const crd_t&) const {
22   mat_t result(dim_, dim_);
23   result(0, 0) = -1;
24   return result;
25 }
26
27 mat_t cloud::d_dx(const crd_t &x, const crd_t &z) const {
28   mat_t result(dim_, x.size());
29   result(0, 0) = rho_ - z.at(0);
30   return result;
31 }
32
33 mat_t cloud::d_dz(const crd_t &x, const crd_t &z) const {
34   mat_t result(dim_, z.size());
35   result(0, 0) = -x.at(0);
36   return result;
37 }
38
39 cloud& cloud::operator+=(const cloud &other) {
40   y_.at(0) += other.y_.at(0);
41   rho_ += other.rho_;
42   return *this;
43 }
44 cloud& cloud::operator-=(const cloud &other) {
45   y_.at(0) -= other.y_.at(0);
46   rho_ -= other.rho_;
47   return *this;
48 }
49
50 cloud& cloud::operator*=(real_t value) {
51   y_.at(0) *= value;
52   rho_ *= value;
53   return *this;
54 }
```

Figure 8.9(e): ground.h

```
1  #ifndef _H_GROUND_
2  #define _H_GROUND_
3
4  #include "mat.h"
5
6  class ground {
7  public:
8    ground(real_t y, real_t rho);
9
10   const crd_t& coordinate() const;
11   real_t beta() const;
12   ground d_dt(const crd_t&, const crd_t&) const;
13   mat_t d_dGround(const crd_t&, const crd_t&) const;
14   mat_t d_dx(const crd_t&, const crd_t&) const;
15   mat_t d_dy(const crd_t&, const crd_t&) const;
16   ground& operator+=(const ground&);
17   ground& operator-=(const ground&);
```

```
18    ground& operator*=(real_t);
19
20    inline dim_t dimensions() const { return dim_t(dim_, dim_); }
21
22  private:
23    static const int dim_;
24    crd_t z_;
25    real_t beta_;
26  };
27
28  #endif //!_H_GROUND_
```

Figure 8.9(f): ground.cpp

```
1   #include "ground.h"
2
3   const int ground::dim_ = 1;
4
5   ground::ground(real_t z, real_t beta) :
6     z_(1, z), beta_(beta)
7   {}
8
9   const crd_t& ground::coordinate() const {
10    return z_;
11  }
12
13  real_t ground::beta() const {
14    return beta_;
15  }
16
17  ground ground::d_dt(const crd_t& x, const crd_t& y) const {
18    return ground((x.at(0) * y.at(0) - beta_ * z_.at(0)), 0);
19  }
20
21  mat_t ground::d_dGround(const crd_t&, const crd_t&) const {
22    mat_t result(dim_, dim_);
23    result(0, 0) = -beta_;
24    return result;
25  }
26
27  mat_t ground::d_dx(const crd_t &x, const crd_t &y) const {
28    mat_t result(dim_, x.size());
29    result(0, 0) = y.at(0);
30    return result;
31  }
32
33  mat_t ground::d_dy(const crd_t &x, const crd_t &y) const {
34    mat_t result(dim_, y.size());
35    result(0, 0) = x.at(0);
36    return result;
37  }
38
39  ground& ground::operator+=(const ground &other) {
40    z_.at(0) += other.z_.at(0);
41    beta_ += other.beta_;
42    return *this;
```

```
43  }
44
45  ground& ground::operator-=(const ground& other) {
46    z_.at(0) -= other.z_.at(0);
47    beta_ -= other.beta_;
48    return *this;
49  }
50
51  ground& ground::operator*=(real_t value) {
52    z_.at(0) *= value;
53    beta_ *= value;
54    return *this;
55  }
```

Figure 8.9. Puppet class definitions in C++. (a)–(b) Class definition of air. (c)–(d) Class definition of cloud. (e)–(f) Class defintion of ground.

Figure 8.10: main.cpp

```
1   #include "atmosphere.h"
2   #include "fmt.h"
3
4   typedef integrand::ptr_t ptr_t;
5
6   int main() {
7     const int num_steps=1000;
8     const real_t x=1., y=1., z=1., sigma=10., rho=28, beta=8./3., dt=0.02;
9
10    air sky(x, sigma);
11    cloud puff(y, rho);
12    ground earth(z, beta);
13    ptr_t boundary_layer = ptr_t(new atmosphere(sky, puff, earth));
14
15    real_t t = 0.;
16    std::cout << fmt(t,5,2) << " "
17              << fmt(boundary_layer->state_vector()) << "\n";
18    for(int step = 1; step <= num_steps; ++step) {
19      integrate(boundary_layer, dt);
20      t += dt;
21      std::cout << fmt(t,5,2) << " "
22                << fmt(boundary_layer->state_vector()) << "\n";
23    }
24  }
```

Figure 8.10. C++ implementation of puppeteer pattern applied to an atmosphere system following the same design as that by Fortran 2003.

8.3 The Consequences

The chief consequence of the PUPPETEER pattern derives from its separation of concerns. Domain experts can construct ADTs that encapsulate widely disparate physics without knowing the implementation details of the other ADTs. In the atmospheric modeling example, a fluid dynamicist might build an Air abstraction that solves the Navier-Stokes equations for wind velocities and pressures, while a chemist might build a Cloud abstraction that predicts acid rain by modeling chemical species adsorption at droplet surfaces. The PUPPETEER would first request droplet locations from the Cloud instance, then request species concentrations at the droplet location from the Air instance, and finally pass these concentrations to the Cloud in the process of constructing $\vec{R}$ and $\partial\vec{R}/\partial\vec{V}$.

A second important consequence derives from the aforementioned containment relationships. Implemented using Fortran's allocatable components, the PUPPETEER can be viewed as holding only references to its puppets (similar to the assignment of a smart pointer in C++, Fortran's `move_alloc` transfers the Air, Cloud, and Ground to the corresponding puppets during the constructor call in 8.4). The actual construction of puppets' data and their lifetime, therefore, can be independent from the existence of the PUPPETEER (see `main` in 8.6). This opens the possibility of varying the physical models dynamically mid-simulation. In the atmospheric boundary layer model, the cloud model would not have to be included until the atmospheric conditions became ripe for cloud formation. Of course, when an absent object would otherwise supply information required by another object, the PUPPETEER must substitute default values. Whether this substitution happens inside the PUPPETEER or inside each ADT (via optional arguments) is implementation-dependent.

The cost of separating concerns and varying the physics at runtime lies in the conceptual work of discerning who does what, where they do it, and in which format. Consider the following lines from the Fortran implementation of the trapezoidal time integration algorithm. This algorithm is implemented in the `integrate()` procedure of 8.6(a):

```
dRHS_dState    = integrand_estimate%dRHS_dV()
...
jacobian = identity - 0.5*dt*dRHS_dState
residual = integrand_estimate - &
  ( integrand + (integrand%t() + integrand_estimate%t())*(0.5*dt))

integrand_estimate = integrand_estimate - &
  (jacobian .inverseTimes. residual)
```

where the ellipses indicate deleted lines.[3] The first line represents an abstract calculation of $\partial\vec{R}/\partial\vec{V}$. All calculations, including the assignment, happen inside the PUPPETEER. This design decision stems from the fact that the details of the PUPPETEER (and the objects it contains) determine $\vec{R}$, so the information-hiding philosophy of OOP precludes exposing these to `integrate()`. By contrast, the formula for calculating the Jacobian in the next line depends on the chosen time-integration algorithm, so while `dt*dRHS_dState` represents an overloaded operation implemented inside the PUPPETEER, that operation returns a Fortran array of intrinsic type (C++ *primitive type*), and the remainder of the line represents arithmetic on intrinsic entities. Thus, the identity matrix and the Jacobian are simple floating-point arrays. Finally, the subsequent two lines would be the same for any algorithm that employs implicit

[3] The name **inverseTimes** is intended to be suggestive of the ultimate result. The sample code in 8.4 employs Gaussian elimination rather than computing and premultiplying the inverse Jacobian.

advancement of nonlinear equations, so those lines represent overloaded arithmetic carried out wholly inside the PUPPETEER.

Ultimately, we believe the forethought that goes into deciding what gets calculated where and by whom pays off in keeping the code flexible. There is no hardwiring of algorithm-specific details of the Jacobian calculation into individual physics abstractions. Nor is there any hardwiring of physics-specific details of the residual calculation into the time integrator or the nonlinear solver. Each can be reused if the implementation of the other changes in fundamental ways.

The puppeteer provides an elegant solution to a common problem in managing complexity in multiscale, multiphysics applications. For efficiency and numerical stability, individual model components often run in nonidentical dimensions (e.g., when coupling two- and three-dimensional simulation models) or incompatible (e.g., reduced) units. All such considerations can be delegated to the puppeteer, greatly easing the reuse of existing simulation software for the individual single-physics models.

In any multiphysics package employing the ABSTRACT CALCULUS pattern, the `integrand` abstract type would be highly stable because it depends on no other ADTs. Hence, the afferent couplings vanish ($Ce = 0$). Also, the associated integrate procedure depends only on the dynamical system being integrated, which will typically be a single puppeteer in a multiphysics simulation.

8.4 Related Patterns

Other parts of this book detail the similarities and close relationships between a PUPPETEER and other design pattern. Section 8.1 describes the ways in which a PUPPETEER serves a role similar to that of a MEDIATOR. Section 6.4 details the close relationship between a PUPPETEER and an ABSTRACT CALCULUS in the context of time integration. In this context, a PUPPETEER might bear an equally close relationship with a STRATEGY if the latter supplies numerical quadrature methods used by an ABSTRACT CALCULUS for time integration. A PUPPETEER can serve as the concrete implementation of an abstract FACADE defined by an ABSTRACT CALCULUS or a STRATEGY.

EXERCISES

1. Develop your own PUPPET pattern that abstracts the notion of an individual subsystem to be manipulated by a PUPPETEER. Shield the PUPPETEER from knowledge of the concrete PUPPETS by having the PUPPETEER manipulate a collection of an abstract `puppet` objects, each one having a private state vector along with public methods for accepting and providing coupling terms. Write an `atmosphere` PUPPETEER with an three-element `puppet` array and pass it puppets corresponding to each of the Lorenz equations (4.5). Advance your PUPPETEER by an explicit time advancement algorithm such as 2nd-order Runge Kutta.
2. Consider the number of possible interpretations of the time integration line (46) in Figure 6.2(b) with each interpretation corresponding to a unique dynamic type that extends "integrand" type. Calculate the growth in information entropy as each new single-physics abstraction is added to a multiphysics package.

9 Factory Patterns

"Separate the physics from the data."
Jaideep Ray

9.1 The Problem

The patterns discussed in Chapters 6–7 adhere closely to the GoF philosophy of designing to an interface rather than an implementation. That maxim inspired the use of abstract classes in defining an ABSTRACT CALCULUS, a STRATEGY, and a SURROGATE. The PUPPETEER definition in Chapter 8 represented the one setting in which client code manipulated concrete implementations directly – although an exercise at the end of that chapter describes a way to liberate the PUPPETEER from the tyranny of implementations also.

Independent of whether the classes comprising the pattern demonstrations are abstract, the client codes (`main` programs in the cases considered) exploit knowledge of the concrete type of each object constructed in Chapters 6–8. Although we were able to write polymorphic procedures such as `integrate` in Figures 6.2(b) and 6.3(b)–(c), in each case, the actual arguments passed to these procedures were references to concrete objects the client constructed. One can write even more flexible code by freeing clients of even this minimal knowledge of concrete types. This poses the dilemma of where object construction happens. Put simply, how does one construct an object without directly invoking its constructor? Or from another angle, how does an object come to be if the client knows only the interface defined by its abstract parent, but that parent's abstract nature precludes definition of a constructor? By definition, no instance of an abstract type can be instantiated.

The desire for any reference to the object to be abstract from its inception conflicts with the need for that inception to result in a concrete instantiation. To place this dilemma in a scientific context, consider the Burgers equation (4.9):

$$u_t = \nu u_{xx} - u u_x$$

which we might consider advancing from t_n to $t_{n+1} \equiv t_n + \Delta t$ according to the explicit Euler marching algorithm:

$$u^{n+1} = u^n + \Delta t \left(\nu u^n_{xx} - u^n u^n_x \right)$$

where u^n and u^{n+1} are the solution at t_n and t_{n+1}, respectively. Given an interface, `field`, to a family of 1D scalar field implementations, we desire to write Fortran client code statements on a `field` reference `u` of the form:

```
u = u + dt*(nu*u%xx() - u*u%x())
```

or equivalent C++ client statements of the form:

```
u += dt*(nu*u.xx() - u*u.x());
```

where the Fortran and C++ `field` classes are both `abstract`, and `u` can dynamically (at runtime) take on any type that implements the `field` interface.

The quandary surrounding the *where* of object instantiation poses related questions about the *when* and the *how* of instantiation. Eliminating the client's control over the instantiation's spatial position in the code offers the potential to also eliminate the client's control over its temporal position and its path to completion. Flexibility in the timing matters most when an object's construction occupies significant resources such as memory or execution time. In such cases, it might be desirable for a constructor to return a lightweight reference or an identifying tag but to delay the actual resource allocation until the time of the object's first use. Flexibility in the steps taken to complete the instantiation matters most when the process involves constructing multiple parts that must fit coherently into a complex whole.

The solutions discussed in Section 9.2 focus primarily on the *where* question, moving the object constructor invocation outside the client code and into a class hierarchy parallel to the one in question. Section 9.4 briefly discusses some of the options for addressing the *when* and *how* questions.

Give Me Liberty and Give Me Birth!

The freedom associated with keeping clients' references to an object abstract can be realized throughout the object's lifetime if these references are abstract from moment of the object's creation. This precludes letting a client invoke the constructors defined in the object's concrete implementation and creates a quandary regarding how, where, and when the object's instantiation occurs.

9.2 The Solution

The GoF define several creational design patterns that resolve the dilemmas presented in Section 9.1. This section details two: the ABSTRACT FACTORY and FACTORY METHOD patterns. The GoF describe the intent of an ABSTRACT FACTORY as follows:

> Provide an interface for creating families of related or dependent objects without specifying their concrete classes.

The GoF demonstrated the construction of an ABSTRACT FACTORY for manufacturing mazes for computer games. A maze comprised a set of rooms, walls, and doors. In an ABSTRACT FACTORY, one defines a method for each product manufactured. Thus, the GoF C++ example comprises a `MazeFactory` class containing `MakeMaze`, `MakeWall`, `MakeRoom`, and `MakeDoor` virtual member functions with default implementations that return pointers to new product instances. Because their `MazeFactory` was not `abstract`, it served as both the ABSTRACT FACTORY and the concrete factory. The GoF define two concrete `MazeFactory` subclasses: `EnhantedMaze` and `BombedMaze`. These overload the aforementioned virtual member functions to produce spellbound doors and boobytrapped walls, respectively.

In most designs, an ABSTRACT FACTORY contains one or more FACTORY METHOD implementations, the intent of which the GoF define as follows:

> Define an interface for creating an object, but let subclasses decide which class to instantiate. Factory method lets a class defer instantiation to subclasses.

In `MazeFactory`, the `MakeMaze`, `MakeWall`, `MakeRoom`, and `MakeDoor` product-manufacturing methods are FACTORY METHODS. Whereas the `MazeFactory` base class defines default implementations for these procedures, subclasses can override that behavior to provide more specific implementations.

The GoF present an alternative based on parameterizing the FACTORY METHOD. In this case, one defines a single FACTORY METHOD for the entire product family. The value of a parameter passed to the FACTORY METHOD determines which family member it constructs and returns.

By analogy with the `MazeFactory`, an ABSTRACT FACTORY that could support the Burgers equation client code of Section 9.1 might construct a product family in which each member abstracts one aspect of a 1D scalar field. These might include Boundary, Differentiator, and Field classes. The Boundary abstraction could store values associated with boundary conditions of various types. The Differentiator could provide methods for computing derivatives based on various numerical approximation schemes. The Field class could store values internal to the problem domain and aggregate a Boundary object and a `Differentiator` object. Alternatively, all of these classes could be abstract with default implementations of their methods or with no method implementations, leaving to subclasses the responsibility for providing implementations.

By analogy with the `EnchantedFactory` and `BombedFactory`, one might define additional product families by subclassing each of the classes described in the previous paragraph. For example, a Periodic6thOrder subclass of Field might aggregate a PeriodicBoundary object and a PeriodicPade6th object. The PeriodicBoundary subclass of Boundary might or might not store data, considering that the boundary values are redundant in the periodic case. The PeriodicPade6th subclass of Differentiator could provide algorithms for a sixth-order-accurate Padé finite difference scheme as described in Appendix A.

Figure 9.1 depicts a minimal class model for supporting the Burgers equation client code from Section 9.1. The ABSTRACT FACTORY FieldFactory in that figure publishes a FACTORY METHOD *create()*. The concrete Periodic6thFactory implements the FieldFactory interface and therefore provides a concrete create() method. That method constructs a Periodic6thOrder object, but returns a Field reference, freeing the client code to manipulate only this reference without any knowledge of the class of the actual object returned. The greatest benefit of this freedom accrues from the ability to switch between Periodic6thOrder and another Field implementation without changing the client code. An exercise at the end of this chapter explores this possibility.

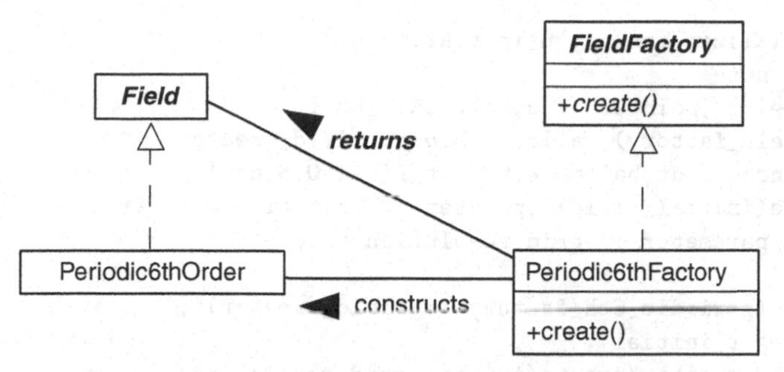

Figure 9.1. UML class diagram of an ABSTRACT FACTORY implementation.

Periodic6thOrder constructs a Field product family that has only one member, Periodic6thOrder, under the assumption that any additional dependencies on other products are either eliminated by subsuming the necessary logic into Periodic6thOrder or delegated to Periodic6thOrder for satisfying internally. The code examples in Sections 9.2.1–9.2.2 take the former approach. An example at the end of this chapter asks reader to design and implement the second approach.

Figure 9.1 debuts the UML diagrammatic technique for adorning relationships. Adornments indicate the nature of adjacent relationships. The arrows indicate the implied order for reading the adornment: "a Periodic6thFactory object constructs a Periodic6thOrder object and returns a Field reference."

9.2.1 Fortran Implementation

Figure 9.2 presents a Fortran implementation of Figure 9.1. Figure 9.2(a) shows a main program that time advances the Burgers equation via 2nd-order Runge-Kutta, adapting the time step continually to maintain stability. Each module in Figure 9.2(b)–(e) corresponds to one class in the UML diagram of Figure 9.1.

Figure 9.2(a): main.f03

```
1   module initializer
2     use kind_parameters ,only : rkind
3     implicit none
4   contains
5     real(rkind) pure function u_initial(x)
6       real(rkind) ,intent(in) :: x
7       u_initial = 10._rkind*sin(x)
8     end function
9     real(rkind) pure function zero(x)
10      real(rkind) ,intent(in) :: x
11      zero = 0.
12    end function
13  end module
14
15  program main
16    use field_module ,only : field,initial_field
17    use field_factory_module ,only : field_factory
18    use periodic_6th_factory_module ,only : periodic_6th_factory
19    use kind_parameters ,only : rkind, ikind
```

```
20   use initializer ,only : u_initial,zero
21   implicit none
22   class(field) ,pointer :: u,half_uu,u_half
23   class(field_factory), allocatable :: field_creator
24   real(rkind) :: dt,half=0.5,t=0.,t_final=0.6,nu=1.
25   procedure(initial_field) ,pointer :: initial
26   integer, parameter :: grid_resolution = 16
27
28   allocate (periodic_6th_factory :: field_creator)
29   initial => u_initial
30   u => field_creator%create(initial,grid_resolution)
31   initial => zero
32   half_uu => field_creator%create(initial,grid_resolution)
33   u_half => field_creator%create(initial,grid_resolution)
34
35   do while (t<t_final) ! 2nd-order Runge-Kutta:
36     dt = u%runge_kutta_2nd_step(nu ,grid_resolution)
37     half_uu = u*u*half
38     u_half = u + (u%xx()*nu - half_uu%x())*dt*half ! first substep
39     half_uu = u_half*u_half*half
40     u  = u + (u_half%xx()*nu - half_uu%x())*dt ! second substep
41     t = t + dt
42   end do
43   print *,'u at t=',t
44   call u%output()
45 end program
```

Figure 9.2(b): field.f03

```
1  module field_module
2    use kind_parameters ,only : rkind, ikind
3    implicit none
4    private
5    public :: field
6    public :: initial_field
7    type ,abstract :: field
8    contains
9      procedure(field_op_field) ,deferred :: add
10     procedure(field_eq_field) ,deferred :: assign
11     procedure(field_op_field) ,deferred :: subtract
12     procedure(field_op_field) ,deferred :: multiply_field
13     procedure(field_op_real)  ,deferred :: multiply_real
14     procedure(real_to_real)  ,deferred :: runge_kutta_2nd_step
15     procedure(derivative) ,deferred :: x  ! 1st derivative
16     procedure(derivative) ,deferred :: xx ! 2nd derivative
17     procedure(output_interface) ,deferred  :: output
18     generic :: operator(+)   => add
19     generic :: operator(-)   => subtract
20     generic :: operator(*)   => multiply_real,multiply_field
21     generic :: assignment(=) => assign
22   end type
23
24   abstract interface
25     real(rkind) pure function initial_field(x)
26       import :: rkind
27       real(rkind) ,intent(in) :: x
```

```
28      end function
29      function field_op_field(lhs,rhs)
30        import :: field
31        class(field) ,intent(in) :: lhs,rhs
32        class(field) ,allocatable :: field_op_field
33      end function
34      function field_op_real(lhs,rhs)
35        import :: field,rkind
36        class(field) ,intent(in)  :: lhs
37        real(rkind) ,intent(in) :: rhs
38        class(field) ,allocatable :: field_op_real
39      end function
40      real(rkind) function real_to_real(this,nu,grid_resolution)
41        import :: field,rkind,ikind
42        class(field) ,intent(in)  :: this
43        real(rkind) ,intent(in) :: nu
44        integer(ikind),intent(in) :: grid_resolution
45      end function
46      function derivative(this)
47        import :: field
48        class(field) ,intent(in)  :: this
49        class(field) ,allocatable :: derivative
50      end function
51      subroutine field_eq_field(lhs,rhs)
52        import :: field
53        class(field) ,intent(inout) :: lhs
54        class(field) ,intent(in) :: rhs
55      end subroutine
56      subroutine output_interface(this)
57        import :: field
58        class(field) ,intent(in) :: this
59      end subroutine
60    end interface
61  end module
```

Figure 9.2(c): periodic_6th_order.f03

```
1   module periodic_6th_order_module
2     use field_module ,only : field,initial_field
3     use kind_parameters ,only : rkind,ikind
4     use matrix_module       ,only : matrix, new_matrix
5     implicit none
6     private
7     public :: periodic_6th_order
8     type ,extends(field) :: periodic_6th_order
9       private
10      real(rkind) ,dimension(:) ,allocatable :: f
11    contains
12      procedure :: add => total
13      procedure :: assign => copy
14      procedure :: subtract => difference
15      procedure :: multiply_field => product
16      procedure :: multiply_real => multiple
17      procedure :: runge_kutta_2nd_step => rk2_dt
18      procedure :: x  => df_dx   ! 1st derivative w.r.t. x
19      procedure :: xx => d2f_dx2 ! 2nd derivative w.r.t. x
```

```
20      procedure :: output
21    end type
22
23    real(rkind) ,dimension(:) ,allocatable :: x_node
24
25    interface periodic_6th_order
26      procedure constructor
27    end interface
28
29    real(rkind) ,parameter :: pi=acos(-1._rkind)
30    real(rkind) ,parameter :: first_coeff_6th(5)= (/1.0_rkind/3.0_rkind,&
31      0.0_rkind,14.0_rkind/9.0_rkind,1.0_rkind/9.0_rkind,0.0_rkind/)
32    real(rkind) ,parameter :: second_coeff_6th(5)=(/2.0_rkind/11.0_rkind,&
33      0.0_rkind,12.0_rkind/11.0_rkind,3.0_rkind/11.0_rkind,0.0_rkind/)
34
35  contains
36    function constructor(initial,grid_resolution)
37      type(periodic_6th_order) ,pointer :: constructor
38      procedure(initial_field) ,pointer :: initial
39      integer(ikind) ,intent(in) :: grid_resolution
40      integer :: i
41      allocate(constructor)
42      allocate(constructor%f(grid_resolution))
43      if (.not. allocated(x_node)) x_node = grid()
44      forall (i=1:size(x_node)) constructor%f(i)=initial(x_node(i))
45    contains
46      pure function grid()
47        integer(ikind) :: i
48        real(rkind) ,dimension(:) ,allocatable :: grid
49        allocate(grid(grid_resolution))
50        forall(i=1:grid_resolution) &
51          grid(i)  = 2.*pi*real(i-1,rkind)/real(grid_resolution,rkind)
52      end function
53    end function
54
55    real(rkind) function rk2_dt(this,nu, grid_resolution)
56      class(periodic_6th_order) ,intent(in) :: this
57      real(rkind) ,intent(in) :: nu
58      integer(ikind) ,intent(in) :: grid_resolution
59      real(rkind)              :: dx, CFL, k_max
60      dx=2.0*pi/grid_resolution
61      k_max=grid_resolution/2.0_rkind
62      CFL=2.0/(24.0*(1-cos(k_max*dx))/11.0/(1.0+4.0/11.0*cos(k_max*dx))+ &
63          3.0*(1.0-cos(2.0*k_max*dx))/22.0/(1.0+4.0/11.0*cos(k_max*dx)))
64      rk2_dt = CFL*dx**2/nu
65    end function
66
67    function total(lhs,rhs)
68      class(periodic_6th_order) ,intent(in) :: lhs
69      class(field) ,intent(in) :: rhs
70      class(field) ,allocatable :: total
71      type(periodic_6th_order) ,allocatable :: local_total
72      select type(rhs)
73        class is (periodic_6th_order)
74          allocate(local_total)
```

```
75          local_total%f = lhs%f + rhs%f
76          call move_alloc(local_total,total)
77        class default
78          stop 'periodic_6th_order%total: unsupported rhs class.'
79      end select
80    end function
81
82    function difference(lhs,rhs)
83      class(periodic_6th_order) ,intent(in) :: lhs
84      class(field) ,intent(in) :: rhs
85      class(field) ,allocatable :: difference
86      type(periodic_6th_order) ,allocatable :: local_difference
87      select type(rhs)
88        class is (periodic_6th_order)
89          allocate(local_difference)
90          local_difference%f = lhs%f - rhs%f
91          call move_alloc(local_difference,difference)
92        class default
93          stop 'periodic_6th_order%difference: unsupported rhs class.'
94      end select
95    end function
96
97    function product(lhs,rhs)
98      class(periodic_6th_order) ,intent(in) :: lhs
99      class(field) ,intent(in)  :: rhs
100     class(field) ,allocatable :: product
101     type(periodic_6th_order) ,allocatable :: local_product
102     select type(rhs)
103       class is (periodic_6th_order)
104         allocate(local_product)
105         local_product%f = lhs%f * rhs%f
106         call move_alloc(local_product,product)
107       class default
108         stop 'periodic_6th_order%product: unsupported rhs class.'
109     end select
110   end function
111
112   function multiple(lhs,rhs)
113     class(periodic_6th_order) ,intent(in) :: lhs
114     real(rkind) ,intent(in)  :: rhs
115     class(field) ,allocatable :: multiple
116     type(periodic_6th_order) ,allocatable :: local_multiple
117     allocate(local_multiple)
118     local_multiple%f = lhs%f * rhs
119     call move_alloc(local_multiple,multiple)
120   end function
121
122   subroutine copy(lhs,rhs)
123     class(field) ,intent(in) :: rhs
124     class(periodic_6th_order) ,intent(inout) :: lhs
125     select type(rhs)
126       class is (periodic_6th_order)
127         lhs%f = rhs%f
128       class default
129         stop 'periodic_6th_order%copy: unsupported copy class.'
```

```
130       end select
131     end subroutine
132 
133     function df_dx(this)
134       class(periodic_6th_order) ,intent(in) :: this
135       class(field) ,allocatable  :: df_dx
136       class(matrix),allocatable,save  :: lu_matrix
137       integer(ikind) :: i,nx, x_east, x_west
138       integer(ikind) :: x_east_plus1,x_east_plus2
139       integer(ikind) :: x_west_minus1,x_west_minus2
140       real(rkind) ,dimension(:,:) ,allocatable :: A
141       real(rkind) ,dimension(:)   ,allocatable :: b,coeff
142       real(rkind) :: dx
143       class(periodic_6th_order) ,allocatable :: df_dx_local
144 
145       nx=size(x_node)
146       dx=2.*pi/real(nx,rkind)
147       coeff = first_coeff_6th
148       if (.NOT. allocated(lu_matrix)) allocate(lu_matrix)
149       if (.NOT. lu_matrix%is_built()) then
150         allocate(A(nx,nx))
151         !__________Initialize coeffecient matrix A _____
152         A=0.0_rkind
153         do i=1, nx
154           x_east = mod(i,nx)+1
155           x_west = nx-mod(nx+1-i,nx)
156           if (i==2) then
157             x_east_plus1=x_east+1; x_west_minus1=nx
158             x_east_plus2=x_east+2; x_west_minus2=nx-1
159           else if (i==3) then
160             x_east_plus1=x_east+1; x_west_minus1=1
161             x_east_plus2=x_east+2; x_west_minus2=nx
162           else if (i==nx-1) then
163             x_east_plus1=1; x_west_minus1=x_west-1
164             x_east_plus2=2; x_west_minus2=x_west-2
165           else if (i==nx-2) then
166             x_east_plus1=nx; x_west_minus1=x_west-1
167             x_east_plus2=1; x_west_minus2=x_west-2
168           else
169             x_east_plus1=x_east+1; x_west_minus1=x_west-1
170             x_east_plus2=x_east+2; x_west_minus2=x_west-2
171           end if
172           A(i,x_west_minus1) =coeff(2)
173           A(i,x_west)    =coeff(1)
174           A(i,i)         =1.0_rkind
175           A(i,x_east)    =coeff(1)
176           A(i,x_east_plus1) =coeff(2)
177         end do
178         lu_matrix=new_matrix(A)
179         deallocate(A)
180       end if
181       allocate(b(nx))
182       b=0.0
183       do i=1,nx
184         x_east = mod(i,nx)+1
```

```
185        x_west = nx-mod(nx+1-i,nx)
186        if (i==2) then
187          x_east_plus1=x_east+1; x_west_minus1=nx
188          x_east_plus2=x_east+2; x_west_minus2=nx-1
189        else if (i==3) then
190          x_east_plus1=x_east+1; x_west_minus1=1
191          x_east_plus2=x_east+2; x_west_minus2=nx
192        else if (i==nx-1) then
193          x_east_plus1=1; x_west_minus1=x_west-1
194          x_east_plus2=2; x_west_minus2=x_west-2
195        else if (i==nx-2) then
196          x_east_plus1=nx; x_west_minus1=x_west-1
197          x_east_plus2=1; x_west_minus2=x_west-2
198        else
199          x_east_plus1=x_east+1; x_west_minus1=x_west-1
200          x_east_plus2=x_east+2; x_west_minus2=x_west-2
201        end if
202
203        b(i)=(0.25*coeff(4)*(this%f(x_east_plus1)-this%f(x_west_minus1))+&
204            0.5*coeff(3)*(this%f(x_east)-this%f(x_west))+ &
205            coeff(5)/6.0*(this%f(x_east_plus2)-this%f(x_west_minus2)))/dx
206      end do
207      allocate(df_dx_local)
208      df_dx_local%f=lu_matrix .inverseTimes. b
209      call move_alloc(df_dx_local, df_dx)
210    end function
211
212    function d2f_dx2(this)
213      class(periodic_6th_order)  ,intent(in)   :: this
214      class(field) ,allocatable :: d2f_dx2
215      class(matrix),allocatable, save      :: lu_matrix
216      integer(ikind) :: i,nx,x_east,x_west
217      integer(ikind) :: x_east_plus1,x_east_plus2
218      integer(ikind) :: x_west_minus1,x_west_minus2
219      real(rkind) ,dimension(:,:) ,allocatable :: A
220      real(rkind) ,dimension(:)   ,allocatable :: coeff,b
221      real(rkind)                              :: dx
222      class(periodic_6th_order)  ,allocatable  :: d2f_dx2_local
223
224      nx=size(this%f)
225      dx=2.*pi/real(nx,rkind)
226      coeff = second_coeff_6th
227      if (.NOT. allocated(lu_matrix)) allocate(lu_matrix)
228      if (.NOT. lu_matrix%is_built()) then
229        allocate(A(nx,nx))
230
231        !__________Initialize coeffecient matrix A _____
232        A=0.0_rkind
233        do i=1, nx
234          x_east = mod(i,nx)+1
235          x_west = nx-mod(nx+1-i,nx)
236          if (i==2) then
237            x_east_plus1=x_east+1; x_west_minus1=nx
238            x_east_plus2=x_east+2; x_west_minus2=nx-1
239          else if (i==3) then
```

```
240           x_east_plus1=x_east+1; x_west_minus1=1
241           x_east_plus2=x_east+2; x_west_minus2=nx
242         else if (i==nx-1) then
243           x_east_plus1=1; x_west_minus1=x_west-1
244           x_east_plus2=2; x_west_minus2=x_west-2
245         else if (i==nx-2) then
246           x_east_plus1=nx; x_west_minus1=x_west-1
247           x_east_plus2=1; x_west_minus2=x_west-2
248         else
249           x_east_plus1=x_east+1; x_west_minus1=x_west-1
250           x_east_plus2=x_east+2; x_west_minus2=x_west-2
251         end if
252         A(i,x_west_minus1) =coeff(2)
253         A(i,x_west)        =coeff(1)
254         A(i,i)             =1.0_rkind
255         A(i,x_east)        =coeff(1)
256         A(i,x_east_plus1)  =coeff(2)
257       end do
258       lu_matrix=new_matrix(A)
259       deallocate(A)
260     end if
261     allocate(b(nx))
262     do i=1, nx
263       x_east = mod(i,nx)+1
264       x_west = nx-mod(nx+1-i,nx)
265       if (i==2) then
266         x_east_plus1=x_east+1; x_west_minus1=nx
267         x_east_plus2=x_east+2; x_west_minus2=nx-1
268       else if (i==3) then
269         x_east_plus1=x_east+1; x_west_minus1=1
270         x_east_plus2=x_east+2; x_west_minus2=nx
271       else if (i==nx-1) then
272         x_east_plus1=1; x_west_minus1=x_west-1
273         x_east_plus2=2; x_west_minus2=x_west-2
274       else if (i==nx-2) then
275         x_east_plus1=nx; x_west_minus1=x_west-1
276         x_east_plus2=1; x_west_minus2=x_west-2
277       else
278         x_east_plus1=x_east+1; x_west_minus1=x_west-1
279         x_east_plus2=x_east+2; x_west_minus2=x_west-2
280       end if
281       b(i)=(0.25*coeff(4)* &
282         (this%f(x_east_plus1)-2.0*this%f(i)+this%f(x_west_minus1))+ &
283         coeff(3)*(this%f(x_east)-2.0*this%f(i)+this%f(x_west))+ &
284         coeff(5)/9.0* &
285         (this%f(x_east_plus2)-this%f(i)+this%f(x_west_minus2)))/dx**2
286     end do
287     allocate(d2f_dx2_local)
288     d2f_dx2_local%f=lu_matrix .inverseTimes. b
289     call move_alloc(d2f_dx2_local, d2f_dx2)
290   end function
291 
292   subroutine output(this)
293     class(periodic_6th_order) ,intent(in) :: this
294     integer(ikind) :: i
```

```
295       do i=1,size(x_node)
296         print *, x_node(i), this%f(i)
297       end do
298     end subroutine
299
300   end module
```

Figure 9.2(d): field_factory.f03

```
1   module field_factory_module
2     use field_module ,only : field,initial_field
3     implicit none
4     private
5     public :: field_factory
6     type, abstract :: field_factory
7     contains
8       procedure(create_interface), deferred :: create
9     end type
10    abstract interface
11      function create_interface(this,initial,grid_resolution)
12        use kind_parameters ,only : ikind
13        import :: field, field_factory ,initial_field
14        class(field_factory), intent(in) :: this
15        class(field) ,pointer :: create_interface
16        procedure(initial_field) ,pointer :: initial
17        integer(ikind) ,intent(in) :: grid_resolution
18      end function
19    end interface
20  end module
```

Figure 9.2(e): periodic_6th_factory.f03

```
1   module periodic_6th_factory_module
2     use field_factory_module! , only : field_factory
3     use field_module ! ,only : field,initial_field
4     implicit none
5     private
6     public :: periodic_6th_factory
7     type, extends(field_factory) :: periodic_6th_factory
8     contains
9       procedure :: create=>new_periodic_6th_order
10    end type
11  contains
12    function new_periodic_6th_order(this,initial,grid_resolution)
13      use periodic_6th_order_module ,only : periodic_6th_order
14      use kind_parameters ,only : ikind
15      class(periodic_6th_factory), intent(in) :: this
16      class(field) ,pointer :: new_periodic_6th_order
17      procedure(initial_field) ,pointer :: initial
18      integer(ikind) ,intent(in) :: grid_resolution
19      new_periodic_6th_order=> periodic_6th_order(initial,grid_resolution)
20    end function
21  end module
```

Figure 9.2(f): matrix.f03

```
1   module matrix_module
2     use kind_parameters ,only : rkind, ikind
```

```
3    implicit none
4    private
5    public :: matrix
6    public :: new_matrix
7    type  :: matrix
8      integer(ikind), allocatable :: pivot(:)
9      real(rkind), allocatable :: lu(:,:)
10   contains
11     procedure  :: back_substitute
12     procedure  :: is_built
13     procedure  :: matrix_eq_matrix
14     generic    :: operator(.inverseTimes.) => back_substitute
15     generic    :: assignment(=) => matrix_eq_matrix
16   end type
17
18   interface new_matrix
19     procedure constructor
20   end interface
21
22   contains
23     logical function is_built(this)
24       class(matrix),intent(in)  :: this
25       is_built = allocated (this%pivot)
26     end function
27
28     function constructor(A) result(new_matrix)
29       class(matrix) ,allocatable                :: new_matrix
30       real(rkind), intent(in) ,dimension(:,:)  :: A
31       integer                                   :: n,info
32
33       n=size(A,1)
34       allocate(new_matrix)
35       allocate(new_matrix%pivot(n), new_matrix%lu(n,n))
36       new_matrix%lu=A
37       call dgetrf(n,n,new_matrix%lu,n,new_matrix%pivot,info)
38     end function
39
40     subroutine matrix_eq_matrix(lhs,rhs)
41       class(matrix) ,intent(in)  :: rhs
42       class(matrix) ,intent(out) :: lhs
43       lhs%pivot = rhs%pivot
44       lhs%lu    = rhs%lu
45     end subroutine
46
47     function back_substitute(this,b) result(x)
48       class(matrix) ,intent(in)                     :: this
49       real(rkind)  ,intent(in)  ,dimension(:)   :: b
50       real(rkind)  ,allocatable ,dimension(:)   :: x, temp_x
51       real(rkind)  ,allocatable ,dimension(:,:) :: lower, upper
52       real(rkind)  ,allocatable ,dimension(:)   :: local_b
53       integer(ikind)                             :: n,i,j
54       real(rkind)                                :: temp
55
56       n=size(this%lu,1)
57       allocate(lower(n,n), upper(n,n))
```

```
58      lower=0.0_rkind
59      upper=0.0_rkind
60      do i=1,n
61        do j=i,n
62          upper(i,j)=this%lu(i,j)
63        end do
64      end do
65      do i=1,n
66        lower(i,i)=1.0_rkind
67      end do
68      do i=2,n
69        do j=1,i-1
70          lower(i,j)=this%lu(i,j)
71        end do
72      end do
73      allocate(local_b(n))
74      local_b=b
75      do i=1,n
76        if (this%pivot(i)/=i) then
77          temp=local_b(i)
78          local_b(i)=local_b(this%pivot(i))
79          local_b(this%pivot(i))=temp
80        end if
81      end do
82      allocate(temp_x(n))
83      temp_x(1)=local_b(1)
84      do i=2,n
85        temp_x(i)=local_b(i)-sum(lower(i,1:i-1)*temp_x(1:i-1))
86      end do
87      allocate(x(n))
88      x(n)=temp_x(n)/upper(n,n)
89      do i=n-1,1,-1
90        x(i)=(temp_x(i)-sum(upper(i,i+1:n)*x(i+1:n)))/upper(i,i)
91      end do
92      do i=n,1,-1
93        if (this%pivot(i)/=i) then
94          temp=x(i)
95          x(i)=x(this%pivot(i))
96          x(this%pivot(i))=temp
97        end if
98      end do
99    end function
100 end module
```

Figure 9.2. Fortran demonstration of an abstract factory: (a) main client, (b) abstract field class, (c) periodic 6th-order Padé field, (d) abstract field factory, (e) periodic 6th-order Padé factory, (f) matrix module.

Fortran constructs appearing for the first time in Figure 9.2 include typed allocation and procedure pointers. Line 28 in the `main` program of Figure 9.2(a) demonstrates how to allocate the dynamic type of a `field_factory` instance to be its child class `periodic_6th_factory`. That allocation yields a concrete object to which results can be assigned and on which type-bound procedures can be invoked.

Line 25 in main demonstrates the declaration of a procedure pointer, initial that conforms to the abstract interface, initial_field, published by the field class at line 6 and subsequently prototyped at line 25–28 of Figure 9.2(b). Lines 29 and 31 in main show pointer assignments of initial to functions intended for initializing u and for initially defaulting other fields to zero, respectively. Lines 30 and 32–33 pass initial to the field factory.

The matrix module shown in Figure 9.2(f) is used to solve the 6th-order first derivatives and second derivatives from Equations A.57 and A.58 using LU decomposition method. LU decomposition method is an efficient procedure for solving a system of linear algebraic equations $\mathbf{Ax} = \mathbf{b}$, the details of LU decomposition method is introduced in A.2.2. Line 37 in matrix module is calling the function dgetrf from Linear Algebra PACKage (LAPACK, http://www.netlib.org/lapack), which performs the LU factorization of a general matrix.

9.2.2 C++ Implementation

Figure 9.3 presents a C++ implementation of Figure 9.1. Figure 9.3(a) shows a main program that time advances the Burgers equation via 2nd-order Runge-Kutta, adapting the time step continually to maintain stability. The class structure implemented in Figure 9.3(b)–(h) mirrors that in the Fortran implementation of Figure 9.2.

Figure 9.3(a): main.cpp

```
1   #include "field.h"
2   #include "field_factory.h"
3   #include "periodic_6th_factory.h"
4   #include "fmt.h"
5   #include "initializer.h"
6
7   int main ()
8   {
9       typedef field::ptr_t field_ptr_t;
10      typedef field_factory::ptr_t field_factory_ptr;
11
12      const real_t t_final = 0.6;
13      const real_t half = 0.5;
14      const real_t nu = 1.0;
15      const int grid_resolution = 128;
16
17      real_t t = 0.0;
18
19      field_factory_ptr field_creator =
20          field_factory_ptr (new periodic_6th_factory());
21
22      field_ptr_t u      = field_creator->create(u_initial,
23                                    grid_resolution);
24
25      field_ptr_t u_half = field_creator->create(zero,
26                                    grid_resolution);
27
28      real_t dt;
```

```
29
30      while (t < t_final) {
31          // use 2nd-order Runge-Kutta
32          dt = u->runge_kutta_2nd_step(nu, grid_resolution);
33
34          // first substep
35          u_half = u + (u->xx()*nu - (u*u*half)->x())*dt*half;
36
37          // second substep
38          u += (u_half->xx()*nu - (u_half*u_half*half)->x())*dt;
39
40          t += dt;
41      }
42
43      std::cout << " u at t = " << fmt(t) << std::endl;
44      u->output();
45
46      return 0;
47  }
```

Figure 9.3(b): field.h

```
1   #ifndef _FIELD_H_
2   #define _FIELD_H_ 1
3
4   #include "RefBase.h"
5   #include "globals.h"
6
7   class field : public virtual RefBase {
8       public:
9           typedef Ref<field> ptr_t;
10
11          virtual ~field() {};
12
13          virtual ptr_t operator+=(ptr_t)                  = 0;
14          virtual ptr_t operator-(ptr_t)             const = 0;
15          virtual ptr_t operator*(real_t)            const = 0;
16          virtual ptr_t operator*(ptr_t)             const = 0;
17
18          virtual ptr_t x()                          const = 0;
19          virtual ptr_t xx()                         const = 0;
20          virtual real_t runge_kutta_2nd_step(real_t,
21                                          int)       const = 0;
22
23          virtual ptr_t clone()                      const = 0;
24          virtual void  output()                     const = 0;
25      protected:
26          field() {};
27  };
28
29
30  //
31  // define operators to be used for field ptr_t
32  //
33  inline field::ptr_t operator+= (field::ptr_t a, field::ptr_t b) {
34      *a += b;
```

```
35      return a;
36  }
37
38  inline field::ptr_t operator+ (field::ptr_t a, field::ptr_t b) {
39      field::ptr_t tmp = a->clone();
40
41      *tmp += b;
42      return tmp;
43  }
44
45  inline field::ptr_t operator- (field::ptr_t a, field::ptr_t b) {
46      field::ptr_t tmp = a->clone();
47
48      tmp = *tmp - b;
49      return tmp;
50  }
51
52
53  inline field::ptr_t operator*(field::ptr_t a, field::ptr_t b) {
54      field::ptr_t tmp = a->clone();
55
56      tmp = *tmp * b;
57      return tmp;
58  }
59
60  inline field::ptr_t operator*(field::ptr_t a, real_t b) {
61      field::ptr_t tmp = a->clone();
62
63      tmp = *tmp * b;
64      return tmp;
65  }
66
67  #endif
```

Figure 9.3(c): field_factory.h

```
1   #ifndef _FIELD_FACTORY_H_
2   #define _FIELD_FACTORY_H_ 1
3
4   #include "RefBase.h"
5   #include "field.h"
6
7   class field_factory : public virtual RefBase {
8       public:
9           typedef Ref<field_factory> ptr_t;
10
11          virtual ~field_factory() {};
12          virtual field::ptr_t create(real_t (*initial) (real_t x),
13                                      int grid_resolution) = 0;
14      protected:
15          field_factory() {};
16  };
17
18  #endif
```

Figure 9.3(e): periodic_6th_factory.h

```
#ifndef PERIODIC_6TH_FACTORY_H_
#define PERIODIC_6TH_FACTORY_H_

#include "field.h"
#include "field_factory.h"

class periodic_6th_factory : public field_factory {
    public:
        periodic_6th_factory ();
        virtual ~periodic_6th_factory();

        virtual field::ptr_t create(real_t (*initial) (real_t x),
                    int grid_resolution);
};
#endif
```

Figure 9.3(f): periodic_6th_factory.cpp

```
#include "periodic_6th_factory.h"
#include "periodic_6th_order.h"

periodic_6th_factory::periodic_6th_factory ()
{ }

periodic_6th_factory::~periodic_6th_factory ()
{ }


field::ptr_t periodic_6th_factory::create(real_t (*initial) (real_t x),
        int grid_resolution)
{
    return field::ptr_t
        (new periodic_6th_order(initial, grid_resolution));
}
```

Figure 9.3(g): periodic_6th_order.h

```
#ifndef _PERIODIC_6TH_ORDER_H_
#define _PERIODIC_6TH_ORDER_H_ 1

#include <vector>
#include "field.h"
#include "globals.h"

class periodic_6th_order : public field {
    public:
        typedef field::ptr_t ptr_t;

        periodic_6th_order(real_t (*initial) (real_t x),
                    int grid_resolution);

        periodic_6th_order(const crd_t& other);
        virtual ~periodic_6th_order();

        virtual ptr_t operator+=(ptr_t);
        virtual ptr_t operator-(ptr_t)  const;
```

```
20	        virtual ptr_t operator*(real_t) const;
21	        virtual ptr_t operator*(ptr_t)  const;
22	
23	        virtual ptr_t x()                                     const;
24	        virtual ptr_t xx()                                    const;
25	        virtual real_t runge_kutta_2nd_step(real_t nu,
26	                                  int grid_resolution)  const;
27	
28	        virtual ptr_t clone()   const;
29	        virtual void output()   const;
30	
31	        static void set_grid(int num_grid_pts);
32	        static const crd_t & get_grid();
33	        static int get_grid_size();
34	
35	    private:
36	        crd_t f_;
37	        static crd_t x_node_;
38	};
39	
40	#endif
```

Figure 9.3(h): periodic_6th_order.cpp

```
1	#include <cmath>
2	#include <exception>
3	#include "periodic_6th_order.h"
4	#include "fmt.h"
5	#include "mat.h"
6	#include "gaussian_elimination.h"
7	
8	struct periodic_6th_order_error : public std::exception {
9	    virtual ~periodic_6th_order_error() throw () {};
10	};
11	
12	
13	const real_t pi = 3.14159265f;
14	
15	const real_t first_coeff_6th[5] =
16	      {1.0/3.0, 0.0, 14.0/9.0, 1.0/9.0, 0.0};
17	
18	const real_t second_coeff_6th[5] =
19	      {2.0/11.0, 0.0, 12.0/11.0, 3.0/11.0, 0.0};
20	
21	
22	typedef field::ptr_t ptr_t;
23	
24	crd_t periodic_6th_order::x_node_ = crd_t();
25	
26	void periodic_6th_order::set_grid(int grid_resolution)
27	{
28	    if (x_node_.empty()) {
29	        for (int i=0; i< grid_resolution;++i) {
30	            x_node_.push_back (2.0*pi*(i)/real_t(grid_resolution));
31	        }
32	
```

```
33      }
34  }
35
36  int periodic_6th_order::get_grid_size() const {
37      return (x_node_.size());
38
39  }
40
41  const crd_t & periodic_6th_order::get_grid () const {
42      return x_node_;
43  }
44
45
46  periodic_6th_order::periodic_6th_order(real_t (*initial) (real_t x),
47                  int grid_resolution) {
48      set_grid(grid_resolution);
49
50      f_ = get_grid();
51
52      for (int i = 0; i < f_.size(); ++i) {
53          f_.at(i) = initial(f_.at(i));
54      }
55  }
56
57  periodic_6th_order::periodic_6th_order(const crd_t & other)
58      : f_(other)
59  {}
60
61
62  periodic_6th_order::~periodic_6th_order() {
63  }
64
65
66  ptr_t periodic_6th_order::operator+=(ptr_t rhs) {
67
68      Ref<periodic_6th_order> other = cast<periodic_6th_order> (rhs);
69
70      if ((other == NULL) || (f_.size() != other->f_.size())) {
71          std::cerr << "periodic_6th_order::operator+= " <<
72              "rhs is invalid" << std::endl;
73
74          throw periodic_6th_order_error();
75      }
76
77      for (int i = 0; i < f_.size(); ++i) {
78          f_.at(i) += other->f_.at(i);
79      }
80
81      return ptr_t(this);
82  }
83
84
85  ptr_t periodic_6th_order::operator-(ptr_t rhs)  const {
86      Ref<periodic_6th_order> other = cast<periodic_6th_order> (rhs);
87
```

```
88      if ((other == NULL) || (f_.size() != other->f_.size())) {
89          std::cerr << "periodic_6th_order::operator- " <<
90              "rhs is invalid" << std::endl;
91
92          throw periodic_6th_order_error();
93      }
94
95      Ref<periodic_6th_order> result =
96          Ref<periodic_6th_order>(new periodic_6th_order(*this));
97
98      for (int i = 0; i < f_.size(); ++i) {
99          result->f_.at(i) -= other->f_.at(i);
100     }
101
102     return result;
103 }
104
105
106 ptr_t periodic_6th_order::operator*(real_t rhs) const {
107     Ref<periodic_6th_order> result =
108         Ref<periodic_6th_order>(new periodic_6th_order(*this));
109
110     for (int i = 0; i < f_.size(); ++i) {
111         result->f_.at(i) *= rhs;
112     }
113
114     return result;
115 }
116
117
118 ptr_t periodic_6th_order::operator*(ptr_t rhs)  const {
119     Ref<periodic_6th_order> other = cast<periodic_6th_order> (rhs);
120
121     if ((other == NULL) || (f_.size() != other->f_.size())) {
122         std::cerr << "periodic_6th_order::operator* " <<
123             "rhs is invalid" << std::endl;
124
125         throw periodic_6th_order_error();
126     }
127
128     Ref<periodic_6th_order> result =
129         Ref<periodic_6th_order>(new periodic_6th_order(*this));
130
131     for (int i = 0; i < f_.size(); ++i) {
132         result->f_.at(i) *= other->f_.at(i);
133     }
134
135     return result;
136 }
137
138
139 ptr_t periodic_6th_order::clone()  const {
140     return ptr_t (new periodic_6th_order(*this));
141 }
142
```

```
143
144  void periodic_6th_order::output() const {
145      for (int i = 0; i < f_.size(); ++i) {
146          std::cout << fmt(x_node_.at(i)) << " " << fmt(f_.at(i))
147              << std::endl;
148      }
149  }
150
151  ptr_t periodic_6th_order:: x()  const {
152      int nx = get_grid_size();
153      real_t dx = 2.0 * pi /real_t(nx);
154
155      // __Initialize coeffecient matrix A and right handside b__
156      mat_t A;
157      A.clear_resize(nx, nx, 0.0);
158
159      crd_t b = crd_t(nx, 0.0);
160
161      crd_t coeff;
162
163      for (int i = 0; i < 5; ++i)
164          coeff.push_back(first_coeff_6th[i]);
165
166      int x_east, x_west, x_east_plus1,
167          x_east_plus2,x_west_minus1,x_west_minus2;
168
169      for (int i = 0; i < nx; ++i) {
170          x_east = (i+1)%nx;
171          x_west = nx -1 - (nx-i)%nx;
172
173          if (i == 1) {
174              x_east_plus1=x_east+1;
175              x_east_plus2=x_east+2;
176              x_west_minus1=nx-1;
177              x_west_minus2=nx-2;
178          }
179          else if (i == 2) {
180              x_east_plus1=x_east+1;
181              x_east_plus2=x_east+2;
182              x_west_minus1=0;
183              x_west_minus2=nx-1;
184          }
185          else if (i == nx-2) {
186              x_east_plus1=0;
187              x_east_plus2=1;
188              x_west_minus1=x_west-1;
189              x_west_minus2=x_west-2;
190          }
191          else if (i == nx-3) {
192              x_east_plus1=nx-1;
193              x_east_plus2=0;
194              x_west_minus1=x_west-1;
195              x_west_minus2=x_west-2;
196          }
197          else {
```

```
198                 x_east_plus1=x_east+1;
199                 x_east_plus2=x_east+2;
200                 x_west_minus1=x_west-1;
201                 x_west_minus2=x_west-2;
202             }
203 
204             A(i,x_west_minus1) =coeff.at(1);
205             A(i,x_west)        =coeff.at(0);
206             A(i,i)             =1.0;
207             A(i,x_east)        =coeff.at(0);
208             A(i,x_east_plus1)  =coeff.at(1);
209 
210             b.at(i) = (0.25*coeff.at(3)*
211                   (f_.at(x_east_plus1)-f_.at(x_west_minus1))+
212                   0.5*coeff.at(2)*(f_.at(x_east) - f_.at(x_west)) +
213                   coeff.at(4)/6.0 * (f_.at(x_east_plus2)-
214                   f_.at(x_west_minus2)))/dx;
215         }
216 
217         return ptr_t(new periodic_6th_order(gaussian_elimination(A,b)));
218 }
219 
220 
221 ptr_t periodic_6th_order:: xx()  const {
222         int nx = get_grid_size();
223         real_t dx = 2.0 * pi /real_t(nx);
224 
225         // __Initialize coeffecient matrix A and right handside b__
226         mat_t A;
227         A.clear_resize(nx, nx, 0.0);
228 
229         crd_t b = crd_t(nx, 0.0);
230 
231         crd_t coeff;
232 
233         for (int i = 0; i < 5; ++i)
234             coeff.push_back(second_coeff_6th[i]);
235 
236         int x_east, x_west, x_east_plus1,
237             x_east_plus2,x_west_minus1,x_west_minus2;
238 
239         for (int i = 0; i < nx; ++i) {
240             x_east = (i+1)%nx;
241             x_west = nx -1 - (nx-i)%nx;
242 
243             if (i == 1) {
244                 x_east_plus1=x_east+1;
245                 x_east_plus2=x_east+2;
246                 x_west_minus1=nx-1;
247                 x_west_minus2=nx-2;
248             }
249             else if (i == 2) {
250                 x_east_plus1=x_east+1;
251                 x_east_plus2=x_east+2;
252                 x_west_minus1=0;
```

```
253            x_west_minus2=nx-1;
254        }
255        else if (i == nx-2) {
256            x_east_plus1=0;
257            x_east_plus2=1;
258            x_west_minus1=x_west-1;
259            x_west_minus2=x_west-2;
260        }
261        else if (i == nx-3) {
262            x_east_plus1=nx-1;
263            x_east_plus2=0;
264            x_west_minus1=x_west-1;
265            x_west_minus2=x_west-2;
266        }
267        else {
268            x_east_plus1=x_east+1;
269            x_east_plus2=x_east+2;
270            x_west_minus1=x_west-1;
271            x_west_minus2=x_west-2;
272        }
273
274        A(i,x_west_minus1) =coeff.at(1);
275        A(i,x_west)        =coeff.at(0);
276        A(i,i)             =1.0;
277        A(i,x_east)        =coeff.at(0);
278        A(i,x_east_plus1)  =coeff.at(1);
279
280        b.at(i) = (0.25*coeff.at(3)*
281             (f_.at(x_east_plus1)-2.0*f_.at(i)+f_.at(x_west_minus1))+
282             coeff.at(2)*(f_.at(x_east)-2.0*f_.at(i)+f_.at(x_west)) +
283             coeff.at(4)/9.0 * (f_.at(x_east_plus2)-
284                 f_.at(i)+f_.at(x_west_minus2)))/(dx*dx);
285    }
286
287    return ptr_t(new periodic_6th_order(gaussian_elimination(A,b)));
288 }
289
290 real_t periodic_6th_order:: runge_kutta_2nd_step(real_t nu,
291        int grid_resolution) const
292 {
293    real_t dx, CFL, k_max;
294
295    dx=2.0*pi/grid_resolution;
296
297    k_max=grid_resolution*0.5;
298
299    CFL=2.0/(24.0*(1-cos(k_max*dx))/11.0/(1.0+4.0/11.0*cos(k_max*dx))+
300           3.0*(1.0-cos(2.0*k_max*dx))/22.0/(1.0+4.0/11.0*cos(k_max*dx)));
301    return (CFL*dx*dx/nu);
302
303 }
```

Figure 9.3(i): initializer.h

```
1 #ifndef INITIALIZER_H_
2 #define INITIALIZER_H_
```

```
3
4   #include "globals.h"
5
6   real_t u_initial (real_t x);
7   real_t zero      (real_t x);
8
9   #endif
```

Figure 9.3(j): initializer.cpp

```
1   #include "initializer.h"
2   #include <cmath>
3
4   real_t u_initial (real_t x)
5   {
6       return (10.0 * sin (x));
7   }
8
9
10  real_t zero (real_t x)
11  {
12      return 0.0;
13  }
```

Figure 9.3. Abstract factory demonstration in C++: (a) main client, (b) field interface, (c) field factory header, (d) abstract field factory, (e) concrete periodic 6th-order Padé field factory, (f) abstract field factory, (g) periodic 6th-order Padé factory, (i) field initializing routine interfaces, (j) field initializing routines.

9.3 The Consequences

The GoF list four consequences of using an ABSTRACT FACTORY: its isolation of concrete classes, its easing of exchanging product families, its promotion of product consistency, and its potential hindering of extending support to new classes. The first two of these reiterate the primary purpose of an ABSTRACT FACTORY. The third prevents mistakenly mixing products from different product families. Centralizing product creation and hiding implementations prevents, for example, a sixth-order Differentiator instance from being used with an eighth-order Padé field instance. The fourth consequence crops up generally in all uses of abstract classes as interfaces: Any changes to the interface cascade down through every implementation. Adding a product to the family necessitates adding corresponding factory methods to every implementation.

The FACTORY METHOD consequences overlap with those of the ABSTRACT FACTORY in which it plays a role. A FACTORY METHOD accomplishes the aforementioned concrete product isolation. Additionally, the GoF note that a FACTORY METHOD "provides hooks for subclasses" and "connects parallel class hierarchies." In the first of these roles, the FACTORY METHOD defined in a base class can provide a default implementation. Client code that links to the base class retains this fully functional hook for use when extended implementations are unavailable or unnecessary. In the second role, the FACTORY METHOD enables one class to delegate responsibilities to another without exposing implementation details. The parallel class then takes on some aspects of the client when it needs to construct its own instances of its sister class. It then benefits from the same flexibility a FACTORY METHOD affords clients.

Another consequence of this chapter's two patterns lies in the resulting proliferation of classes. Were the Burgers equation solver architect willing to expose Periodic6thOrder to the client, she could eliminate the other three classes. Obviously, the flexibility afforded by not exposing it must be balanced against the complexity associated with hiding it.

The latter paragraph strikes a theme that recurs throughout this text. Flexibility often comes at the cost of increased complexity. Engendering flexibility in an application requires increasing the sophistication of the infrastructure that supports it. This proved true in Part I of the text, wherein we argued that the sophistication of OOP matters most as a software application grows larger. A similar statement holds for the design patterns in Part II. They prove more worthwhile with increasing scale in OOP projects. A successful designer allows for growth while consciously avoiding slicing butter with a machete.

Up to this point, we have emphasized the role of factory patterns in determining where object construction occurs: inside factories. In the context of private data, *where* also determines *who*. To the extent factories hide concrete classes from clients, they assume sole responsibility for object construction.

Other creational patterns also manage the *how* of object construction. The GoF BUILDER pattern, for example, adds a Director abstraction between the client a Builder interface. Clients direct the Director to construct a product. The concrete Builder steps through the construction of the parts required to piece together a complex object. This isolates the client from even knowing the construction steps, in contrast to the MazeFactory example in which the client individually invokes methods to manufacture the maze's constituent parts.

Lastly, the GoF SINGLETON pattern address one aspect of *when* construction happens: upon the first request and only then. A SINGLETON instance ensures that only one instance of a class gets instantiated. This provides an OOD alternative to global data, ensuring all clients access the same instance.

9.4 Related Patterns

ABSTRACT FACTORY and FACTORY METHOD represent the smallest of this book's patterns. Whereas OBJECT potentially impacts every ADT in a package, ABSTRACT FACTORY and FACTORY METHOD impact only those for which it proves valuable to vary the class of the object instantiated at a given point in the code. This could apply to all classes in a given project or none. Furthermore, these two patterns always play a supporting role: They exist only to support the instantiation of another class. In these senses, these patterns sit at the end of, and complete, a long process in multiphysics model construction. That process starts with the universally useful basic services in OBJECT, progresses through the expression of the mathematical problem in an ABSTRACT CALCULUS, the enabling of adaptive, semidiscrete time integration via a STRATEGY, the aggregation of individual submodels into a PUPPETEER, and finally the flexible object construction in an ABSTRACT FACTORY containing one or more FACTORY METHODS.

In a different sense, the ABSTRACT FACTORY could have a large footprint if a substantial number of ADTs need the flexibility it affords at object construction time. Unlike some of the other patterns for which the corresponding implementation sits in one node or one node hierarchy of a class diagram, the ABSTRACT FACTORY and FACTORY METHOD patterns might be replicated at multiple locations across a package wherever their services find useful application.

The role an ABSTRACT FACTORY plays also relates closely to that of a STRATEGY pattern. One might use a STRATEGY to vary the finite difference schemes discussed in this chapter instead of using the ABSTRACT FACTORY approach. One distinction lies in the timing and duration, with an ABSTRACT FACTORY acting only at the time of instantiation to vary the scheme by choosing the corresponding object's identity; whereas a STRATEGY acts throughout the duration of the scheme's use and thereby allows for additional variation beyond the point of instantiation.

The viewpoint and expected knowledge of the user harbors a more important distinction. The STRATEGY client code in Figure 7.2(a), for example, exploits explicit knowledge of which time-integration scheme is being used. The ABSTRACT FACTORY code in the current chapter, by contrast, remains agnostic with respect to the concrete type of `field` it uses.

The GoF listed several other relationships between this chapter's two patterns and patterns not discussed in this text. The SINGLETON pattern, for example, ensures that only one instance of a class gets instantiated. Because applications typically require only a single factory, an ABSTRACT FACTORY can be implemented profitably as a SINGLETON. A TEMPLATE METHOD defines a procedure abstractly by building up statements (e.g., expressions in an ABSTRACT CALCULUS) from procedures specified in an interface. Since a TEMPLATE METHOD works only with interfaces, its only option for object instantiation is via FACTORY METHODS. See (Gamma et al. 1995) for additional pattern relationships.

EXERCISES

1. Augment the UML class diagram of Figure 9.1 by including interfaces and corresponding implementations of Boundary and Differentiator abstractions. Modify the Fortran code in Figure 9.2 or the C++ code in Figure 9.3 so that the Periodic6thOrder constructor instantiates and stores Boundary and Differentiator abstractions.
2. Refactor the code from Exercise 2 so that the Periodic6thOrder stores a pointer to a DifferentationStrategy using the STRATEGY pattern of Chapter 7. Supply an additional DifferentiatorStrategy8th eighth-order-accurate Padé finite difference scheme as defined in Appendix A.

PART III

GUMBO SOOP

10 Formal Constraints

"Learn the rules so you know how to break them properly."
Tenzin Gyatso, The 14th Dalai Lama

10.1 Why Be Formal?

Whereas Parts I and II focused on canonical examples, Part III marches toward complete applications, resplendent with runtime considerations. The current chapter addresses code and compiler correctness. Chapter 11 discusses language interoperability. Chapter 12 addresses scalability and weaves elements of the entire book into a vision for multiphysics framework design.

Formal methods form an important branch of software engineering that has apparently been applied to the design of only a small percentage of scientific simulation programs (Bientinesi and van de Geijn 2005; van Engelen and Cats 1997). Two pillars of formalization are specification and verification – that is, specifying mathematically what a program must do and verifying the correctness of an algorithm with respect to the specification. The numerical aspects of scientific programming are already formal. The mathematical equations one wishes to solve in a given scientific simulation provide a formal specification, whereas a proof of numerical convergence provides a formal verification. Hence, formal methods developers often cite a motivation of seeking correctness standards for non-scientific codes as rigorous as those for scientific codes (Oliveria 1997). This ignores, however, the nonnumerical aspects of scientific programs that could benefit from greater rigor. One such aspect is memory management. The current chapter specifies formal constraints on memory allocations in the Fortran implementation of the Burgers equation solver from Chapter 9.

There have been long-standing calls for increased use of formal methods in scientific programming to improve reliability. Stevenson (1997) proclaimed a crying need for highly reliable system simulation methodology, far beyond what we have now. He cited formal methods among the strategies that might improve the situation. Part of his basis was Hatton's measurement of an average of 10 statically detectable serious faults per 1,000 lines among millions of lines of commercially released scientific C and Fortran 77 code (Hatton 1997). Hatton defined a fault as a misuse of the language that will very likely fail in some context. Examples include interface inconsistencies and using uninitialized variables. Hatton suggested that formal methods reduce defect density by a factor of three (cited in J3 Fortran Standards Technical Committee 1998). This chapter shows how code inspired by formal constraints can also detect compiler faults.

Pace (2004) expressed considerable pessimism about the prospects for adopting formal methods in scientific simulation because of the requisite mathematical training, which often includes set theory and a predicate calculus. A candidate for adoption must balance such rigor against ease of use. The Object Constraint Language (OCL) strikes such a balance, facilitating expressing of formal statements about software models without using mathematical symbols unfamiliar to nonspecialists. OCL's designers intended for it to be understood by people who are not mathematicians or computer scientists (Warmer and Kleppe 2003).

To attract scientific programmers, any software development strategy must address performance. Fortunately, program specification and verification are part of the software design rather than the implementation. Thus, they need not impact run-time performance. However, run-time checking of assertions, a third pillar of formal methods, is part of the implementation (Clarke and Rosenblum 2006; Dahlgren 2007). Assertions are tool-based or language-based mechanisms for gracefully terminating execution whenever programmer-inserted Boolean expressions fail. When the Boolean expressions are sufficiently simple, they occupy a negligible fraction of the run time compared to the long loops over floating point calculations typical of scientific software.

A final factor influencing adoption of formal methods is the lack of a common approach for describing the structure of traditional scientific codes beyond flow charts. OCL's incorporation into the UML standard (Warmer and Kleppe 2003) suggests that newcomers must simultaneously leap two hurdles: learning OCL and learning UML. Fortunately, increasing interest in object-oriented scientific programming has led to more frequent scientific program structural descriptions resembling UML class diagrams (Akin 2003; Barton and Nackman 1994), and several scientific programming projects use automated documentation tools that produce class diagrams (Heroux et al. 2005).

This chapter details how exposure to formal methods inspired systematic run-time assertion checking in a production research code, and how these assertions enabled the detection of an inscrutable compiler bug. The chapter also explains how applying OCL constraints to a UML software model forced us to think abstractly about an otherwise language-specific construct, pointers, and how this abstraction process inspired a workaround to a dilemma at the intersection of language design and existing compiler technology.

The remaining sections apply constraints on memory use in the Burgers equation solver of Chapter 9. Section 10.2 defines the problem that the constraints address. Section 10.3 describes an aspect of Fortran that poses a relevant dilemma. Section 10.4 specifies a useful set of constraints on a UML class diagram. Section 10.5 describes a useful design pattern that facilitates applying the constraints. Section 10.7 details performance results from a case study culled from a research application.

I'm Not as Think as You Drunk I Am!

Runtime checking of assertions inserted during software implementation can help detect source- and compiled-code faults. Design constraints can motivate the assertions. OCL provides a formal way to specify constraints.

10.2 Problem Statement

OCL constraints are meaningful only in the context of an object-oriented class model (Warmer and Kleppe 2003). Figure 10.1 presents a UML class model for a Burgers equation solver. As we move toward more complex examples in this part of the book, many implementation details will be left to the reader as exercises at the end of each chapter. We provide here a broad description of such an implementation with emphasis on the perspective of client code.

In the sample Burgers solver design of Figure 10.1, the Burgers class sits at the core and provides the primary interface with which most clients would interact. Burgers implements the ABSTRACT CALCULUS defined by the Integrand interface. In doing so, it likely makes itself an eligible argument for a time integration procedure such as that the `integrate` subroutine packaged in Figure 6.1. Clients could declare

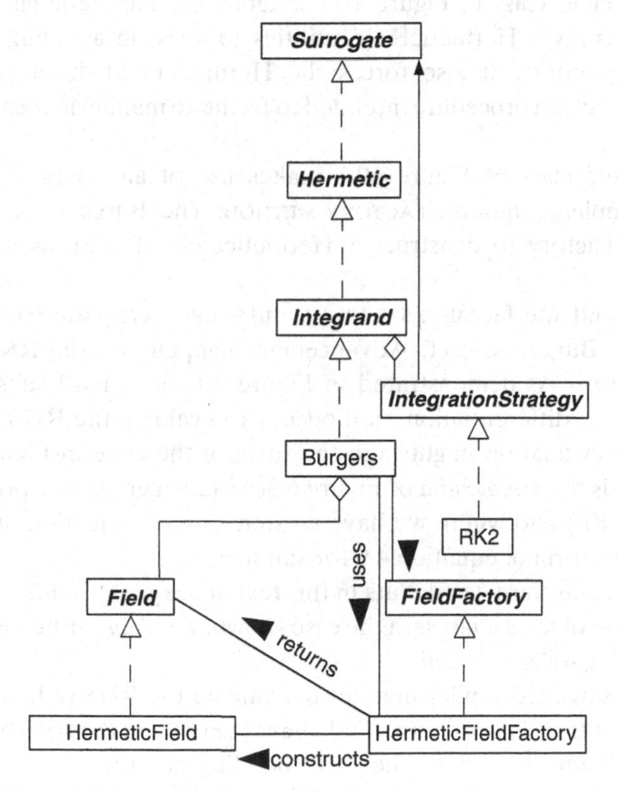

Figure 10.1. UML class diagram of a hermetic Burgers solver.

and construct a Burgers object and pass it repeatedly to `integrate` for advancement one time step per invocation. Advancing from an initial time `t` to a final time `t_final` over time increments `dt` might take the form:

```
type(burgers) :: cosmos
cosmos = burgers()
do while (t<t_final)
  call integrate(cosmos,dt)
  t = t + dt
end do
```

where we have assumed the existence of default component initialization in the overloaded structure constructor `burgers`.

In Figure 10.1, the Burgers class aggregates a reference to a Field component `u`. Since Field is abstract, this reference must be a `pointer` or `allocatable` component in Fortran. The structure constructor would use this component to dynamically allocate a Field instance. The closed diamond on the Burgers end of the Burgers/Field relationship indicates that this instance's lifetime coincides with that of the Burgers object. It gets constructed when the Burgers object is constructed and finalized when the Burgers object gets finalized.

Because this chapter focuses on situations where one needs to monitor compilers' compliance with the memory management stipulations in the Fortran 2003 standard, the Field class in Figure 10.1 extends the Hermetic class described in Chapter 5. This gives HermeticField utilities to assist in avoiding memory leaks and dangling pointers. It also forces the HermeticField developer to define a `force_finalization` procedure intended to facilitate manual invocation of the final subroutine.

The Burgers class of Figure 10.1 makes use of an ABSTRACT FACTORY and its concrete implementation's FACTORY METHOD. The Burgers constructor uses a HeremticFieldFactory to construct a HermeticField. It returns a pointer to an abstract Field.

The Integrand interface uses STRATEGY and SURROGATE pattern implementations to advance the Burgers object. Advancement happens via the RK2 algorithm in the design shown. As demonstrated in Figure 9.1, each RK2 substep involves a invoking the time differentiation method `t()` to evaluate the RHS of the Burgers equation. This evaluation might take the form of the code in Figure 10.2, where `burgers` extends the `integrand` of Figure 6.2(b) and aggregates a pointer to a Field from Figure 9.2(b) and where we have written Burgers equation at line 22 in the nonconservative form of equation 4.9 for simplicity.

The `associate` construct debuts in this text in `burgers_module` at line 23. It allows the creation of local aliases. In the case shown, `v` and `dv_dt` become a shorthand for `this%u` and `d_dt%u`, respectively.

The procedures a compiler invokes to evaluate the RHS of line 24 include the type-bound operators `operator(*)` and `operator(-)` and the type-bound functions `xx()` and `x()`. Figure 10.3 shows the corresponding call tree.

The object `dv_dt` in Figure 10.3 ultimately becomes the result of the `t()` type-bound procedure call in a time integration procedure such as `integrate` in

```
1  module burgers_module
2    use integrand_module ,only : integrand
3    implicit none
4    private
5    public :: burgers
6    type ,extends(integrand) :: burgers
7      class(field) ,allocatable :: u
8    contains
9      procedure :: t
10     procedure :: add => total
11     procedure :: assign
12     procedure :: multiply => product
13     generic :: operator(*) => product
14     generic :: assignment(=) => assign
15   end type
16 contains
17   function t(this) result(d_dt)
18     class(burgers) :: this
19     type(burgers) :: d_dt
20     class(field) ,pointer :: v, dv_dt
21     if (.not. allocated(this%u)) stop 'invalid argument inside t()'
22     allocate(d_dt%u,source=this%u)
23     associate( v => this%u , dv_dt =>d_dt%u)
24       dv_dt = v%xx()*nu - v*v%x()
25     end associate
26   end function
27   ! (additional type-bound procedures omitted)
28 end module burgers_module
```

Figure 10.2. Partial implementation of a Burgers solver class.

Figure 7.2(h). Line 27 of that figure reads:

```
this_half = this + this%t()*(0.5*dt) ! predictor step
```

which results in the additional cascade of calls detailed in Figure 10.4.

Each function call in Figure 10.3 produces a `field` result that is no longer needed once the procedure or operator of lower precedence (lower on the call tree) completes. Each function call in Figure 10.4 likewise produces an equally transient `integrand` containing a `field`. Thus, the combined Figures 10.3–10.4 represent eight intermediate instantiations. These correspond to the results of `u%xx()`, `u%x()`, `operator(-)`, and two invocations of `operator(*)` in Figure 10.3 along with the results of `this%t()`, `operator(*)`, and `operator(+)` in Figure 10.4.

Judicious handling of the memory allocations and deallocations associated with intermediate results proves critical when they occupy sufficient memory to crash a run in the case of a memory leak. In 3D simulations of fluid turbulence, for example, direct calculation of even modest laboratory conditions requires storing discrete fields with roughly 512^3 grid point values (de Bruyn Kops and Riley 1998). Given typical 4-byte precision per value, each 512^3-point `field` would occupy 0.5 gigabytes. State-of-the-art simulations on the largest supercomputers involve distributed memory allocations one to two orders of magnitude larger. Furthermore, coupling

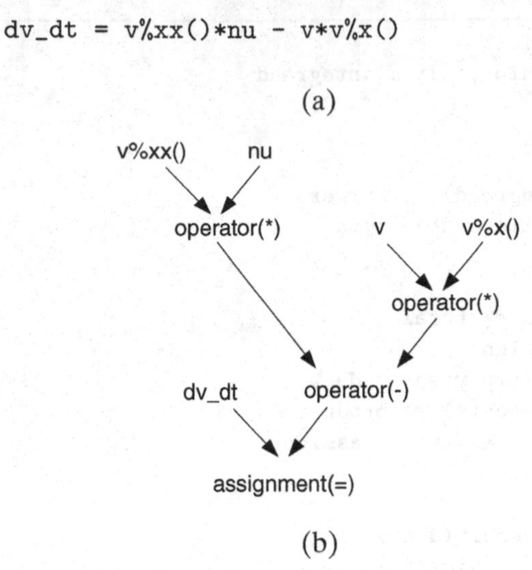

Figure 10.3. Burgers equation: (a) abstract calculus and (b) the resultant call tree.

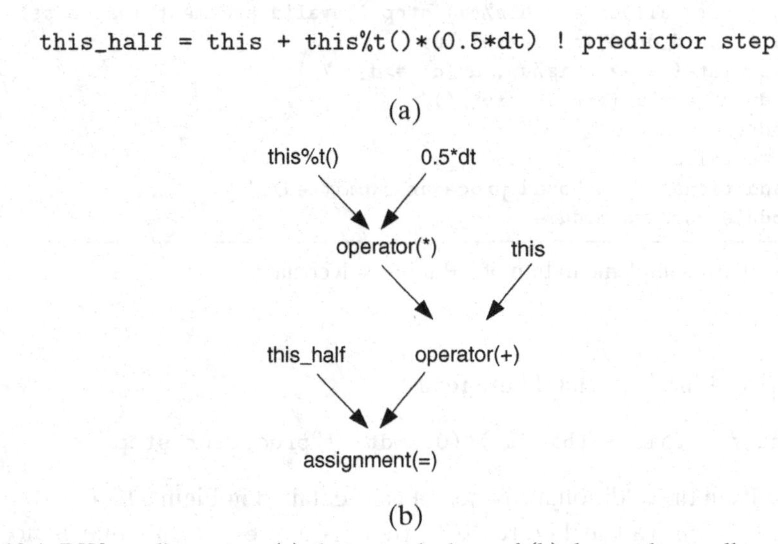

Figure 10.4. RK2 predictor step: (a) abstract calculus and (b) the resultant call tree.

in additional physics such as magnetic or quantum effects multiplies the memory footprint several-fold more. If compilers do not free the intermediate results, the burden falls on the programmer to do so. When the final assignment at the bottom of each call tree executes, the names associated with intermediate allocations are out of scope, and it would be difficult or impossible for the programmer to free the associated memory at that point.

Rouson, Morris, and Xu (2005) and Markus (2003) independently developed essentially equivalent rules for freeing the intermediate memory allocated for results produced in the process of evaluating ABSTRACT CALCULUS expressions. Rouson, Xu, and Morris (2006) formalized these rules in a set of OCL constraints. Here we update those constraints, applying the memory deallocation strategy of Stewart (2003), which handles a broader set of cases and allows us to leverage the OBJECT pattern implementation of Chapter 5.

For the remainder of Chapter 10, we refer to all functions called in ABSTRACT CALCULUS expressions as "operators." In practice, some of these would be generic type-bound operators and defined assignments, for example, `operator(*)` and `assignment(=)` in Figure 10.3(b). Others would be functions that mimic integral or differential mathematical operators such as `xx()`. With this nomenclature, we define intermediate results as temporary according to the following:

Definition 10.1. Any object is defined as temporary if it can be safely finalized at the termination of the first operator to which it is passed as an argument.

Corollary 10.1. *All objects that are not temporary are persistent.*

In the OBJECT implementation of Figure 5.2, any object that extends `hermetic` starts life with a `temporary` component pointer (in its `hermetic` parent) initialized to `null()`. This defaults all objects to being persistent. With this definition and its corollary in hand, one can summarize the Stewart memory management scheme in five rules:

Rule 10.1. Mark all operator results as temporary.

Rule 10.2. Mark all left-hand arguments to defined assignments as persistent.

Rule 10.3. Guard all temporary operator arguments upon receipt.

Rule 10.4. Finalize all temporary operator arguments that were received unguarded.

Rule 10.5. Finalize persistent objects inside the procedure that instantiated them.

These rules constrain code behavior. The challenge for this chapter lies in expressing these constraints in OCL and incorporating the resulting formalisms into the UML class model.

It Beats Flipping Burgers

The continual turnover of Burgers object allocations illustrates the necessity for disciplined memory management by the compiler or the developer. Sorting out when to free an object's memory can be difficult or impossible in the absence of rules that distinguish temporary from persistent objects.

10.3 Side Effects in Abstract Calculus

OCL constraints can be satisfied by code in any language at development time. Nonetheless, several Fortran-specific issues motivate the content of this chapter. This section outlines these issues.

In cases where the `field` component inside a `burgers` object is `allocatable`, the Fortran 2003 standard obligates the compiler to free the associated memory after the object goes out of scope. In cases where the `field` component is a `pointer`, the most straightforward way is to free the memory inside a `final` subroutine. The developer can then rely on the compiler to invoke the `final` subroutine after the object goes out of scope. As explained in Section 5.2, the `allocatable`-component approach has proved problematic with recent versions of several compilers. That section also explains that the `pointer`-component approach remains infeasible for most compilers. Even when either approach works, developers might desire to control the memory utilization more closely in order to recycle allocated space or to free it sooner than the compiler might.

Fortran's requirement that type-bound operators and the procedures they call be free of side effects greatly complicates programmers' ability to free memory allocated for operator arguments. The `intent(in)` attribute enforces the requirement but also poses a dilemma. In nested invocations of operators, an argument passed to one operator might be a result calculated by another operator of higher precedence. That higher-precedence operator is likely to have allocated memory for its result just as the lower-precedence operator might allocate memory for its result. The easiest and most efficient place for the programmer to release memory that was dynamically allocated inside the result of one operator is inside the operator to which this result is passed. However, the operator receiving the result cannot modify it. A similar dilemma relates to defined assignments such as the `assignment(=)` binding in Figure 10.2, wherein there frequently arises a need to free memory associated with allocatable components of the RHS argument if it is the result of an expression evaluation.

Even compilers that do the proper deallocations might not do so economically. For example, the first author has analyzed intermediate code received from one vendor and found that this vendor's compiler carries along all the memory allocated at intermediate steps in the call tree, performing deallocations only after the final assignment at the top of the tree (Rouson et al. 2006). In complicated expressions written on large objects, that vendor's approach could make ABSTRACT CALCULUS infeasible. Hence, in addition to correcting for compiler nonconformance with the Fortran standard's memory management stipulations, the rules and resulting constraints in this chapter economize memory usage in ways the standard does not require.

Warning: Calculus May Cause Memory Loss, Bloating, and Insomnia

Fortran's prohibition against side effects in type-bound operators complicates the process of explicitly deallocating objects before they go out of scope and can lead to unrestrained growth in memory usage. Even compilers that correctly automate deallocation or finalization might not do so economically, resulting in developers losing sleep strategizing to free memory sooner than would the compiler.

10.4 Formal Specification

10.4.1 Modeling Arrays with OCL

Although the `field` class component array in our Burgers solver is 1D, the lion's share of scientific research problems are more naturally expressed with multidimensional arrays. Modeling these arrays in OCL therefore bears some discussion. Although OCL has a "sequence" type that could naturally be used to model 1D arrays, the language does not contain an intrinsic multidimensional array type.

Because OCL is a subset of UML, however, types defined by a UML model are considered OCL model types. We can therefore incorporate the Array type in Figure 2.3 into models for use in OCL. We do so in the remainder of this chapter for the 1D case and leave the multidimensional case as a modeling exercise for the reader.

10.4.2 Hermeticity

We can now specify the constraints that preclude a category of memory leaks in composite operator calls. We refer to leak-free execution as hermetic memory management, or simply hermeticity. The memory of interest is associated with the allocatable array components inside data structures passed into, through, and out of call trees of the form of Figures 10.3–10.4. We assume no other large memory allocations inside operators. This assumption is discussed further vis-á-vis economy in Section 10.4.3.

Figure 10.5 provides a partial class diagram, including several ABSTRACT CALCULUS operators: the unary operator `x()`, binary operator add, and a defined assignment `assign`. Our primary task is to constrain the behavior of such operators using the memory management rules listed in Section 10.2. We model Definition 10.1 and Corollary 10.1 simply by including in the HermeticField class a "temporary" attribute, modeled as an instance of the Boolean model type described in Figure 2.3. The value of this attribute classifies the object in the set of objects defined by the Definition or in the complementary set defined by the Corollary. The use of the Boolean "temporary" attribute demonstrates several differences between a software model and a software implementation. For example, an implementation of HermeticField would likely extend Hermetic as in Figure 10.1, thereby inheriting the "temporary" and "depth" attributes as well as the force_finalization() method. That level of detail is unnecessary for purposes of writing the intended constraints. Along this vein, to consider the `hermetic` derived type component `temporary` in Figure 5.2 to be an implementation of the HermeticField temporary attribute, one has to map the `integer` value `0` for the component to the value True for the Boolean type in Figure 2.3, and map all nonzero values to False. Finally, even the last artifice misses the detail that the `temporary` component in the `hermetic` implementation Figure 5.2 is a pointer. UML does not contain a pointer type or attribute, but the modeling of pointers in UML plays a central role in Section 10.4.3, so we defer further discussion to that section.

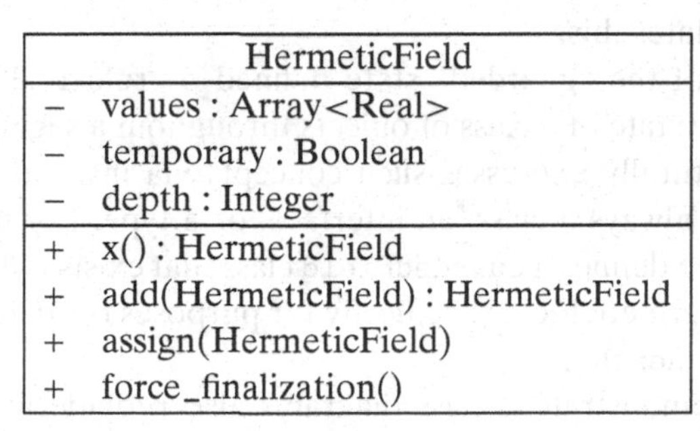

HermeticField
– values : Array<Real>
– temporary : Boolean
– depth : Integer
+ x() : HermeticField
+ add(HermeticField) : HermeticField
+ assign(HermeticField)
+ force_finalization()

Figure 10.5. HermeticField partial class diagram.

In formal methods, constraints take three forms: preconditions that must be true before a procedure commences; postconditions that must be true after it terminates; and invariants that must be true throughout its execution. OCL preconditions bear the prefix "`pre:`", whereas "`post:`" precedes postconditions and "`inv:`" precedes invariants. The contexts for OCL pre- and postconditions are always operations or methods. When writing OCL constraints as standalone statements, a software designer writes the constraint context above the constraint. Hence, Rule 10.1 applied to the Hermetic method `SetTemp()` in Figure 10.6 could be written:

```
context : Hermetic :: SetTemp()
post: temporary = true
```

Alternatively, the designer indicates the context by linking the constraint to the corresponding operations visually in a UML diagram. The visual link might involve writing a constraint just after the method it constrains in a class diagram or by drawing connectors between the constraint and the method as in Figure 10.6. When using connectors, one commonly encloses the constraint in a UML annotation symbol as Figure 10.6 depicts.

The argument and return value declarations in Figure 10.6 reflect OCL conventions along with a desire for succinct presentation. For both of these reasons, Figure 10.6 omits the implicit passing of the object on which the method is invoked. In C++ and in many of the Fortran examples in this book, that object is referenced by the name `this`. In OCL, that object is named `self` and can be omitted in the interest of brevity. The postconditions applied to the Hermetic in Figure 10.6 omit `self` because it is the only object in play. Several postconditions applied to HermeticField, however, include it to explicitly distinguish its attributes from those of the other the objects received or returned by the methods.

We formally specify Rule 10.1 through postconditions on the arithmetic and differential operators `x()` and `add()`, respectively, and Rule 10.2 through a postcondition on the defined assignment `assign()`. Specifically, two postconditions in Figure 10.6 specify that the results of `x()` and `add()` must be temporary. Another postcondition specifies that the LHS argument (`self`) passed implicitly to `assign` must be persistent.

Rule 10.3 governs the treatment of the subset of temporary objects that must be guarded against premature finalization. According to the memory management scheme of Stewart (2003), the GuardTemp() procedure guards a temporary object by incrementing its depth count. The postcondition applied to the like-named method in Figure 10.6 stipulates this.

Formalization of the "guarded" state defined by rule 10.3 requires making a statement about the state of a class of objects throughout a segment of the program execution. One naturally expresses such concepts via invariants. In OCL invariants, the context is always a class, an interface, or a type. For modeling purposes, Figure 10.6 therefore defines a `GuardedField` class that exists only to provide a well-defined subset of HermeticField specifically for purposes of specifying the invariant that subset satisfies: `depth>1`.

As the GoF demonstrated, class diagrams also provide an effective medium through which to describe conforming implementations. The note linked to the

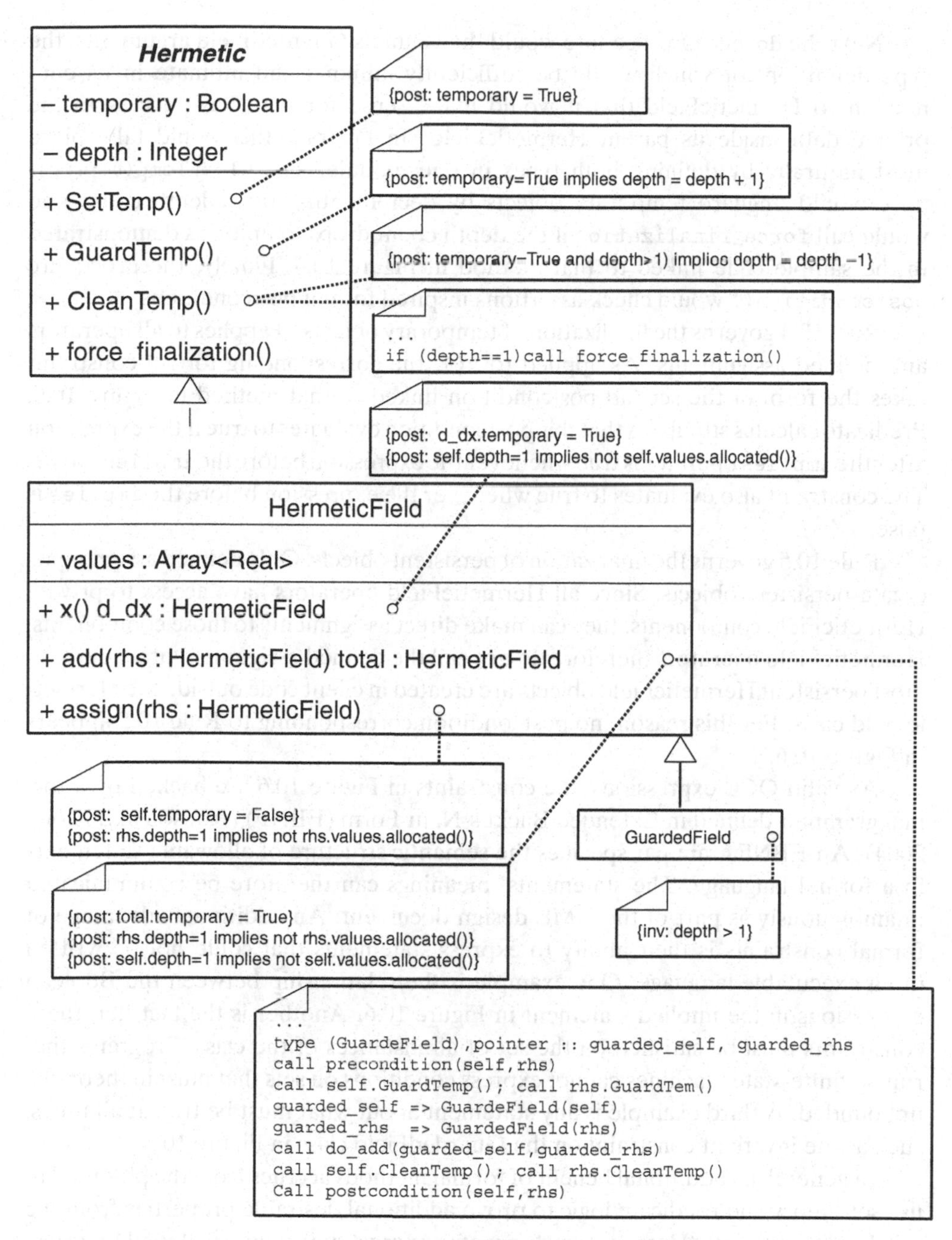

Figure 10.6. UML class diagram of a hermetic Burgers solver.

`add()` operation in Figure 10.6 displays a skeletal implementation. In a complete implementation, the depicted `precondition()` procedure would likely check a set of assertations inspired by formal preconditions. Then invoking `GuardTemp()` on `self` and `rhs` would guard these objects by increasing their depth counts if they are temporary. Next the `GuardedField` constructor calls might verify that the depth count exceeds unity before returning pointers to the HermeticField objects, giving them the dynamic type GuardedField.

Next the `do_add()` procedure would then sum its GuardedField arguments, the type definition for which would be sufficiently minimal and intimate in its connection to HermeticField that it would make sense for it to have access to the private data inside its parent HermeticField. In Fortran, this would take place most naturally by defining both types in one module. The `CleanTemp()` procedure would unguard temporary objects by decrementing their depth count and would call `force_finalization` if the depth count drops to unity, as demonstrated in the sample code linked to that method in Figure 10.6. Finally, the procedure `postcondition()` would check assertions inspired formal postconditions.

Rule 10.4 governs the finalization of temporary objects. It applies to all operators and defined assignments. As applied to `x()`, the corresponding formal constraint takes the form of the second postcondition linked to that method in Figure 10.6. Predicate calculus stipulates that this postcondition evaluates to true if the expression after the `implies` operator is true whenever the expression before the `implies` is true. The constraint also evaluates to true whenever the expression before the `implies` is false.

Rule 10.5 governs the finalization of persistent objects. Only defined assignments create persistent objects. Since all HermeticField operators have access to private HermeticField components, they can make direct assignments to those components. HermeticField operators therefore do not call the defined assignment procedure, so most persistent HermeticField objects are created in client code outside the HermeticField class. For this reason, no postcondition corresponding to Rule 10.5 appears in Figure 10.6.

As valid OCL expressions, the constraints in Figure 10.6 are backed by a formal grammar defined in Extended Backus-Naur Form (EBNF) (Rouson and Xiong 2004). An EBNF grammar specifies the semantic structure of allowable statements in a formal language. The statements' meanings can therefore be communicated unambiguously as part of the UML design document. An additional advantage of formal constraints is their ability to express statements that could not be written in an executable language. One example is the relationship between the Boolean expressions in the implied statement in Figure 10.6. Another is the fact that these constraints must be satisfied on the set of all instances of the class. Programs that run on finite-state machines cannot express conditions on sets that must in theory be unbounded. A third example is any statement about what must be true at all times, such as the invariant constraint on the GuardedField class in Figure 10.6.

In general, an additional benefit of formal methods accrues from the potential to use set theory and predicate logic to prove additional desirable properties from the specified constraints. However, such proofs are most naturally developed by translating the constraints into a mathematical notation that is opaque to those who are not formal methods specialists. Since OCL targets non-specialists, proving additional properties mathematically would detract from the main purpose of simply clarifying design intent and software behavior.

10.4.3 Economy

Refining Rules 10.1–10.5 can lead to more economical memory usage. Although the memory reductions are modest in some instances, we hope to demonstrate an important tangential benefit from the way OCL forces designers to think more abstractly about otherwise language-specific constructs. For this purpose, consider again the function `add()`, which in the model of Figure 10.6 takes the implicit argument `self` and the explicit argument `rhs` and returns `total`. We can ensure economical memory usage by specifying that temporary memory be recycled. To facilitate this, the scalars `self`, `rhs`, and `total` must have the `pointer` or `allocatable` attribute. In many respects, the `allocatable` approach is more robust: It obligates the compiler to free the memory when it goes out of scope and the `move_alloc()` intrinsic facilates moving the memory allocation from one object to another without costly deallocation and reallocation (see Section 6.2.1). However, because `allocatable` scalars are a Fortran 2003 construct without universal compiler support, the `pointer` approach is desirable from the standpoint of portability. We consider that approach for the remainder of this chapter.

To pass pointers in Fortran, one writes the `add()` function signature as:

```
function add(self,rhs) result(total)
  class(HermeticField), pointer, intent(in) :: self
  type(HermeticField), pointer, intent(in) :: rhs
  type(HermeticField), pointer :: total
```

Since OCL does not have a pointer type, however, we model pointers as a UML association between two classes. Specifically, we model `self`, `rhs`, and `total` as instances of a HermeticFieldPointer class that exists solely for its association with the HermeticField class. We further assume HermeticFieldPointer implements the services of Fortran pointers, including an `associated()` method that returns a Boolean value specifying whether the object on which it is invoked is associated with the object passed explicitly as an argument.

Figure 10.7 diagrams the HermeticFieldPointer/HermeticField relationship. A diagrammatic technique appearing for the first time in that figure is the adornment of either end of the relationship with multiplicities (see Section 2.5) describing the allowable number of instances of the adjacent classes. The adornments in Figure 10.7 indicate that one or more HermeticFieldPointer instances can be associated with any one HermeticField instance.

In Figure 10.7, we apply the label "target" to the HermeticField end of the association. From a HermeticFieldPointer object, this association can be navigated through the OCL component selection operator ".", so an economical postcondition might take the form shown in the annotation linked to HermeticFieldPointer, wherein the `implied` conditions stipulate that `total` must be associated with one of

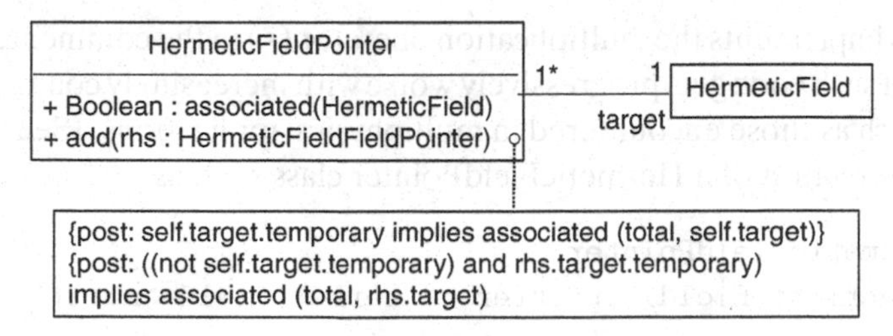

Figure 10.7. HermeticField and HermeticFieldPointer class association.

the temporary targets. In writing the constraint, we arbitrarily chose to overwrite `self` in the case in which both `self` and `rhs` are recyclable.

We advocate reusing the entire temporary object while giving its array component, if any, the `allocatable` attribute, rather than the `pointer` attribute. In addition to the aforementioned convenience of `allocatable` arrays, they offer performance advantages over pointers. An `allocatable` array references contiguous memory that can be accessed with unit stride, whereas a `pointer` can target an array subsection that might not be contiguous (Metcalf, Reid, and Cohen 2004). Any resulting lack of spatial locality could retard cache performance.

10.5 The Shell Pattern

From a programmability standpoint, the greatest benefit of developing the economizing constraints in Figure 10.7 stems from the conceptual leap required to model pointers in OCL. Fortran's prohibition against side effects in defined operators necessitates the ability to specify `pointer` arguments `intent` – another technology for which compiler support lagged the publication of the Fortran 2003 standard by several years. Developing the HermeticFieldPointer class model in UML naturally inspires the notion of constructing an actual class analogous HermeticFieldPointer. Doing so facilitates emulating pointer intent without compiler support for it.

The construction of a class analogous to HermeticFieldPointer served as the critical linchpin that enabled the first ADT calculus syntax in Fortran (Rouson, Xu, and Morris 2006). Without it, it would have been impossible to satisfy the hermeticity constraints in Figure 10.6 with recent compiler versions. In the absence of correct, automatic compiler deallocation of memory, the programmer's hands are tied for the reasons described in Section 10.3. For this reason, earlier attempts at ADT calculus in Fortran used an awkward syntax, evaluating derived type expressions via explicit procedure invocation with an adjacent comment displaying the desired operator syntax:

```
! total = lhs + rhs
call assign(total,add(lhs,rhs))
```

Although this approach worked well for simple expressions, more complicated examples render code nearly indecipherable on first inspection. Consider adopting a similar approach to writing the 3D heat equation in a form patterned after Table 3.1:

```
! dT_dt = alpha *(T%xx() + T%yy() + T%zz())
call assign(dT_dt,times(alpha,(addT%xx(),add(T%yy(),T%zz()))))
```

where `times` implements the multiplication operator (`*`) in the commented code. The readability of such code gets progressively worse with increasingly complicated sets of equations such as those encountered in multiphysics applications. (See Chapter 12.)

The construction of a HermeticFieldPointer class such as:

```
type HermeticFieldPointer
  type(HermeticField) ,pointer :: target_field=>null()
end type
```

along with a corresponding constructor and various defined operators such as:

```
function add(self,rhs) result(total)
  class(HermeticFieldPointer) ,intent(in) :: self
  type(HermeticFieldPointer)  ,intent(in) :: rhs
  type(HermeticFieldPointer)              :: total
end function
```

facilitates satisfying the Fortran's `intent(in)` requirement and restricts us from changing with what target object the `target_field` is associated while allowing us to deallocate any allocatable entities inside the target with compilers that fail to do so.

Since HermeticFieldPointer exists only to wrap a HermeticField instance, one might refer to the former as a shell and the latter as a kernel. Together they comprise a SHELL pattern. Chapter 11 describes similar instances in which a lightweight class serves as a wrapper or identification tag for another class. An exercise at the end of the current chapter asks the reader to write a SHELL pattern description using the three-part rule of Section 4.1.

10.6 An Assertion Utility for Fortran

Formal constraints make statements that must always be true for an application to conform to the given specification. Relative to this level of certitude, run-time assertion checking must necessarily be considered incomplete: Each run tests only one set of inputs. Nonetheless, the regularity of many scientific simulations mitigates the impact of this limitation. In the Burgers solver, for example, the memory allocations, reallocations, and deallocations are the same at every time step independent of the initial data.

Ideally, constraints could be translated into program statements automatically. Tools exist for generating Java code from OCL. Although some of the earliest work on automated assertion insertion targeted Fortran 77 (Stucki and Foshee 1971), we know of no such efforts for Fortran 2003. Fortran lacks even a language mechanism intended specifically for writing assertions as exists in C, C++, and Java. In the absence of a language construct, the options remain to write ad hoc conditional (`if-then`) statements or to link to an assertion toolkit.

Press et al. (1996) published assertion tools similar to those in Figure 10.8. The two subroutines in that figure terminate execution after printing a message, if provided one, to `error_unit`, which on Unix-like systems corresponds to printing to the “standard error” stream. The subroutine `assert()` traverses an `logical` array, each element of which would likely be a Boolean expression. Execution terminates gracefully (without a platform-dependent, system-generated error message) if any element of the array evaluates to `.false.`. The subroutine `assert_identical()` traverses an `integer` array, checking whether each element matches the first one. If not, the execution halts gracefully after reporting the mismatches. As suggested by Press and colleagues, a common use of the latter utility involves passing a list of array sizes. The first executable line of `assert()` invokes `assert_identical()` for this purpose.

To be viable in production settings, an assertion facility must have negligible impact on run time. Languages with built-in assertion capabilities typically allow the developers to disable assertions at compile-time. In such cases, the developers can turn assertion checking on and off, using assertions when debugging but eliminating them during production runs. In Fortran, one can emulate this capability by providing a global, compile-time, `logical` constant `assertions` value and conditioning all assertion checking on this value in the manner:

```
logical ,parameter :: assertions=.false.
if (assertions) call assert(sanity_checks,insanity_warnings)
```

where `sanity_checks` and `insanity_warnings` would contain arrays of `logical` expressions and corresponding error messages, respectively. With this approach, most compilers would eliminate the assertion calls during a dead-code-removal optimization phase because it is clear that the call can never be made. (Section 1.7 described a similar scenario for a global `debugging` parameter.) This behavior would likely depend on the optimization level used; however, most production scientific codes use the highest optimization levels available.

```
1  module assertion_utility
2    use iso_fortran_env ,only : error_unit
3    implicit none
4    private
5    public :: error_message,assert,assert_identical
6    type error_message
7      character(:) ,allocatable :: string
8    end type
9  contains
10   subroutine assert(assertion,text)
11     logical ,dimension(:) ,intent(in) :: assertion
12     type(error_message) ,dimension(:) ,intent(in) :: text
13     integer :: i
14     logical :: any_failures
15     call assert_identical( [size(assertion),size(text)] )
16     any_failures=.false.
17     do i=1,size(assertion)
18       if (.not. assertion(i)) then
19         any_failures=.true.
20         write(error_unit,*) 'Assertion failed with message: '
21         if (allocated(text(i)%string)) then
22           write(error_unit,*) text(i)%string
23         else
24           write(error_unit,*) '(no message provided).'
25         end if
26       end if
27     end do
28     if (any_failures) stop 'Execution halted on failed assertion(s)!'
29   end subroutine
30   subroutine assert_identical(integers)
31     integer ,dimension(:) ,intent(in) :: integers
32     integer :: i
33     logical :: any_mismatches
34     any_mismatches = .false.
35     do i=2,size(integers)
```

```
36        if (integers(i) /= integers(1)) then
37          any_mismatches = .true.
38          write(error_unit,*) &
39          'Value ',i,' does not match expected value ',integers(1)
40        end if
41      end do
42      if (any_mismatches) stop 'Execution halted on failed assertion!'
43    end subroutine
44  end module
```

Figure 10.8. Fortran assertions utility.

10.7 Case Study: A Fluid Turbulence Solver

This section describes results associated with a production research code that employed run-time assertions based on a set of rules and constraints that were simpler than Rules 10.1–10.5 and Figure 10.6. The code solves the Navier-Stokes equations for fluid dynamics for a canonical yet computationally demanding turbulent flow. Rouson et al. (2008a) gave a comprehensive description of the software architecture and the numerical methods employed. Rouson et al. (2008b) and Morris et al. (2008) described new science illuminated via multiphysics applications of the software. Here we focus on the performance impact of checking run-time assertions inspired by OCL constraints.

The rules employed in this case inspired the inclusion of a Boolean temporary flag in the derived type definition but no depth count. Given the role the depth count plays in guarding temporary objects from premature deletion, the lack of a depth count necessitates dropping Rule 10.3 and modifying 10.4. The new set of rules follow:

Rule 10.6. Mark all operator results as temporary.

Rule 10.7. Mark all left-hand arguments to defined assignments as persistent.

Rule 10.8. Deallocate all memory associated with temporary operator arguments.

Rule 10.9. Deallocate all memory associated with persistent objects inside the procedure that instantiated them.

where the meanings of "temporary" and "persistent" remain based on Definition 10.1 and Corollary 10.1. These are equivalent to the rules put forth by Markus (2003) and Rouson, Morris, and Xu (2005). Rules 10.6–10.9 use "deallocate" instead of "finalize" to address situations in which it is necessary or desirable for the programmer to explicitly deallocate memory rather than to rely on a `final` binding.

In the absence of a depth count, Rules 10.6–10.9 can cause problems in nested invocations. For illustrative purposes, consider the following case adapted from Stewart (2003):

```
1      type(field) :: a, b, c
2      a = field()
3      b = field()
4      call do_something(a + b, c)
```

where we assume the `field` type contains one or more `allocatable` components and a `logical` component that identifies it as temporary or persistent. We further assume the `field()` constructor allocates memory for `allocatable` components inside `a` and `b` and gives those components default values. In this scenario, the addition operator invoked inside the first argument slot at line 4 would mark its result temporary based on Rule 10.6.

Next consider the following definition of the above `do_something` procedure:

```
1  subroutine do_something(x, y)
2    type(field) ,intent(in) :: x
3    type(field) ,intent(inout) :: y
4    y =x
5    y = x + y
6  end subroutine
```

where the temporary object now referenced by `x` inside `do_something()` would have its `allocatable` components deallocated by the defined assignment at line 4 in the latter code in accordance with Rule 10.6. That deallocation would in turn preclude proper execution of the addition operator in the subsequent line. Nonetheless, the current case study involved no such scenarios and the simplified Rules 10.6–10.9 sufficed.

Figure 10.9 presents `precondition()` and `postcondition()` subroutines adapted from Rouson et al. (2006) using the assertion utility of Figure 10.8. In the subject turbulence code, each `field` instance contains two `allocatable` array components: One stores a physical-space representation and another stores the corresponding Fourier coefficients. Since one can always compute Fourier coefficients from the physical-space samples that generated those coefficients, and vice versa, having both allocated is either redundant or inconsistent. It is therefore never valid for both to be allocated except momentarily in the routine that transforms one to the other. This condition is checked in the Figure 10.9.

The aforementioned turbulence simulations employed analogous routines. The code called a `precondition()` procedure at the beginning of each `field` method. It called a `postcondition()` procedure at the end of each method.

Besides monitoring the source code behavior, these assertions proved eminently useful in checking for compiler bugs. Using these routines, Rouson et al. (2006) discovered that one vendor's compiler was not always allocating arrays as requested. Despite repeated attempts to reproduce this problem in a simple code to submit in a bug report, the simplest demonstration of the error was a 4,500-line package in which the error only occurred after several time steps inside a deeply nested call tree. The related saga cost several weeks of effort plus a wait of more than a year for a compiler fix. The disciplined approach to checking assertions at the beginning and end of each procedure, as inspired by the OCL pre- and postconditions, paid its greatest dividend in motivating a switch to a different compiler.

Figure 10.10 presents a procedure that verifies economical memory recycling based on a SHELL class named `field_pointer`. Rouson et al. (2006) used analogous code to check whether memory allocated for temporary arguments had been recycled by associating that memory with the function result.

```
1  subroutine precondition(this,constructor)
2    use assertion_utility ,only : assert
3    implicit none
4    type(field) ,intent(in) :: this
5    logical ,intent(in) :: constructor
6    logical :: both_allocated ,at_least_one_allocated
7    both_allocated= allocated(this%fourier) .and. allocated(this%physical)
8    call assert( [.not. both_allocated] &
9    ,[error_message('redundant or inconsistent argument')])
10   at_least_one_allocated = &
11   allocated(this%fourier) .or. allocated(this%physical)
12   if (constructor) then
13     call assert( [.not. at_least_one_allocated] &
14     ,[error_message('constructor argument pre-allocated')])
15   else; call assert( [at_least_one_allocated] &
16     ,[error_message('argument data missing')])
17   end if
18 end subroutine
19
20 subroutine postcondition(this,public_operator,deletable)
21   use assertion_utility ,only : assert
22   implicit none
23   type(field) ,intent(in) :: this
24   logical ,intent(in) :: public_operator,deletable
25   logical :: both_allocated,at_least_one_allocated
26   both_allocated= allocated(this%fourier) .and. allocated(this%physical)
27   call assert( [.not. both_allocated] &
28   ,[error_message('redundant or inconsistent result')])
29   at_least_one_allocated = &
30   allocated(this%fourier) .or. allocated(this%physical)
31   if (public_operator .and. deletable) then
32      if (this%temporary) then
33        call assert([.not. at_least_one_allocated] &
34        ,[error_message('temporary argument persisting']))
35      else; call assert([at_least_one_allocated] &
36        ,[error_message('persistent argument deleted')])
37      end if
38   else
39      call assert([at_least_one_allocated] &
40      ,[error_message('invalid result']))
41   end if
42 end subroutine
```

Figure 10.9. Memory allocation assertions in a Fourier-spectral turbulence code.

Since our pre- and postcondition subroutines contain simple Boolean expressions, they require insignificant amounts of execution time. Table 10.1 shows that the most costly procedure for the case studied by Rouson et al. (2006) the `transform()`, which contained 3D FFT calls. That procedure accounted for 34 percent of the processor time. By contrast, `precondition()` and `postcondition()` occupied immeasurably low percentages of execution time even though the number of calls to these routines exceeds the calls to `transform()` by roughly an order of magnitude.

```
1  subroutine economical_postcondition(lhs,rhs,result)
2    use assertions ,only : assert
3    implicit none
4    type(field_pointer) ,intent(in) :: lhs,rhs,result
5    if (lhs%target%temporary) then
6      call assert( [associated(result%target,lhs%target)] &
7       ,[error_message('lhs not recycled')])
8    else if (right%target%temorary) then
9      call assert( &
10     [associated(result%target,rhs%target)] &
11     ,[error_message('rhs not recycled')])
12   end if
13 end subroutine
```

Figure 10.10. Memory economy assertions in a Fourier-spectral turbulence code.

Table 10.1. *Procedural run-time distribution*

Function Name	Number of Calls	% of Total Runtime
transform	108	34.42
assign	167	19.67
xx	156	16.18
dealias	36	7.70
add	75	7.43
times	75	5.03
...	...	...
precondition	1477	0
postcondition	885	0

Thus, the capability to turn assertions off appears to be nonessential in the chosen application. This observation is likely to hold for the majority of demanding, leading-edge scientific applications.

EXERCISES

1. Write a set of preconditions prescribing desirable memory management behavior for the methods in Figure 10.6.
2. Using the three-part rule of Section 4.1, write a more complete description of the SHELL pattern.
3. Modify the assertion utility of Figure 10.8 so that the `error_message` derived type has private data along with a `error_message` generic procedure that overloads the intrinsic structure constructor and supports the constructor call syntax in Figure 10.9.

11 Mixed-Language Programming

"But if thought corrupts language, language can also corrupt thought."
George Orwell

11.1 Automated Interoperability Tools

In human cultures, membership in a community often implies sharing a common language. This holds as much for computer programming languages as for natural human tongues. Certain languages dominate certain research communities. As a result, the modern drive towards multidisciplinary studies inevitably leads to mixed-language development. As noted in the preface, survey evidence suggests that approximately 85% of high-performance computing users write in some flavor of C/C++/C#, whereas 60% write in Fortran. These data imply that at least 45% write in both Fortran and C/C++/C#, so the union of these two language families likely contains the most important pairings in mixed-language scientific programming.

In many settings, the language version matters as much as the language identity. For example, interfacing other languages with object-oriented Fortran 2003 poses a much broader set of challenges than does interfacing with procedural Fortran 77 or even object-based Fortran 95. It also matters whether one language invokes code in a second language, the second invokes code in the first, or both. As suggested by Rasmussen et al. (2006), one can account for invocation directionality by considering ordered pairs of languages, where the order determines which language is the caller and which is the callee.

Consider a set $\{A, B, C, \ldots\}$ of N dialects, where each member is a language or language version distinct from each other member. In the worst case, when each member comes from a different language or when language versions do not maintain perfect backward compatibility, a project that involves code in each dialect invoking code in each other dialect generates $N(N-1) = O(N^2)$ ordered dialect pairings. Most such pairings require glue code (also called "bridging code") that ensures type, variable, and procedure compatibility. Supporting language interoperability in such a setting may be cost-prohibitive.

Broadly, two categories of solutions reduce the cost of supporting language pairings. One class of tools reduces the programming overhead associated with each bilingual pairing by generating the relevant glue code automatically. A more ambitious class of tools eliminates the quadratic growth in ordered pairings by inserting an intermediary through which an object declared in one language communicates with objects in all other languages without either side knowing the identity of the other side. The intermediary plays a role analogous to the MEDIATOR mentioned in Sections 4.2.3 and 8.1. Such tools generally work with a common interface representation: an Interface Definition Language (IDL). The remainder of this section discusses tools that follow both approaches.

The Simplified Wrapper and Interface Generator (SWIG) automates the interface construction process. SWIG interfaces C and C++ code with codes written in several other languages, including Python, Perl, Ruby, PHP, Java, and C#[1]. To access C functions, for example, from another language, one writes a SWIG interface file. This file specifies the function prototypes to be exported to other languages. From an interface file, SWIG generates a C wrapper file that, when linked with the rest of the project code, can be used to build a dynamically loadable extension that is callable from the desired language. Often, the developer need not write the interface file line by line and can instead incorporate a C/C++ header file in a SWIG interface file via a single `#include` directive.

SWIG does not provide direct support for interfacing to Fortran. Since SWIG interfaces to Python, one workaround involves using the PyFort tool[2] that enables the creation of Python language extensions from Fortran routines as suggested by Pierrehumbert[3]. PyFort supports Fortran 77 with plans for supporting Fortran 90, but Pierrehumbert found it easier to manually construct C wrappers for Fortran routines.

The Chasm language interoperability toolkit targets scientific programmers. Chasm enables bidirectional invocations between statically typed languages – that is languages in which type information can be discovered in the compiled code prior to runtime – including Fortran, C, C++, and Java. (Chasm also enables unidirectional invocations from dynamically typed languages such as Ruby and Python to statically typed languages but not the reverse.) Because many scientific applications are written in Fortran and many libraries in C++, the most oft-cited discussion of Chasm focuses on Fortran/C++ pairings (Rasmussen et al. 2006). In this setting, Chasm facilitates invoking methods on C++ objects from Fortran 95 and vice versa. Chasm aims for the interoperability "holy grail" articulated by Barrett (1998):

> Informally, a polylingual system is a collection of software components, written in diverse languages, that communicate with one another transparently. More specifically, in a polylingual system it is not possible to tell, by examining the source code of a polylingual software component, that it is accessing or being accessed by components of other languages. All method calls, for example, appear to be within a component's language, even if some are interlanguage calls. In addition, no IDL or other intermediate, foreign type model is visible to developers, who may create types in their native languages and need not translate them into a foreign type system.

Chasm therefore eschews the use of a specially tailored IDL. Chasm instead automatically generates an Extensible Markup Language (XML) representation of

[1] http://www.swig.org
[2] http://sourceforge.net/projects/pyfortran
[3] http://geosci.uchicago.edu/ rtp1/itr/Python.html

procedure interfaces. Chasm does so by parsing a compiler's intermediate language representation of a code's abstract syntax tree, which is a simplified description of the code's syntactical structure.

A complete multilanguage build process with Chasm takes three steps:

1. Static analysis,
2. Bridging-code generation, and
3. Compilation and linking.

Step 1 produces the XML description of the procedure interfaces found in the compiler's intermediate-language representation. With the XML output from Step 1 as input, Step 2 outputs stub and skeleton interfaces as illustrated in Figure 11.1. Step 3 takes whatever form the user chooses. For the smallest packages, manually invoking the compiler and linker might suffice. For moderately larger packages on Windows and POSIX-compliant (Unix) platforms, building might involve writing a Makefile that specifies file dependencies and compilation and linking syntax. For the largest collections of packages, Makefile construction itself might be automated with tools such as Gnu[4] Autotools or CMake[5] (Martin and Hoffman 2008).

In the schema of Figure 11.1, a calling procedure in language *A* invokes a language-*A* stub. Stubs exemplify the GoF ADAPTER pattern. The GoF describes the following goal of an ADAPTER: "Convert the interface of a class into another interface clients expect." A language-*A* stub provides a natural interface in *A*, hiding compiler-specific conventions the language-*B* compiler uses to generate symbol names in object files and libraries. Section 11.2 details some of these conventions.

Continuing with Figure 11.1, the language-*A* stub invokes the language-*B* skeleton, which addresses additional compiler-specific handling of argument and return-value data representations, specifically the layout of Fortran-derived types, assumed-shape arrays, and pointers thereto. For some argument and return types, no skeleton need be constructed. Furthermore, in some cases, the compiler can inline the stub, incorporating the associated executable code directly into that associated with the calling routine. With an inlined stub and no skeleton, the overhead associated with *A*-*B* interoperability is minimal.

When language syntax and semantics lead to ambiguities in transferring data to another language, Chasm requires the user to tweak some of the generated XML

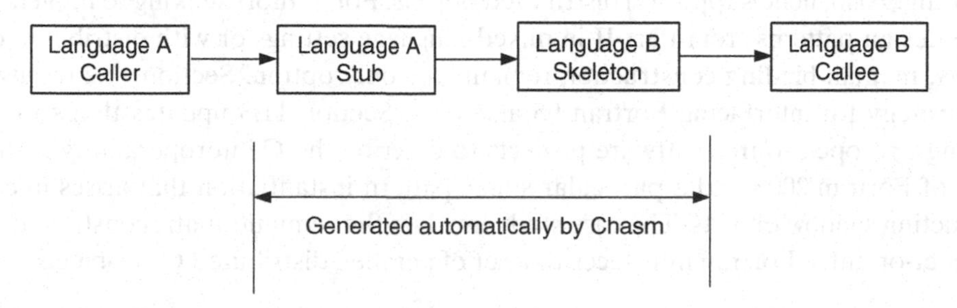

Figure 11.1. Chasm stub and skeleton interfaces for language A procedure invoking language B procedure.

[4] http://www.gnu.org
[5] http://www.cmake.org

manually. An important case involves pointers in C and C++. An argument listed as a pointer to a C/C++ value, for example, an `int*` or `double*`, might resolve to a single value passed by reference in Fortran or to a 1D array. Knowledge about the development conventions of a particular project can assist in resolving such ambiguities, as Section 11.3 discusses.

According to Rasmussen et al. (2006), Chasm does not support distributed objects. This limits its utility for the high-performance computing applications that dominate much of scientific programming. Furthermore, its most recently published version precedes the December 2008 release of the first fully compliant Fortran 2003 compiler by at least two years. Chasm is therefore unlikely to support the OOP constructs of Fortran 2003.

The Babel tool described by Allan et al. (2006) takes the second of the aforementioned two paths to reducing interoperability complexity: eliminating the quadratic explosion of dialect-pairings by inserting a MEDIATOR. A Babel user writes an interface file in a specification language: the Scientific Interface Definition Language (SIDL) described by Dahlgren et al. (2009). Babel uses a single SIDL interface to generate language-specific bindings for C, C++, Fortran 77, Fortran 90, Python, and Java pairings.

As its name implies, SIDL is tailored to scientific programming. For example, it supports a rich set of mathematical constructs, including, for example, dynamically, allocated, multidimensional, complex arrays of non-unit stride. A desire to support scientific computing also sets the agenda for future Babel development. The Babel team cites support for massively parallel, distributed objects as a priority for future development.

Babel mediates communications between languages via an intermediate object representation (IOR) written in ANSI C. Babel enables OOP in each supported language, including languages that do not provide native support for OOP. In each dialect, Babel enables inheritance, polymorphism, encapsulation, and object identity. For example, Babel implements object identity via struct pointers in C, via 64-bit integers in Fortran 77, and via derived types with 64-bit integer components in Fortran 90/95. With Babel, one could implement each level of an inheritance hierarchy in a different language.

Owing to the recent advent of compiler support for Fortran 2003, none of the above tools supported object-oriented Fortran 2003 constructs as of the time of this writing. Also, none supported distributed objects. For authors seeking to implement the design patterns from Part II in mixed-language settings or with distributed objects, manual binding construction remains the only option. Section 11.2 discusses a strategy for interfacing Fortran 95 and C++. Section 11.3 updates that strategy using two open-source software projects to describe the C interoperability feature set of Fortran 2003 and a particular SHELL pattern instantiation that arises in constructing shadow objects. The case study involves the semiautomatic construction of object-oriented Fortran interfaces to a set of parallel, distributed C++ objects.

11.2 Manual Interoperability: C++/Fortran 95

The remainder of this chapter outlines a strategy for wrapping object-oriented C++ with object-oriented Fortran 2003 using the C interoperability constructs in Fortran

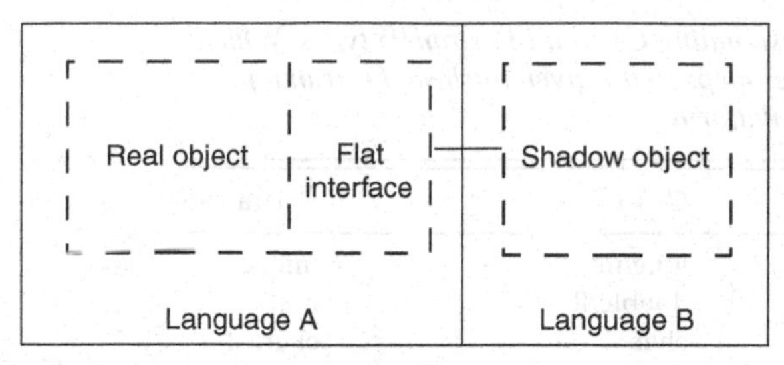

Figure 11.2. A flat interface exported to a shadow object in language B by a real object in language A.

2003. The strategy builds on and updates the approach Gray et al. (1999) outlined for interfacing object-based Fortran 95 with object-oriented C++.Gray and colleagues aimed to make user-defined types in Fortran available in C++ and vice versa. In their approach, a real object in one language exports its behavior to the other language via a flat interface consisting of intrinsic types and 1D arrays thereof. The original object exports its identity and state to the second language via a "shadow" object that presents a "logical" interface. Figure 11.2 illustrates this architecture. Through use of the logical interface, the shadow object may appear as a native object to a user in language B.

Gray and colleagues distinguish an object's logical interface from its physical interface. In their schema, constructing a physical interface requires three steps:

1. Unmangling procedure names,
2. Flattening interfaces with interoperable built-in data types,
3. Ensuring code initialization.

We explain these steps next.

Unmangling reverses the name-mangling practices compilers use to map global identifiers in source code to unique identifiers in object code. The root cause for name mangling stems from the C++ and Fortran languages' allowing certain global identifiers to have the same name. For example, C++ allows two procedures to have the same name while being distinguishable by only their signatures, and Fortran allows different modules to have the same module procedure names. Compilers generate these global names and pass them to linkers as entries in a symbol table. Linkers generally reject duplicate symbols in producing the object or executable code. Thus, a mangling scheme enables successful linking by eliminating symbol duplication.

C++ compilers mangle all procedure names by default. In Fortran, the most common global identifiers are procedure names (external procedures or module procedures)[6]. However, Fortran restricts the name conflicts on global identifiers used in a program so that only module procedures require name mangling.

Name-mangling rules are not standardized, which results in different name-mangling schemes used in different compilers. Therefore, object codes produced by different compilers are not linkable even on the same platform. C++ provides

[6] Internal procedure names in Fortran are not global identifiers. Also generic names in Fortran are not global identifiers.

Table 11.1. *Corresponding C++ and Fortran 95 types: Which primitive C++ types maps to the given intrinsic Fortran type depending on the platform*

Type	C++	Fortran 95
Integer	long/int	integer
Real	double/float	real
Character	char	character

the `extern "C"` construct to unmangle the procedure name, assuming there is no function overloading in the program, thus enabling interfacing to C, which disallows function overloading. In Fortran, one can use external procedures to avoid name mangling. Since Fortran is case-insensitive, it further requires C++ to use lower case in procedure names. Gray and colleagues took this approach to unmangle the procedure names: Their physical interfaces consisted of a set of C++ `extern "C"` procedures in lower case and Fortran 95 external procedures. Since the Fortran 2003 standard introduced features to enhance interoperability with C, we can update the Gray and colleagues unmangling scheme by requiring C bindings on Fortran procedures in the physical interface. Section 11.3.1 describes the mechanics of this process.

Flattening (interfaces), the second step in the Gray et al. strategy, is required for physical interfacing through nonmember functions in C++ or external procedures in Fortran 95. Flattening requires passing only intrinsic data types that are common to C++ and Fortran 95. Gray et al. suggest that the first types listed in Table 11.1 match Institute of Electrical and Electronics Engineers (IEEE) standards on most platforms, even though the C++ and Fortran 95 standards do not guarantee matching representations. The second types listed in the C++ column, however, might match the corresponding type in the Fortran column on other platforms. Section 11.3.1 eliminates this ambiguity by leveraging the Fortran 2003 standard's interoperable data types.

In addition, Gray et al. limit array arguments to be contiguous, 1D arrays in the flattened physical interface because of the semantic disparity in treating arrays between the two languages. In Fortran 95, a multidimensional array has its rigid shape and its elements can only be accessed via array subscripts. In C++, a multidimensional array is treated the same as an 1D array, and the elements can be accessed via a pointer using pointer arithmetic. Furthermore, pointer arrays and assumed-shape arrays in Fortran can represent noncontiguous portions of a data array. This concept is not naturally supported in C++.[7] Thus it is reasonable to the restrict the array arguments to the contiguous, 1D arrays.

Since Fortran 95 passes all arguments by reference, Gray et al. further require all C++ arguments to be passed as pointers or references. With this approach, all parameters passed by value in the C++ logical interface must be passed by reference in the physical interface. One benefit of this calling convention is that it can be extended to support code that conforms to earlier Fortran standards, for example

[7] Currently Fortran language committee, J3, is working on a TR to provide utilities for C programmers to access Fortran's assumed-shape and pointer arrays, and so on.

Fortran 77. Nevertheless, one can eliminate the need for this convention in Fortran 2003 by leveraging that standard's provision for passing by value. Section 11.3.1 explains further.

In applications, interface flattening may expose safety and portability issues. One issue Gray et al. have to address is the `this` pointer passed implicitly to all nonstatic member functions. By the rules of interface flattening, the `this` pointer has to be passed using a reference to a `long` by C++ and an `integer` by Fortran 95 in the physical interface. This raises concerns as to its safety and portability. To address these issues, Gray et al. chose to use an opaque pointer in encapsulating the details of casting between the `this` pointer and a `long` or `integer` formal parameter in the calls. In addition, Gray et al. strongly recommended automating the conversion process from the logical interface seen by end-users to its corresponding physical interface, which is normally hidden from the same users.

Initialization of data, the last step in the Gray et al. strategy, relates to the underlying operating systems (OS) and has been a long-standing portability issue for programs written in C, C++, and Fortran. The issue almost exclusively concerns the `extern const` and `static` objects in C/C++, and objects with the `save` attribute in Fortran. In C++ `extern const` and `static` objects can be explicitly initialized. Similarly, a `save` object in Fortran can be explicitly initialized either by its declaration statement or by a separate `data` statement. These initialized data objects are stored in a `.data` segment and are all initialized before `main` program starts executing. However neither C++ nor Fortran stipulate the explicit initializations for these objects, that is, `extern const`, `static`, or `save` objects can be declared without being initialized in their native languages. These uninitialized objects are inserted into a `.bss` segment, whose initialization is dependent on the OS. These `.bss` objects may cause porting issues. For example, a programmer who solely develops on an `ELF`-based OS may comfortably leave all the `static` variables uninitialized because the operating system mandates data in `.bss` be filled up with zeros at startup time. However, on other systems, such as `XCOFF`-based AIX, these objects are left with uninitialized states.

In their strategy, Gray et al. built code using shared libraries on `ELF`-based systems. They reported success with this technique on Linux, Sun, and SGI platforms but not on IBM's AIX machines.

In the Gray et al. approach, the logical interface uses the the physical interfacing techniques to enable the shadow-object relationship of Figure 11.2. The logical interface consists of C++ class definitions or Fortran 95-derived type definitions in a module that a programmer would see naturally in her native language. We illustrate this approach by using the `astronaut` example from Section 2.2. Let's assume the `astronaut` class in Figure 2.4(b) has to be implemented in Fortran 95. Thus the logical interfaces in Fortran 95 for `astronaut` class would be the following:

```
module astronaut_class
  implicit none
  public :: astronaut
  type astronaut
     private
     character(maxLen), pointer :: greeting
```

```
end type

  interface astronaut
    module procedure constructor
  end interface

  interface destruct
    module procedure free_greeting
  end interface

  contains

  function constructor(new_greeting) result(new_astronaut)
    type(astronaut) :: new_astronaut
    character(len=*), intent(in) :: new_greeting
    ! ...
  end function

  subroutine greet(this)
    type(astronaut), intent(in) :: this
    ! ...
  end subroutine

  subroutine free_greeting (this)
    type(astronaut), intent(inout) :: this
    ! ...
  end subroutine
end module
```

To enable accessing the astronaut class in C++, one might construct the following logical interface in C++:

```
class astronaut
{
  public:
    astronaut (const string &);
    void greet() const;
    ~astronaut();
  private:
    string greeting;
};
```

The physical interface consists of a set of procedures with flat interfaces seen by compilers but not necessarily by application programmers. The Fortran 95 physical interface corresponding to the C++ logical interface comprises the following external procedures:

```
  subroutine astronaut_construct(this, new_greeting)
    integer :: this                     ! this pointer
```

```
    integer, intent(in) :: new_greeting ! points to message
    ! code omitted, use opaque pointers to cast this pointer ...
  end function

  subroutine astronaut_greet(this)
    integer, intent(in) :: this         ! this pointer
    ! code omitted, use opaque pointers to cast this pointer ...
  end function

  subroutine astronaut_free_greeting (this)
    type(astronaut), intent(inout) :: this ! this pointer
    ! code omitted, use opaque pointers to cast this pointer ...
  end subroutine
```

The following lists the C++ physical interface for the `astronaut` class:

```
// in flat interfaces parameter name "self" is used because
// "this" is a reserved keyword in C++
extern "C"
{
    void astronaut_construct(long & self, long & new_greeting);
    void astronaut_greet (long & self);
    void astronaut_free_greeting (long & self);
}
```

Gray et al. has demonstrated this approach can be successfully applied to export the behaviors of a real object from C++ to Fortran 95, and vice versa. As with any other techniques, there are also pitfalls in the strategy. As admitted by Gray et al., this set of physical interfaces using `long` or `integer` to represent C pointers is not type-safe. The outcome could be disastrous if an invalid value is passed in the call. Gray et al. mitigated this safety risk by using an opaque pointer technique, which insulates users from the need to do direct type casting. Today, the data safety issue can be sufficiently addressed by C-interoperability features introduced in Fortran 2003, as Section 11.3.1 discusses.

A separate issue omitted by Gray et al. relates to their method requiring special "glue-code" or "wrapper code" by both C++ and Fortran 95 to wrap the logical interface and generate calls to the physical interfaces. The amount of work required might limit the utility of applying their approach to the large, preexisting C++ and Fortran 95 programs that are already well tested for use within their language of origin.

11.3 Case Study: ForTrilinos and CTrilinos

This section uses examples from the ForTrilinos and CTrilinos open-source software projects to present an updated approach to interfacing object-oriented Fortran and C++. ForTrilinos provides Fortran interfaces to the C++ packages published by the Trilinos project[8]. CTrilinos exports C interfaces to Trilinos packages and serves as an intermediary between ForTrilinos and other Trilinos packages.

[8] http://trilinos.sandia.gov

Heroux et al. (2005) described the goals, philosophy, and practices of the Trilinos project:

> The Trilinos Project is an effort to facilitate the design, development, integration, and ongoing support of mathematical software libraries within an object-oriented framework for the solution of large-scale, complex multiphysics engineering and scientific problems. Trilinos addresses two fundamental issues of developing software for these problems: (i) providing a streamlined process and set of tools for development of new algorithmic implementations and (ii) promoting interoperability of independently developed software.
>
> Trilinos uses a two-level software structure designed around collections of packages. A Trilinos package is an integral unit usually developed by a small team of experts in a particular algorithms area such as algebraic preconditioners, nonlinear solvers, etc. Packages exist underneath the Trilinos top level, which provides a common look-and-feel, including configuration, documentation, licensing, and bug-tracking.

With more than 50 packages in its most recent release, Trilinos might best be described as a meta-project aimed at encouraging consistently professional software engineering practices in scalable, numerical algorithm development projects. Object orientation plays a central role in the Trilinos design philosophy. Hence, when requests came in for Fortran interfaces to Trilinos, which is mostly written in C++, the Trilinos developers assigned a high value to writing object-oriented interfaces while using idioms that feel natural to Fortran programmers.

Figure 11.3 shows the global architecture of a ForTrilinos-enabled application. The top layer comprises application code a user writes by instantiating Fortran-derived type instances and invoking type-bound procedures on those instances. These instances are lightweight in that they hold only identifying information about the underlying Trilinos C++ objects.

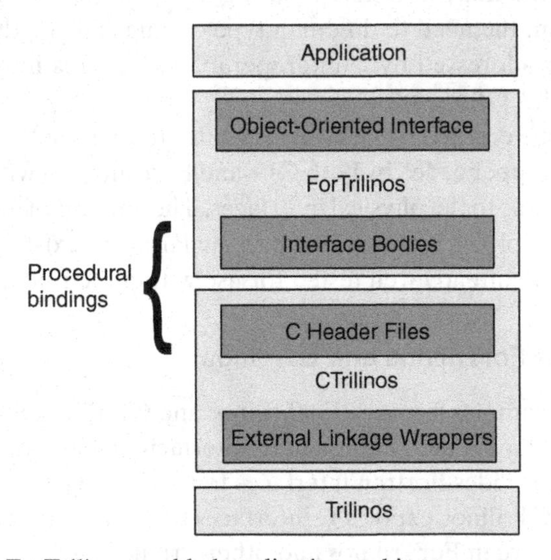

Figure 11.3. Sample ForTrilinos-enabled application architecture.

The next layer down, the ForTrilinos layer, supports a user's application code by defining public, extensible derived types arranged in hierarchies that mirror the hierarchies in the underlying Trilinos packages. As such, the Fortran layer reintroduces properties like polymorphism, inheritance, function overloading, and operator overloading, each of which gets sacrificed internally in the CTrilinos layer due to the C language's lack of support for OOP. That internal sacrifice ensures portability by exploiting the C interoperability constructs in the Fortran 2003 standard to guarantee type compatibility, procedural name consistency, and compatible argument passing conventions between Fortran and C.

Each ForTrilinos method wraps one or more CTrilinos functions and invokes those functions via procedural bindings consisting of Fortran interface bodies and corresponding C header files. CTrilinos produces the ForTrilinos procedural bindings automatically via scripts that parse the underlying C++ and simultaneously generate the CTrilinos header files and the ForTrilinos interface bodies. Although the CTrilinos scripts output C function prototypes for the header files, the actual implementation of each function is written in C++. Embedding the C++ in a `extern "C"{}` constructs provides external linkage – that is, the ability to link with any C procedures that do not use function overloading.

In the bottom layer of Figure 11.3 lie the C++ class implementations defined by various Trilinos packages. This layered design approach relieves the developers of Trilinos numerical packages from any burden of designing for interoperability. In this sense, the ForTrilinos/CTrilinos effort is noninvasive.

Sections 11.3.1–11.3.4 adapt material from Morris et al. (2010), presenting code snippets and class diagrams from ForTrilinos and CTrilinos to demonstrate the C interoperability features of Fortran 2003 in 11.3.1, method invocation in 11.3.2, object construction and destruction in 11.3.3, and mirroring C++ inheritance hierarchies in Fortran in 11.3.4.

11.3.1 C Interoperability in Fortran 2003

The C interoperability constructs of Fortran 2003 ensure the interoperability of types, variables, and procedures. These constructs provide for referencing global data and procedures for access by a "companion processor." That companion processor need not be a C compiler and might even be a Fortran compiler. In the case of a procedure, for example, the Fortran standard merely constrains the procedure to be describable by a C function prototype.

A ForTrilinos/CTrilinos procedural binding example provides a broad sampling of the Fortran 2003 C interoperability features. For that purpose, we consider here one of the Trilinos Petra (Greek for "foundation") packages: Epetra (essential petra). Epetra provides the objects and functionality needed for the development of linear and nonlinear solvers in a parallel or serial computational environment. All Trilinos packages build atop one of the Petra packages, most commonly Epetra, so enabling Fortran developers to build Epetra objects is an essential first step toward enabling their use of other Trilinos packages.

The Epetra MultiVector class facilitates constructing and using dense multi-vectors, vectors, and matrices on parallel, distributed-memory platforms. The original C++ prototype for an `Epetra_MultiVector` method that fills the multi-vector with random elements has the following prototype:

```
class Epetra_MultiVector {
  int Random();
}
```

which CTrilinos scripts parse and automatically export the corresponding C wrapper prototype:

```
int Epetra_MultiVector_Random (CT_Epetra_MultiVector_ID_t selfID);
```

as well as the corresponding ForTrilinos interface block:

```
1 interface
2   integer(c_int) function Epetra_MultiVector_Random ( selfID ) &
3     bind(C,name='Epetra_MultiVector_Random')
4     import :: c_int ,FT_Epetra_MultiVector_ID_t
5     type(FT_Epetra_MultiVector_ID_t), intent(in), value :: selfID
6   end function
7 end interface
```

where the `Epetra_Multivector_Random` interface body demonstrates several useful C interoperability constructs. An examination of this example from left to right and top to bottom drives much of the remainder of this section.

The interface block beginning and ending at lines 1 and 7, respectively, contains one interface body. Most Trilinos interface blocks contain many more. Each Trilinos package contains multiple classes, each of which exposes multiple methods. CTrilinos typically generates one interface block per package. Each block contains the interface bodies for all methods on all classes in the package. The interface body count includes at least one per method and often more if an overloaded method name presents multiple possible combinations of argument types, ranks, and kind parameters. The ForTrilinos Epetra procedural bindings, for example, contain over 800 interface bodies.

The `Epetra_MultiVector_Random` interface body starts at line 2 in the example. That line specifies an `integer` return type with a `c_int` kind parameter that ensures a bit representation matching the `int` type provided by the companion C compiler. Fortran 2003 compilers provide an intrinsic `iso_c_binding` module that contains the `c_int` parameter along with more than 30 other interoperable kind parameters. See Metcalf et al. (2004) for a complete listing. Table 11.2 updates Table 11.1 with types provided by `iso_c_binding`. When a companion compiler fails to supply a C type corresponding to one of the interoperable kind parameters, the parameters take on negative values.

ForTrilinos supplies the `valid_kind_parameters()` function in Figure 11.4 to test several kind parameters and report the meanings of any negative values found. The function returns `.true.` if the tested parameters are all positive and `.false.` otherwise. ForTrilinos also provides the test code in Figure 11.5 to report on the

Table 11.2. *Fortran 2003 kind parameters and the corresponding interoperable C types*

Fortran Type	Fortran 2003 Kind Parameter	C type
integer	c_int	int
integer	c_long	long
real	c_double	double
real	c_float	float
character	c_char	char

`valid_kind_parameters()` result. Trilinos uses the open-source CMake[9] build system's `ctest` facility to run tests such as that in Figure 11.5 in an automated manner, search the results, and report successes and failures.

The statement at line 2 of the `Epetra_MultiVector_Random` interface body continues on line 3, where the `bind` attribute specifies that the referenced procedure is describable by an interoperable C prototype. The `name=` parameter enforces C's case sensitivity during link-time procedure name resolution. Because it is possible for the linked procedure to be implemented in Fortran, this also provides a mechanism for making Fortran case sensitive insofar as procedure names are concerned. The `generalize_all()` procedure in Figure 11.4 provides an example of a Fortran implementation of a `bind(C)` interface body defined elsewhere in ForTrilinos. It also demonstrates a C pointer argument received as a `c_ptr` derived type and converted to a Fortran pointer by the `c_f_pointer()` subroutine. The `iso_c_binding` module provides `c_ptr` and `c_f_pointer()`.

The `Epetra_MultiVector_Random` interface body also ensures the default C behavior of passing of its `FT_Epetra_MultiVector_ID_t` argument.[10] Elsewhere in ForTrilinos appears that type's definition:

```
type, bind(C) :: FT_Epetra_MultiVector_ID_t
  private
  integer(ForTrilinos_Table_ID_t) :: table
  integer(c_int) :: index
  integer(FT_boolean_t) :: is_const
end type
```

where the `bind(C)` attribute ensures interoperability with the CTrilinos struct:

```
typedef struct {
  CTrilinos_Table_ID_t table; /* Table of object references */
  int index;                  /* Array index of the object  */
  boolean is_const;     /* Whether object was declared const */
} CT_Epetra_MultiVector_ID_t;
```

[9] http://www.cmake.org

[10] Although not specifically mandated by any Fortran standard, the default behavior of most Fortran compilers resembles passing arguments by reference, which provides an object's address to the receiving procedure, thereby allowing the receiving procedure to modify the object directly. An example of an alternative approach that simulates the same behavior is the copy-in/copy-out technique that allows for modifying the passed entity without passing its address.

Figure 11.4

```
1  module fortrilinos_utils
2    implicit none
3    private
4    public :: length
5    public :: generalize_all
6    public :: valid_kind_parameters
7  contains
8
9    ! Fortran-style strings take two forms: one with no C counterpart and
10   ! a second that is an array of character values.  A C string is inter-
11   ! operable with the second type of Fortran string with the primary
12   ! distinction being that Fortran strings carry their own string-length
13   ! information, whereas the length of C strings is indicated by a
14   ! terminating null character. The "count" procedure calculates C
15   ! string length by searching for the null terminator.
16
17   integer(c_int) function length(c_string) bind(C,name="forLinkingOnly")
18     use ,intrinsic :: iso_c_binding ,only: c_char ,c_int  ,c_null_char
19     character(kind=c_char) :: c_string(*)
20     length = 0
21     do
22        if(c_string(length+1) == c_null_char) return
23        length = length + 1
24     end do
25   end function
26
27   ! This is a Fortran implementation of the functionality in the
28   ! Epetra_*_Abstract procedures in each CTrilinos/src/CEpetra* file.
29   ! It effectively casts any given Epetra derived type to a general type
30   ! that can represent any of the Epetra derived types in
31   ! ForTrilinos_enums.F90.
32
33    type(ForTrilinos_Universal_ID_t) function generalize_all(object_id) &
34      bind(C,name="for_linking_only")
35      use ForTrilinos_enums ,only : ForTrilinos_Object_ID_t
36      use ,intrinsic :: iso_c_binding ,only: c_ptr,c_f_pointer
37      type(c_ptr) ,value :: object_id
38      type(ForTrilinos_Universal_ID_t), pointer :: local_ptr
39
40      call c_f_pointer (object_id, local_ptr)
41      generalize_all = local_ptr
42    end function
43
44   ! This procedure checks the values of parameters required to interop-
45   ! erate with CTrilinos.  The Fortran 2003 standard requires that these
46   ! parameters be defined in the intrinsic module iso_c_binding with
47   ! values that communicate meanings specified in the standard.  This
48   ! procedure's quoted interpretations of these values are largely ex-
49   ! cerpted from the standard.  This procedure returns true if all of
50   ! the interoperating Fortran kind  parameters required by ForTrilinos
51   ! have a corresponding C type defined by the companion C processor.
52   ! Otherwise, it returns false.  (For purposes of ForTrilinos, the
53   ! Fortran standard's use of the word 'processor' is interpreted as
54   ! denoting the combination of a compiler, an operating system, and
```

```
55  ! a hardware architecture.)
56
57  logical function valid_kind_parameters(verbose)
58    use ,intrinsic :: iso_c_binding ,only : &
59      c_int,c_char,c_double,c_ptr,c_long,c_bool,c_null_char
60    use ,intrinsic :: iso_fortran_env ,only : error_unit ,output_unit
61    logical ,optional :: verbose
62    logical           :: verbose_output
63
64    character(len=*) ,parameter :: no_fortran_kind= &
65      'The companion C processor defines the corresponding C type, &
66      but there is no interoperating Fortran processor kind.'
67    character(len=*) ,parameter :: no_c_type= &
68      'The C processor does not define the corresponding C type.'
69    character(len=*) ,parameter :: interoperable = &
70      'This interoperating Fortran kind has a corresponding C type.'
71    character(len=*) ,parameter :: imprecise= &
72      'The C processor type does not have a precision equal to the &
73      precision of any of the Fortran processor real kinds.'
74    character(len=*) ,parameter :: limited= &
75      'The C processor type does not have a range equal to the range &
76      of any of the Fortran processor real kinds.'
77    character(len=*) ,parameter :: limited_and_imprecise= &
78      'The C processor type has neither the precision nor range of any &
79      of the Fortran processor real kinds.'
80    character(len=*) ,parameter :: not_interoperable_nonspecific = &
81      'There is no interoperating Fortran processor kind for &
82      unspecified reasons.'
83
84    valid_kind_parameters = .true. ! default return value
85
86    if (present(verbose)) then
87      verbose_output=verbose
88    else
89      verbose_output=.false.
90    end if
91
92    select case(c_long)
93      case(-1)
94        write(error_unit ,fmt='(2a)') 'c_long error: ',no_fortran_kind
95        valid_kind_parameters = .false.
96      case(-2)
97        write(error_unit ,fmt='(2a)') 'c_long error: ',no_c_type
98        valid_kind_parameters = .false.
99      case default
100       if (verbose_output) write(output_unit,fmt='(2a)') &
101         'c_long: ',interoperable
102   end select
103
104   select case(c_double)
105     case(-1)
106       write(error_unit ,fmt='(2a)') 'c_double error: ',imprecise
107       valid_kind_parameters = .false.
108     case(-2)
109       write(error_unit ,fmt='(2a)') 'c_double error: ',limited
```

```
110          valid_kind_parameters = .false.
111        case(-3)
112          write(error_unit ,fmt='(2a)') &
113          'c_double error: ',limited_and_imprecise
114          valid_kind_parameters = .false.
115        case(-4)
116          write(error_unit ,fmt='(2a)') &
117          'c_double error: ',not_interoperable_nonspecific
118          valid_kind_parameters = .false.
119        case default
120          if (verbose_output) write(output_unit,fmt='(2a)') &
121          'c_double: ',interoperable
122      end select
123
124      select case(c_bool)
125        case(-1)
126          write(error_unit ,fmt='(a)') 'c_bool error: invalid value for &
127          a logical kind parameter on the processor.'
128          valid_kind_parameters = .false.
129        case default
130          if (verbose_output) write(output_unit ,fmt='(a)') 'c_bool: &
131            valid value for a logical kind parameter on the processor.'
132      end select
133
134      select case(c_char)
135        case(-1)
136          write(error_unit ,fmt='(a)') 'c_char error: invalid value for &
137          a character kind parameter on the processor.'
138          valid_kind_parameters = .false.
139        case default
140          if (verbose_output) write(output_unit ,fmt='(a)') 'c_char: &
141            valid value for a character kind parameter on the processor.'
142      end select
143    end function
144  end module fortrilinos_utils
```

Figure 11.4. ForTrilinos utilities.

```
1   program main
2     use fortrilinos_utils ,only : valid_kind_parameters
3     use iso_fortran_env   ,only : error_unit ,output_unit
4     implicit none
5     if (valid_kind_parameters(verbose=.true.)) then
6       write(output_unit,*)
7       write(output_unit,fmt='(a)') "End Result: TEST PASSED"
8     else
9       write(error_unit,*)
10      write(error_unit,fmt='(a)') "End Result: TEST FAILED"
11    end if
12  end program
```

Figure 11.5. ForTrilinos interoperability test.

where table identifies which of several tables holds a reference to the underlying Trilinos C++ object and index specifies which entry in the table refers to that object. CTrilinos constructs one table for each class in the packages it wraps, and it stores all instances of that class in the corresponding table.

In the CT_Epetra_MultiVector_ID_t struct, the table member is defined as an enumerated type of the form:

```
typedef enum {
  CT_Invalid_ID,         /*does not reference a valid table entry*/
  /* lines omitted */
  CT_Epetra_MultiVector_ID/*references Epetra_MultiVector entry*/
  /* lines omitted */
} CTrilinos_Table_ID_t;
```

which aliases the int type and defines a list of int values ordered in unit increments from CT_Invalid_ID=0. Likewise, ForTrilinos exploits the Fortran 2003 facility for defining an interoperable list of enumerated integers according to:

```
enum ,bind(C)
  enumerator ::                   &
    FT_Invalid_ID,                &
    ! lines omitted
    FT_Epetra_MultiVector_ID, &
    ! lines omitted
end enum
```

which defines the list but does not alias the associated type. Since Fortran provides no mechanism aliasing intrinsic types, ForTrilinos employs the nearest Fortran counterpart: an alias for the relevant kind parameter:

```
   integer(kind(c_int)) ,parameter :: ForTrilinos_Table_ID_t=c_int
```

which takes on the same kind and value as the c_int parameter itself.

The final FT_Epetra_MultiVector_ID_t component is_const, a Boolean value, allows for marking structs that cannot be modified because they were declared as const. Since C's support for a Boolean type only entered the language in its most recent standard (C99), CTrilinos defines its own integer Boolean type for use with earlier C compilers. ForTrilinos defines a corresponding FT_boolean_t integer kind parameter and employs it in defining its multi-vector type.

CTrilinos reserves exclusive rights to modify its struct members even though they have an interoperable representation in ForTrilinos. ForTrilinos honors these rights by giving the FT_Epetra_MultiVector_ID_t components private scope within a module that contains only type definitions and parameter initializations.

Finally, line 4 of the Epetra_MultiVector_Random interface body imports entities defined in the host module scoping unit. The Fortran standard does not make names from a host scoping unit available within an interface body. Hence, although each ForTrilinos procedural binding module contains use statements of the form:

```
  use ,iso_c_binding ,only : c_int
  use ,ForTrilinos_enums ,only : FT_Epetra_MultiVector_ID_t
```

Table 11.3. *Dummy array declaration interoperability (N/A=none allowed): the interoperability restrictions apply to the dummy (received) array, not to the actual (passed) array*

Array category	Fortran declaration	Interoperable C declaration
Explicit-shape	real(c_float) :: a(10,15)	float a[15][10];
Assumed-size	real(c_float) :: b(10,*)	float b[ ][10];
Allocatable	real(c_float),allocatable:: c(:,:)	N/A
Pointer	real(c_float),pointer:: d(:,:)	N/A
Assumed-shape	real(c_float) :: e(:,:)	N/A

the entities these statements make accessible to the surrounding `module` are not automatically accessible within the interface bodies in that `module`. The `import` statement brings the named entities into the interface body scoping unit. This completes the discussion of the `Epetra_MultiVector_Random` binding.

An important interoperability scenario not considered up to this point involves the interoperability of array arguments. Whereas Gray et al. (1999) mapped all array arguments to their flattened, 1D equivalents, Fortran 2003 facilitates passing multidimensional arrays. Standard-compliant compilers must copy passed data into and out of `bind(C)` procedures in a manner that ensures contiguous memory on the receiving side even when, for example, the actual argument passed is an array pointer that references a noncontiguous subsection of a larger array. The restrictions in doing so relate primarily to how the interface body for receiving procedure must declare its array arguments. Only arrays declared on the receiving end as explicit-shape or assumed-size are interoperable. Table 11.3 delineates the interoperable and noninteroperable declarations.

An important special case concern character array arguments. Metcalf et al. (2004) state:

> In the case of default character type, agreement of character length is not required ... In Fortran 2003, the actual argument corresponding to a default character (or with kind [`c_char`]) may be any scalar, not just an array element or substring thereof; this case is treated as if the actual argument were an array of length 1 ...

where the bracketed text corrects a typographical error in the original text. Passing a `character` actual argument represents a special case of what Fortran 2003 terms *sequence association*, which is an exception to the array shape rules that otherwise apply. The standard allows sequence association only when the dummy argument (declared in the receiving procedure interface) references contiguous memory – that is an explicit-shape or assumed-size arrays. In such cases, the actual (passed) argument need not have a rank and a shape matching those of the dummy (received) argument. The actual argument rank and shape only need to provide at least as many elements as required by the dummy argument.

As an extreme scenario, an actual argument can even be an array element, in which case the dummy takes the rest of the array as input. Although originally intended to provide backward compatibility with Fortran 90, which introduced assumed-shape arrays, this feature now finds new legs in a C-interoperability world.

Figure 11.4 defines a `length()` procedure that makes use of the aforementioned rank and shape flexibility. That procedure returns the character count of null-terminated C strings. The calling procedure need not pass an actual argument with a rank and a shape matching those of the dummy argument `c_string`.

11.3.2 Method Invocation

Trilinos exploits several C++ capabilities not supported by C. Among these are object method invocation; inheritance; polymorphism; overloaded function names and operators; default argument values; class and function templates; string and bool types; exception handling; and safe type-casting. Consequently, the C interface to Trilinos proves rather awkward from a C++ viewpoint. The developers pay this price for standards conformance but hide the inelegant features in the intermediate layers of Figure 11.3. We describe next how CTrilinos copes with the missing C++ capabilities and ForTrilinos restores similar capabilities using natural Fortran forms.

Within C++, one can dereference a pointer to an object to invoke a method on the object directly. A pointer `mv` to an `Epetra_MultiVector`, for example, can be used to invoke the Random() method as in the following pseudocode:

```
Epetra_MultiVector *mv = new Epetra_MultiVector(...);
mv->Random();                 // arguments omitted above
```

wherein the ellipses stand in for arguments not shown in this snippet. In implementing CEpetra, the CTrilinos wrapper for Epetra, however, a dilemma arises from C's ability to receive, pass, and return pointers to C++ classes but C's inability to directly invoke methods on the object the pointer references. One could resolve this dilemma by wrapping the above code in C++ helper functions and embedding such functions in an `extern''C''{}` construct. That approach, however, would require manipulating raw pointers throughout CTrilinos, leaving the code vulnerable to a variety of pitfalls ranging from dangling pointers to memory leaks.

To circumvent the pitfalls associated with the direct manipulation of pointers, CTrilinos grants users no direct access to the object pointer. Instead, users reference objects by a handle that the developers refer to as a struct identification tag, or a *struct ID*,. The aforementioned `CT_Epetra_MultiVector_ID_t` exemplifies a struct ID. Struct IDs contain identifying information that users should never directly modify.

Given a struct ID `mv_id`, the multivector randomize method invocation of section 11.3.1 becomes:

```
Epetra_MultiVector_Random(mv_id);
```

Likewise, although the Fortran 2003 standard defines a `c_ptr` derived type specifically for passing C pointers between Fortran and C, ForTrilinos accesses Trilinos via CTrilinos, so Fortran users never directly manipulate the underlying Trilinos object pointers.

ForTrilinos wraps each struct ID in a SHELL, names the shell type after the underlying Trilinos class, and publishes type-bound procedures named after the type. Figure 11.6 illustrates the Fortran Epetra multivector construction code and randomization method call is thus:

```
type(epetra_multivector) :: mv
mv  = epetra_multivector(...)
call mv%Random()
```

which is the most natural Fortran equivalent to the original C++, and where ellipses again mark arguments omitted from this pseudocode.

CTrilinos tabulates each struct a CTrilinos function returns. Recording the struct in a table facilitates passing it into the appropriate destruction function, which deletes the underlying object and its struct ID. Compile-time type checking is enforced by providing each wrapped class with its own custom (but internally identical) struct ID.

Figure 11.6

```
1  module Epetra_MultiVector_module ! ---------- Excerpt --------------
2    use ForTrilinos_enums ,only: FT_Epetra_MultiVector_ID_t,&
3      FT_Epetra_Map_ID_t,ForTrilinos_Universal_ID_t
4    use ForTrilinos_table_man
5    use ForTrilinos_universal,only:universal
6    use ForTrilinos_error
7    use FEpetra_BlockMap  ,only: Epetra_BlockMap
8    use iso_c_binding     ,only: c_int,c_double,c_char
9    use forepetra
10   implicit none
11   private                        ! Hide everything by default
12   public :: Epetra_MultiVector ! Expose type/constructors/methods
13   type ,extends(universal)         :: Epetra_MultiVector !"shell"
14     private
15     type(FT_Epetra_MultiVector_ID_t) :: MultiVector_id
16   contains
17      procedure :: cpp_delete => ctrilinos_delete_EpetraMultiVector
18      procedure         :: get_EpetraMultiVector_ID
19      procedure ,nopass :: alias_EpetraMultiVector_ID
20      procedure         :: generalize
21      procedure         :: Random
22      procedure         :: Norm2
23      procedure         :: NumVectors
24   end type
25
26    interface Epetra_MultiVector ! constructors
27      module procedure from_scratch,duplicate,from_struct
28    end interface
29
30 contains
31
32   type(Epetra_MultiVector) function from_struct(id)
33      type(FT_Epetra_MultiVector_ID_t) ,intent(in) :: id
34      from_struct%MultiVector_id = id
35      call from_struct%register_self
36   end function
37
```

```
38    type(Epetra_MultiVector) function from_scratch( &
39       BlockMap,Num_Vectors,zero &
40    )
41     use ForTrilinos_enums ,only: FT_boolean_t,FT_TRUE,FT_FALSE
42     use iso_c_binding     ,only: c_int
43     class(Epetra_BlockMap) ,intent(in) :: BlockMap
44     integer(c_int)         ,intent(in) :: Num_Vectors
45     logical                ,intent(in) :: zero
46     integer(FT_boolean_t)              :: zero_in
47     type(FT_Epetra_MultiVector_ID_t)   :: from_scratch_id
48     if (zero) zero_in=FT_TRUE
49     if (.not.zero) zero_in=FT_FALSE
50     from_scratch_id = Epetra_MultiVector_Create( &
51       BlockMap%get_EpetraBlockMap_ID(),Num_Vectors,zero_in &
52     )
53     from_scratch = from_struct(from_scratch_id)
54    end function
55
56    type(Epetra_MultiVector) function duplicate(this)
57      type(Epetra_MultiVector) ,intent(in) :: this
58      type(FT_Epetra_MultiVector_ID_t) :: duplicate_id
59      duplicate_id = Epetra_MultiVector_Duplicate(this%MultiVector_id)
60      duplicate = from_struct(duplicate_id)
61    end function
62
63    type(FT_Epetra_MultiVector_ID_t) function get_EpetraMultiVector_ID( &
64       this &
65    )
66      class(Epetra_MultiVector) ,intent(in) :: this
67      get_EpetraMultiVector_ID=this%MultiVector_id
68    end function
69
70    type(FT_Epetra_MultiVector_ID_t) function alias_EpetraMultiVector_ID( &
71      generic_id &
72    )
73      use ForTrilinos_table_man,only: CT_Alias
74      use iso_c_binding        ,only: c_loc,c_int
75      use ForTrilinos_enums    ,only: ForTrilinos_Universal_ID_t, &
76                                      FT_Epetra_MultiVector_ID
77      type(ForTrilinos_Universal_ID_t) ,intent(in) :: generic_id
78      type(ForTrilinos_Universal_ID_t) ,allocatable ,target :: alias_id
79      integer(c_int) :: status
80      type(error) :: ierr
81      allocate(alias_id,source=CT_Alias( &
82        generic_id,FT_Epetra_MultiVector_ID),stat=status &
83      )
84      ierr=error(status,'FEpetra_MultiVector:alias_EpetraMultiVector_ID')
85      call ierr%check_success()
86      alias_EpetraMultiVector_ID=degeneralize_EpetraMultiVector( &
87        c_loc(alias_id) &
88      )
89    end function
90
91    type(ForTrilinos_Universal_ID_t) function generalize(this)
92     use ForTrilinos_utils ,only: generalize_all
```

```
93     use iso_c_binding     ,only : c_loc
94     class(Epetra_MultiVector) ,intent(in) ,target :: this
95     generalize = generalize_all(c_loc(this%MultiVector_id))
96    end function
97
98    type(FT_Epetra_MultiVector_ID_t) function &
99      degeneralize_EpetraMultiVector(generic_id) bind(C)
100     use ForTrilinos_enums ,only : ForTrilinos_Universal_ID_t, &
101                                   FT_Epetra_MultiVector_ID_t
102     use ,intrinsic :: iso_c_binding ,only: c_ptr,c_f_pointer
103     type(c_ptr)                        ,value   :: generic_id
104     type(FT_Epetra_MultiVector_ID_t) ,pointer :: local_ptr=>null()
105     call c_f_pointer (generic_id, local_ptr)
106     degeneralize_EpetraMultiVector = local_ptr
107   end function
108
109   subroutine Random(this,err)
110    class(Epetra_MultiVector) ,intent(inout) :: this
111    type(error)     ,optional ,intent(out)   :: err
112    integer(c_int)                           :: error_out
113    error_out = Epetra_MultiVector_Random (this%MultiVector_id)
114    if (present(err)) err=error(error_out)
115   end subroutine
116
117   function Norm2(this,err) result(Norm2_val)
118     class(Epetra_MultiVector)    ,intent(in) :: this
119     type(error)     ,optional    ,intent(out) :: err
120     real(c_double) ,dimension(:),allocatable :: Norm2_val
121     integer(c_int)                           :: error_out
122     integer(c_int)                           :: status
123     type(error)                              :: ierr
124     if (.not.allocated(Norm2_val)) then
125       allocate(Norm2_val(this%NumVectors()),stat=status)
126       ierr=error(status,'FEpetra_MultiVector:Norm2')
127       call ierr%check_success()
128     endif
129     error_out = Epetra_MultiVector_Norm2(this%MultiVector_id,Norm2_val)
130     if (present(err)) err=error(error_out)
131   end function
132
133   integer(c_int) function NumVectors(this)
134     class(Epetra_MultiVector) ,intent(in) :: this
135     NumVectors=Epetra_MultiVector_NumVectors(this%MultiVector_id)
136   end function
137
138   subroutine ctrilinos_delete_EpetraMultiVector(this)
139     class(Epetra_MultiVector),intent(inout) :: this
140     call Epetra_MultiVector_Destroy( this%MultiVector_id )
141   end subroutine
142
143 end module
144
```

Figure 11.6. ForTrilinos MultiVector class excerpt.

11.3.3 Tabulation, Construction, and Destruction

CTrilinos wraps constructors for each supported class. Every wrapped class is associated with a pair of tables of templated class CTrilinos_Table. Each table entry is a RCP to an object created by the Trilinos C++ code. The reference-counting ability does not extend beyond CTrilinos and cannot be utilized by applications. Application code written in C must use a type-specific struct ID, such as the aforementioned CT_Epetra_MultiVector_ID_t, to track an object via its table entry. Each type-specific struct ID has an identical form to the following generic struct ID:

```
typedef struct {
  CTrilinos_Table_ID_t table;
  int index;
  boolean is_const;
} CTrilinos_Universal_ID_t;
```

where the struct members have the same meanings as in the aforementioned Epetra_MultiVector_ID_t type and where ForTrilinos defines a corresponding interoperable ForTrilinos_Universal_ID_t. Application code written in Fortran would use the extensible, native derived type rather than the interoperable struct ID.

The CTrilinos table design prevents catastrophic runtime errors by catching attempts to access invalid table entries. Although CTrilinos uses RCPs to decide when to free the underlying object, a reference within a table will prevent the reference count from hitting zero; therefore, a CTrilinos user must explicitly remove objects from the tables when they are no longer needed so that the underlying object can be destroyed. CTrilinos assumes ownership of the lifetimes of objects created with wrapped constructors. CTrilinos destroys such an object once no table entries reference the object.

For every class CTrilinos wraps, it publishes a destruction procedure that can be invoked in the manner:

```
Epetra_MultiVector_Destroy(&mv_id);
```

Such a procedure removes the object reference from the specified table entry and then destroys the C++ object if there are no remaining references to it in the table. Since the struct ID is the only way CTrilinos users can access the C++ object, prevention of memory leaks necessitates that explicit invocation of the destruction procedure before the object goes out of scope.

As a direct consequence of supporting object access only through type-specific tables, error-free use of CTrilinos requires that application code never duplicate a struct ID or pass the ID by value to a function that requests the destruction of the underlying object. Were a user to duplicate an ID and later pass it to a destroy function, the one remaining table entry associated with the original object would refer to a nonexistent object. Future use of that table entry would result in unpredictable behavior and potential access violations. Certain procedures, however, may require creating a new ID struct for an object. In such cases, both struct IDs should be deleted once the underlying object has been destroyed.

ForTrilinos automates the CTrilinos destruction call. This automation requires navigating several subtleties arising from the interplay between multiple languages. These subtleties crop up even in the simple construction pseudocode on page 270, for example. In the statement "`mv=Epetra_MultiVector(...)`", the constructor returns a new object that becomes the RHS argument of a defined assignment. When the assignment terminates, its RHS goes out of scope with no remaining references to it, at which time a standard-compliant compiler must invoke the corresponding object's final subroutine. ForTrilinos final subroutines invoke CTrilinos destruction functions, which in turn delete the underlying Trilinos C++ object and any related table entries. This cascade of calls catastrophically impacts future use of the LHS argument into which the assignment would have just copied the RHS information, causing an existential crisis: The construction process destroys the constructed object!

One way to circumvent premature destruction relies upon counting references to ForTrilinos objects so the destruction of Trilinos C++ objects occurs only after all the corresponding ForTrilinos objects are out of scope. Figure 11.7 shows the UML class diagram for a ForTrilinos OBJECT hierarchy that provides such a service. Figure 11.8 shows the implementation. All ForTrilinos types extend the shown Universal class, which aggregates a Ref_Counter named `ref_count`. The `ref_count` component serves a bookkeeping role for all objects that share the same CTrilinos struct ID. The Hermetic class serves as a SURROGATE for Universal when aggregated into the Ref_Counter.

Similar to reference counting in C++, reference counting in ForTrilinos relies on a language capability to override the behaviors of object construction, assignment, and destruction provided by compilers. First, an object requiring a unique struct ID has to be constructed using `ref_count`'s overloaded constructor in order to participate in reference counting. Also, a type-bound assignment overrides the default assignment, so a reference-counted LHS object does not automatically get finalized on object assignment. Finally, the finalizer of this type calls `release()` to decrease the reference count by one.

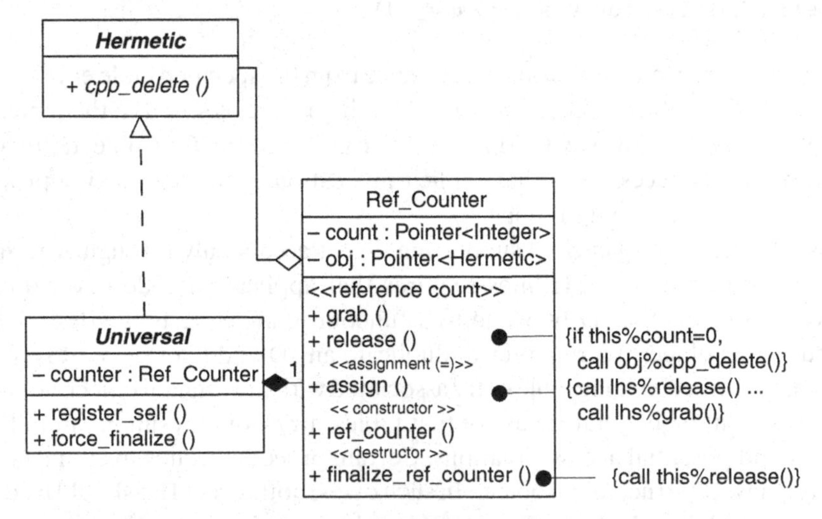

Figure 11.7. ForTrilinos OBJECT hierarchy, including the Hermetic SURROGATE.

The grab() and release() methods respectively increment and decrement the reference count. When the count reaches zero, release() calls the CTrilinos destruction procedure for the underlying C++ object. To achieve this, ref_counter keeps a reference to the object using the obj component, which is of Hermetic type. Hermetic provides a hook for ref_counter to use to invoke the CTrilinos destruction functions. As shown in Figure 11.8(b), the abstract Hermetic class only defines the interface for invoking the CTrilinos destruction procedures. Hermetic delegates to subclasses the implementation of that interface, including the CTrilinos destruction function invocation.

Figure 11.8(c) provides the Universal implementation. All ForTrilinos objects must invoke their inherited register_self() method to register themselves to participate in reference counting before they can be used – that is, after instantiation of a Trilinos C++ object and its ForTrilinos shadow and assignment of the shadow's struct ID, the shadow object must call register_self(). Afterward, the shadow can be used freely in assignments and other operations during which all new references to this object are counted automatically by ref_counter. Universal also provides a force_finalize() method for programmers to manually invoke the release() method of ref_counter. This method is needed *only* for ForTrilinos objects created in the main program or with the save attribute, for which the language does not mandate automatic object finalizations and the programmer must therefore manually release the object's memory.

Fortran provides no facility for overriding its default copying behavior.[11] The aforementioned reference-counting scheme therefore requires no object copying. Thus, ForTrilinos objects cannot be used in places, such as array constructors, where object copying is required.

The need for reference counting in Fortran is unique to the mixed-language programming paradox: An object's creation and deletion in Fortran is constrained by the underlying C++ object's creation and deletion. In the vast majority of cases, developers writing solely in Fortran avoid object creation, deletion, and finalization concerns altogether by using allocatable components instead of pointers and shadows. This practice obligates standard-compliant compilers to automatically deallocate the memory associated with such components when the encompassing object goes out of scope.

Figure 11.8(a):ref_count

```
1  module ref_counter_module
2    use hermetic_module, only : hermetic
3    private
4    public :: ref_counter
5    type ref_counter
6        private
7        integer, pointer :: count => null()
8        class(hermetic), pointer :: obj => null()
9    contains
10       procedure, non_overridable :: grab
```

[11] Fortran 2003 provides no facility analogous to C++ copy constructors.

```
11        procedure, non_overridable :: release
12        procedure :: assign
13        final :: finalize_ref_counter
14        generic :: assignment(=) => assign
15    end type
16
17    interface ref_counter
18        module procedure constructor
19    end interface
20
21 contains
22
23    subroutine grab(this)
24        class(ref_counter), intent(inout) :: this
25        if (associated(this%count)) then
26            this%count = this%count + 1
27        else
28            stop 'Error in grab: count not associated'
29        end if
30    end subroutine
31
32    subroutine release(this)
33        class (ref_counter), intent(inout) :: this
34        if (associated(this%count)) then
35            this%count = this%count - 1
36
37            if (this%count == 0) then
38                call this%obj%cpp_delete
39                deallocate (this%count, this%obj)
40            end if
41        else
42            stop 'Error in release: count not associated'
43        end if
44    end subroutine
45
46    subroutine assign (lhs, rhs)
47        class (ref_counter), intent(inout) :: lhs
48        class (ref_counter), intent(in) :: rhs
49
50        lhs%count => rhs%count
51        lhs%obj => rhs%obj
52        call lhs%grab
53    end subroutine
54
55    recursive subroutine finalize_ref_counter (this)
56        type(ref_counter), intent(inout) :: this
57        if (associated(this%count)) call this%release
58    end subroutine
59
60    function constructor (object)
61        class(hermetic), intent(in) :: object
62        type(ref_counter), allocatable :: constructor
63        allocate (constructor)
64        allocate (constructor%count, source=0)
65        allocate (constructor%obj, source=object)
```

```
66        call constructor%grab
67      end function
68    end module
```

Figure 11.8(b):hermetic

```
1   module hermetic_module
2     private
3     public :: hermetic
4     type, abstract :: hermetic
5     contains
6         procedure(free_memory), deferred :: cpp_delete
7     end type
8
9     abstract interface
10       subroutine free_memory (this)
11          import
12          class(hermetic), intent(inout) :: this
13       end subroutine
14     end interface
15   end module
```

Figure 11.8(c):universal

```
1   module universal_module
2     use hermetic_module ,only : hermetic
3     use ref_counter_module, only : ref_counter
4     implicit none
5
6     type ,abstract ,extends(hermetic) :: universal
7       private
8       type(ref_counter) :: counter
9
10      contains
11      procedure, non_overridable :: force_finalize
12      procedure, non_overridable :: register_self
13    end type
14
15    contains
16
17    subroutine force_finalize (this)
18      class(universal), intent(inout) :: this
19
20      call this%counter%release
21    end subroutine
22
23    subroutine register_self (this)
24      class(universal), intent(inout) :: this
25
26      this%counter = ref_counter(this)
27    end subroutine
28  end module
```

Figure 11.8. Implementation of reference-counted ForTrilinos OBJECT hierarchy.

11.3.4 Polymorphism and Hierarchy Mirroring

Whereas C supports neither polymorphism nor inheritance, Trilinos employs these two technologies ubiquitously. For example, each Trilinos class has one overloaded constructor with an average of six different implementations. In CTrilinos, all wrapped overloaded functions must bear a unique name.

Life without inheritance is nontrivial. Without C support for the inheritance concept, a separate wrapper must be created for each method a child class inherits from its parent classes. For example, a wrapper for the method:

```
int Epetra_MultiVector::Norm2(double * Result) const;
```

will invoke the `Norm2()` method on an `Epetra_MultiVector` instance only. Even though Trilinos publishes an `Epetra_Vector` class that inherits this method, a separate wrapper must be written for `Epetra_Vector` objects.

As another example, a separate wrapper must be created for each combination of argument classes that could result when a child instance is passed where the function prototype specifies a parent instance. Consider the class hierarchy in Figure 11.9 showing that `Epetra_RowMatrix` has four child classes, including `Epetra_BasicRowMatrix`, `Epetra_CrsMatrix`, `Epetra_MsrMatrix`, and `Epetra_VbrMatrix`. Three of these classes also have children. In C++, one can invoke a hypothetical method involving two different `Epetra_RowMatrix` children as follows:

```
Epetra_CrsMatrix *A = new Epetra_CrsMatrix(...);
Epetra_JadMatrix *B = new Epetra_JadMatrix(...);
A->TwoRowMatrixOp(B);
```

A particular method of the `Epetra_RowMatrix` class can take as an argument an instance of any of class that implements the `Epetra_RowMatrix` interface. Therefore, in the specific case of an `Epetra_CrsMatrix` and `Epetra_JadMatrix`, the corresponding

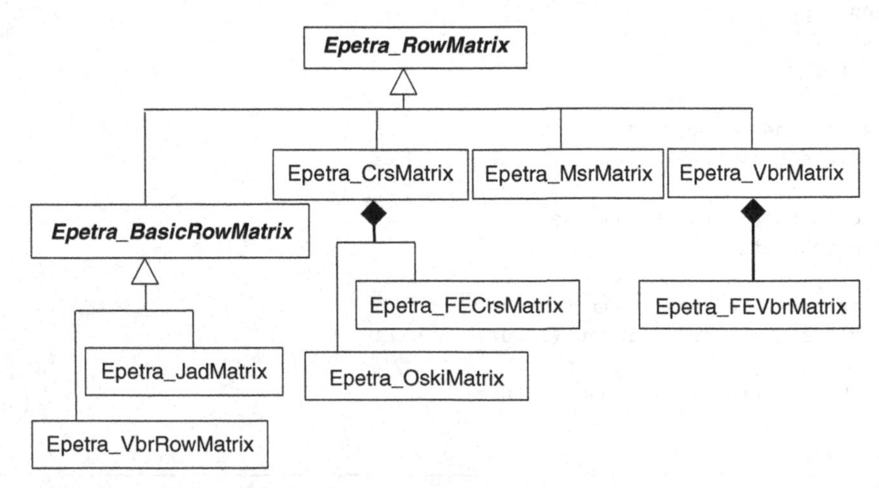

Figure 11.9. Epetra_Row Matrix class hierarchy.

C code would be:

```
CT_Epetra_CrsMatrix_ID_t A;
CT_Epetra_JadMatrix_ID_t B;
A = Epetra_CrsMatrix_Create(...);
B = Epetra_JadMatrix_Create(...);
Epetra_RowMatrix_TwoRowMatrixOp_Crs_Jad(A,B);
```

This simple example requires creating 36 wrappers with unique names in order to allow for all `Epetra_RowMatrix` methods with all possible combinations of types that derive from `Epetra_RowMatrix`. CTrilinos manages inheritance by defining a set of union types corresponding to each of the struct ID types. Within each union, there is a member corresponding to itself, its generic, and each class above it in the inheritance hierarchy. Following this convention, CTrilinos defines the flexible union ID for the `Epetra_CrsMatrix` class as follows:

```
typedef union{
  CTrilinos_Universal_ID_t universal;
  CT_Epetra_CrsMatrix_ID_t Epetra_CrsMatrix;
  CT_Epetra_BLAS_ID_t Epetra_BLAS;
  CT_Epetra_CompObject_ID_t Epetra_CompObject;
  CT_Epetra_DistObject_ID_t Epetra_DistObject;
  CT_Epetra_Object_ID_t Epetra_Object;
  CT_Epetra_Operator_ID_t Epetra_Operator;
  CT_Epetra_RowMatrix_ID_t EPetra_RowMatrix;
  CT_Epetra_SrcDistObject_ID_t Epetra_SrcDistObject;
} CT_Epetra_CrsMatrix_ID_Flex_t;
```

This union type allows CTrilinos to invoke the previously discussed method as:

```
CT_Epetra_CrsMatrix_ID_Flex_t A;
CT_Epetra_JadMatrix_ID_Flex_t B;
A.Epetra_CrsMatrix = Epetra_CrsMatrix_Create(...);
B.Epetra_JadMatrix = Epetra_JadMatrix_Create(...);
Epetra_RowMatrix_TwoRowMatrixOp(
  A.Epetra_RowMatrix,B.Epetra_RowMatrix
);
```

where only one version of the wrapper is needed. This allows safe struct ID type conversion without using explicit type casting in C.

ForTrilinos reintroduces the inheritance and polymorphism that most Trilinos packages employ but CTrilinos cannot. The ForTrilinos `Epetra_Vector` wrapper, for example, extends the `Epetra_MultiVector` derived type and represents the special case of a multivector comprised of one vector. Figure 11.10 excerpts the `Epetra_Vector` implementation.

The `Epetra_Vector` excerpt shows several features common to all ForTrilinos derived types, including the procedures `generalize()`, `degeneralize_*()`, `get_*_ID()`, and `alias_*_ID()`, where the asterisks substitute for derived type names. The object-construction process provides a useful guide to the use of each of these procedures. Consider the following pseudocode:

```
1  program main
2    use Epetra_Vector_module, only : Epetra_Vector
3    implicit none
4    type(Epetra_Vector) :: A
5    A = Epetra_Vector(...)  ! arguments omitted
6  end main
```

where the structure constructor invoked on the RHS of line 5 returns a new `Epetra_Vector` – which has a new Fortran struct ID tag for the underlying C++ object that the ForTrilinos constructor directs CTrilinos to create at lines 55–57 in Figure 11.10. The new struct ID is also passed to the structure constructor of `Epetra_Vector`, `from_struct`, at line 58 to create a `FT_Epetra_MultiVector_ID_t` alias struct ID and construct the new `Epetra_Vector` object's `Epetra_MultiVector` parent component at lines 35–37 in Figure 11.10. Finally, the intrinsic assignment assigns the new `Epetra_Vector` object to LHS and invokes the defined assignment in the `ref_counter` parents component. This completes the construction process.

Creation of the alias struct ID proceeds in three steps. First, starting with the innermost invocation on lines 35–37 of Figure 11.10, `generalize()` converts the `FT_Epetra_Vector_ID_t` struct to the generic `ForTrilinos_Universal_ID_t`. Second, `alias_EpetraMultiVector_ID()` converts the returned generic ID into the parent `FT_Epetra_MultiVector_ID_t`. Third, the `Epetra_MultiVector()` constructor uses the latter ID to construct an `Epetra_MultiVector` that gets assigned to the parent component of the new `Epetra_Vector` at line 35–37.

Lines 70–89 of Figure 11.6 present `alias_EpetraMultiVector_ID()`. At lines 81–83, that procedure passes its `generic_id` argument to the CTrilinos `CT_Alias()` function along with an enumerated value identifying the table of the desired return ID. Lines 86–88 copies that return ID into a newly allocated generic `alias_id`, the C address of which is then passed to the module procedure `degeneralize_EpetraMultiVector()`, which converts `alias_id` into a class-specific struct ID.

Being able to reference the same object by either its own struct ID or that of the parent guarantees the ability to invoke the `Epetra_MultiVector` type-bound procedure `Norm2()` on an `Epetra_Vector` or on an `Epetra_MultiVector`. Thus, the ForTrilinos equivalent of the presented CTrilinos example follows:

```
type(Epetra_CrsMatrix) :: A
type(Epetra_JadMatrix) :: B
A=Epetra_CrsMatrix(...)  ! arguments omitted
B=Epetra_JadMatrix(...)  ! arguments omitted
call A%TwoRowMatrixOp(B)
```

Supporting this syntax requires that `Epetra_RowMatrix_TwoRowMatrixOp()` be wrapped by `TwoRowMatrixOp()`. ForTrilinos uses the type-bound procedure `get_EpetraRowMatrix_ID()` to pass the required arguments to the wrapped function. Consequently, users can pass any `Epetra_RowMatrix` subclass instances to an `Epetra_RowMatrix` method.

In the implementation just described, we have again encountered issues that developers writing solely in Fortran 2003 (or even in C++ in this case) need never confront. The need to explicitly traverse branches of an inheritance hierarchy by converting an instance of a child type to a generic type for subsequent conversion to its parent type arises from using a non-OOP language as an intermediary between two OOP languages. In pure Fortran or C++, one could invoke a method defined in a parent class on an instance of a child class without ever even discovering the instance's dynamic type.

11.3.5 Discussion

Experience with implementing the strategy outlined in this article suggests that it provides a noninvasive approach to wrapping the packages in Trilinos. Developers of individual packages need not concern themselves with designing for interoperability. Several packages have been successfully wrapped without modification to the underlying package.

The strategy has also proven to be less labor intensive and more flexible than prior practices of manually constructing wrappers for individual packages for specific end-users. The labor reduction stems from the possibility of automating portions of the process, namely the generation of procedural bindings consisting of the C header files and the Fortran interface bodies. The increased flexibility is manifold. First, the adherence to language-standard features, designed specifically to promote interoperability, all but guarantees compatible bit representations, linking, and argument-passing conventions. Second, the ability to construct Petra objects, which are used by all Trilinos packages, ensures that wrapping Petra packages immediately decreases the subsequent effort required to wrap new packages. Third, the ability for Fortran programmers to directly control the construction of Epetra objects, in particular, eliminates the need for the developers of individualized wrappers to hardwire assumptions about parameter values into the construction process.

Drawbacks of the above approach include the amount of new code that must be generated and the inability to pass objects directly between Fortran and C++. The desired portability comes at the cost of manipulating only lightweight, Fortran shadows of the underlying C++ objects. Direct manipulation of the objects happens only in C++, and the functionality that supports that manipulation gets exported in C header files, which necessitates sacrificing OOP in the intermediate layer. Fortunately, this sacrifice is purely internal and is not exposed to the end users. Even though there has been some discussion in the Fortran community of adding interoperability with C++ to a future Fortran standard, no formal effort is underway and the standardization of any such effort would likely be many years away.

Figure 11.10

```
1  module Epetra_Vector_module
2    use ForTrilinos_enums    ,only: &
3        FT_Epetra_MultiVector_ID_t,FT_Epetra_Vector_ID_t,&
4        FT_Epetra_BlockMap_ID_t,ForTrilinos_Universal_ID_t,FT_boolean_t
```

```
5   use ForTrilinos_table_man
6   use ForTrilinos_universal
7   use ForTrilinos_error
8   use FEpetra_MultiVector ,only: Epetra_MultiVector
9   use FEpetra_BlockMap    ,only: Epetra_BlockMap
10  use iso_c_binding       ,only: c_int
11  use forepetra
12  implicit none
13  private                      ! Hide everything by default
14  public :: Epetra_Vector ! Expose type/constructors/methods
15  type ,extends(Epetra_MultiVector)      :: Epetra_Vector !"shell"
16    private
17    type(FT_Epetra_Vector_ID_t)  :: vector_id
18  contains
19     procedure          :: cpp_delete => &
20                           ctrilinos_delete_EpetraVector
21     procedure          :: get_EpetraVector_ID
22     procedure ,nopass  :: alias_EpetraVector_ID
23     procedure          :: generalize
24  end type
25
26    interface Epetra_Vector ! constructors
27      module procedure duplicate,from_struct,constructor1
28    end interface
29
30 contains
31
32   type(Epetra_Vector) function from_struct(id)
33     type(FT_Epetra_Vector_ID_t) ,intent(in) :: id
34     from_struct%vector_id = id
35     from_struct%Epetra_MultiVector=Epetra_MultiVector( &
36      from_struct%alias_EpetraMultiVector_ID(from_struct%generalize()) &
37     )
38   end function
39
40   type(Epetra_Vector) function constructor1(BlockMap,zero_initial)
41     use ForTrilinos_enums ,only: &
42       FT_boolean_t,FT_FALSE,FT_TRUE,FT_Epetra_BlockMap_ID_t
43     use FEpetra_BlockMap  ,only: Epetra_BlockMap
44     class(Epetra_BlockMap) ,intent(in) :: BlockMap
45     logical ,optional      ,intent(in) :: zero_initial
46     integer(FT_boolean_t)              :: zero_out
47     type(FT_Epetra_Vector_ID_t)        :: constructor1_id
48     if (.not.present(zero_initial)) then
49      zero_out=FT_FALSE
50     elseif (zero_initial) then
51      zero_out=FT_TRUE
52     else
53      zero_out=FT_FALSE
54     endif
55     constructor1_id = Epetra_Vector_Create( &
56       BlockMap%get_EpetraBlockMap_ID(),zero_out &
57     )
58     constructor1 = from_struct(constructor1_id)
59    end function
```

```
60
61   type(Epetra_Vector) function duplicate(this)
62     type(Epetra_Vector) ,intent(in) :: this
63     type(FT_Epetra_Vector_ID_t) :: duplicate_id
64     duplicate_id = Epetra_Vector_Duplicate(this%vector_id)
65     duplicate = from_struct(duplicate_id)
66   end function
67
68   type(FT_Epetra_Vector_ID_t) function get_EpetraVector_ID(this)
69     class(Epetra_Vector) ,intent(in) :: this
70     get_EpetraVector_ID=this%vector_id
71   end function
72
73   type(FT_Epetra_Vector_ID_t) function alias_EpetraVector_ID(&
74     generic_id &
75   )
76     use iso_c_binding       ,only: c_loc
77     use ForTrilinos_enums   ,only: &
78        ForTrilinos_Universal_ID_t, FT_Epetra_Vector_ID
79     use ForTrilinos_table_man,only: CT_Alias
80     type(ForTrilinos_Universal_ID_t) ,intent(in) :: generic_id
81     type(ForTrilinos_Universal_ID_t) ,allocatable ,target :: alias_id
82     integer(c_int) :: status
83     type(error) :: ierr
84     allocate(alias_id,source= &
85       CT_Alias(generic_id,FT_Epetra_Vector_ID),stat=status &
86     )
87     ierr=error(status,'FEpetra_Vector:alias_Epetra_Vector_ID')
88     call ierr%check_success()
89     alias_EpetraVector_ID=degeneralize_EpetraVector( &
90       c_loc(alias_id) &
91     )
92   end function
93
94   type(ForTrilinos_Universal_ID_t) function generalize(this)
95    use ForTrilinos_utils ,only: generalize_all
96    use iso_c_binding     ,only: c_loc
97    class(Epetra_Vector) ,intent(in) ,target :: this
98    generalize = generalize_all(c_loc(this%vector_id))
99   end function
100
101  type(FT_Epetra_Vector_ID_t) function degeneralize_EpetraVector( &
102    generic_id) bind(C)
103    use ForTrilinos_enums           ,only: &
104       ForTrilinos_Universal_ID_t,FT_Epetra_Vector_ID_t
105    use ,intrinsic :: iso_c_binding ,only: c_ptr,c_f_pointer
106    type(c_ptr)                   ,value   :: generic_id
107    type(FT_Epetra_Vector_ID_t) ,pointer :: local_ptr
108    call c_f_pointer (generic_id, local_ptr)
109    degeneralize_EpetraVector = local_ptr
110  end function
111
112  subroutine ctrilinos_delete_EpetraVector(this)
113    class(Epetra_Vector) ,intent(inout) :: this
114    call Epetra_Vector_Destroy(this%vector_id)
```

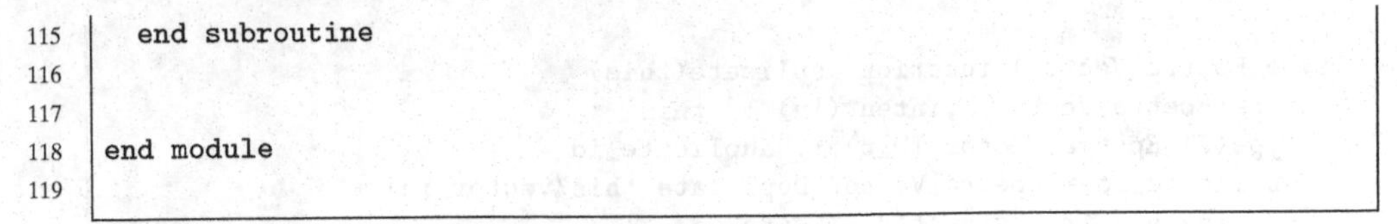

```
115    end subroutine
116
117
118  end module
119
```

Figure 11.10. ForTrilinos Epetra_Vector wrapper excerpt.

EXERCISES

1. Expand the `valid_kind_parameters()` test of Figure 11.4 to include tests for compiler support for the remaining interoperable types in the Fortran 2003 standard. See Metcalf et al. (2004) or the Fortran 2003 standard (a free draft version is available on the Web).
2. Download the Trilinos solver framework from http://trilinos.sandia.gov. Write Fortran 2003 a wrapper module for the `Epetra_VbrMatrix` class.

12 Multiphysics Architectures

"When sorrows come, they come not single spies but in battalions."
William Shakespeare

12.1 Toward a Scalable Abstract Calculus

The canonical contexts sketched in Section 4.3 and employed throughout Part II were intentionally low-complexity problems. Such problems provided venues for fleshing out complete software solutions from their high-level architectural design through their implementation in source code. As demonstrated by the analyses in Chapter 3, however, the issues addressed by OOA, OOD, and OOP grow more important as a software package's complexity grows. Complexity growth inevitably arises when multiple subdisciplines converge into multiphysics models. The attendant increase in the scientific complexity inevitably taxes the hardware resources of any platform employed. Thus, leading-edge research in multiphysics applications must ultimately address how best to exploit the available computing platform.

Recent trends in processor architecture make it clear that fully exploiting the available hardware on even the most modest of computing platforms necessitates mastering parallelism. Even laptop computers now contain multicore processors, and the highest-end machines contain hundreds of thousands of cores. The process of getting a code to run efficiently on parallel computers is referred to as getting a code *to scale*, and code designs that facilitate scaling are termed *scalable*. The fundamental performance question posed by this chapter is whether one can construct a scalable ABSTRACT CALCULUS. The Sundance project (Long 2004) has already answered this question in the affirmative for C++. The Morfeus project (Rouson et al. 2010) aims to provide the first demonstration of a similar capability in Fortran 2003. Whereas this quest remains a subject of ongoing research, the current chapter presents progress to date along with ideas for the path forward.

The holy grail of scalability is automatic parallelization of code without direct programmer intervention. Such automation might happen at the discretion of an optimizing compiler or a linked library. In the absence of automation, parallelization typically happens explicitly at the hands of the programmer. We briefly explore parallelization by automatic means in Section 12.1.2, by compiler directives in Section 12.1.3, by wrapping a library in Section 12.1.4, and by intrinsic language constructs in Section 12.1.5.

One aspect of the holy grail not likely to be addressed in the foreseeable future is algorithmic design. No parallelization tool of which we are aware has the insight to alert a programmer to the availability of a more readily parallelized algorithm than the chosen one. One barrier to fully automated parallelization lies in the difficulty and domain specificity of defining useful high-level abstractions. Building application code atop an ABSTRACT CALCULUS helps in this regard. An ABSTRACT CALCULUS facilitates the specification of a problem's structure at a level high enough to allow great variation in the underlying algorithms. The software expression of the Burgers equation on page 203, for example, conveys the abstract notion that differentiation must be performed without specifying *how* to perform it. The derivative implementation might be serial or parallel with any of several parallelization strategies.

Consider the derivative implementations in Figures 9.2(c) and 9.3(e), which require solving linear systems that yield 6th-order-accurate Padé estimates of the derivatives at each grid point. Portions of the chosen LU factorization and back substitution algorithms are inherently serial. For this reason, we switch to 2nd-order-accurate central differences in Figure 12.1 to take advantage of greater opportunities for parallelism. An exercise at the end of the chapter guides the reader in constructing a new `main` and a `periodic_2nd_factory` to replace the corresponding codes in Figures 9.2(a) and Figures 9.2(e). All other modules from Figures 9.2 can be used untouched with Figure 12.1.

The switch to central differences carries with it a drop in the asymptotic convergence rate and an associated likely drop in accuracy. The accuracy requirements of the problem domain determine the acceptability of any accuracy losses on the given grid. The size of the available hardware determines whether the accuracy can be recovered through grid refinement. The speed of the hardware and efficiency of its utilization determines whether such grid refinement can be accomplished within acceptable solution time constraints. The problem domain thus influences the outcome of efforts to improve scalability.

12.1.1 Amdahl's Law and Parallel Efficiency

As suggested in Section 1.6, we advocate empirically driven optimization. As a useful starting point, one can identify important bottlenecks by measuring the runtime share occupied by each procedure in a representative run of the code. Table 12.1 profiles the runtime share for the dominant procedures in a serial run of the central difference Burgers equation solver compiled without any optimizations. The run goes for 1,000 time steps to render initialization and cleanup costs negligible relative to the cost of marching forward in time.

Figure 12.1

```
1 module periodic_2nd_order_module
2   use field_module ,only : field,initial_field
3   use kind_parameters ,only : rkind,ikind
```

```
4    implicit none
5    private
6    public :: periodic_2nd_order, constructor
7
8    type ,extends(field) :: periodic_2nd_order
9      private
10     real(rkind) ,dimension(:) ,allocatable :: f
11   contains
12     procedure :: add => total
13     procedure :: assign => copy
14     procedure :: subtract => difference
15     procedure :: multiply_field => product
16     procedure :: multiply_real => multiple
17     procedure :: runge_kutta_2nd_step => rk2_dt
18     procedure :: x  => df_dx   ! 1st derivative w.r.t. x
19     procedure :: xx => d2f_dx2 ! 2nd derivative w.r.t. x
20     procedure :: output
21   end type
22
23   real(rkind) ,dimension(:) ,allocatable :: x_node
24
25   interface periodic_2nd_order
26     procedure constructor
27   end interface
28
29   real(rkind) ,parameter :: pi=acos(-1._rkind)
30
31 contains
32   function constructor(initial,num_grid_pts)
33     type(periodic_2nd_order) ,pointer :: constructor
34     procedure(initial_field) ,pointer :: initial
35     integer(ikind) ,intent(in) :: num_grid_pts
36     integer :: i
37     allocate(constructor)
38     allocate(constructor%f(num_grid_pts))
39     if (.not. allocated(x_node)) x_node = grid()
40     do i=1,size(x_node)
41       constructor%f(i)=initial(x_node(i))
42     end do
43   contains
44     pure function grid()
45       integer(ikind) :: i
46       real(rkind) ,dimension(:) ,allocatable :: grid
47       allocate(grid(num_grid_pts))
48       do i=1,num_grid_pts
49         grid(i)  = 2.*pi*real(i-1,rkind)/real(num_grid_pts,rkind)
50       end do
51     end function
52   end function
53
54   real(rkind) function rk2_dt(this,nu, num_grid_pts)
55     class(periodic_2nd_order) ,intent(in) :: this
56     real(rkind) ,intent(in) :: nu
57     integer(ikind) ,intent(in) :: num_grid_pts
58     real(rkind)              :: dx, CFL, k_max
```

```
59      dx=2.0*pi/num_grid_pts
60      k_max=num_grid_pts/2.0_rkind
61      CFL=1.0/(1.0-cos(k_max*dx))
62      rk2_dt = CFL*dx**2/nu
63    end function
64
65    function total(lhs,rhs)
66      class(periodic_2nd_order) ,intent(in) :: lhs
67      class(field) ,intent(in) :: rhs
68      class(field) ,allocatable :: total
69      type(periodic_2nd_order) ,allocatable :: local_total
70      integer                    :: i
71      select type(rhs)
72        class is (periodic_2nd_order)
73          allocate(local_total)
74          allocate(local_total%f(size(lhs%f)))
75          do i=1, size(lhs%f)
76            local_total%f(i) = lhs%f(i) + rhs%f(i)
77          end do
78          call move_alloc(local_total,total)
79        class default
80          stop 'periodic_2nd_order%total: unsupported rhs class.'
81      end select
82    end function
83
84    function difference(lhs,rhs)
85      class(periodic_2nd_order) ,intent(in) :: lhs
86      class(field) ,intent(in)  :: rhs
87      class(field) ,allocatable :: difference
88      type(periodic_2nd_order) ,allocatable :: local_difference
89      integer                    :: i
90      select type(rhs)
91        class is (periodic_2nd_order)
92          allocate(local_difference)
93          allocate(local_difference%f(size(lhs%f)))
94          do i=1, size(lhs%f)
95            local_difference%f(i) = lhs%f(i) - rhs%f(i)
96          end do
97          call move_alloc(local_difference,difference)
98        class default
99          stop 'periodic_2nd_order%difference: unsupported rhs class.'
100     end select
101   end function
102
103   function product(lhs,rhs)
104     class(periodic_2nd_order) ,intent(in) :: lhs
105     class(field) ,intent(in)  :: rhs
106     class(field) ,allocatable :: product
107     type(periodic_2nd_order) ,allocatable :: local_product
108     integer                    :: i
109     select type(rhs)
110       class is (periodic_2nd_order)
111         allocate(local_product)
112         allocate(local_product%f(size(lhs%f)))
113         do i=1, size(lhs%f)
```

```
114            local_product%f(i) = lhs%f(i) * rhs%f(i)
115          end do
116          call move_alloc(local_product,product)
117        class default
118          stop 'periodic_2nd_order%product: unsupported rhs class.'
119      end select
120    end function
121
122    function multiple(lhs,rhs)
123      class(periodic_2nd_order) ,intent(in) :: lhs
124      real(rkind) ,intent(in)  :: rhs
125      class(field) ,allocatable :: multiple
126      type(periodic_2nd_order) ,allocatable :: local_multiple
127      integer                   :: i
128      allocate(local_multiple)
129      allocate(local_multiple%f(size(lhs%f)))
130      do i=1, size(lhs%f)
131        local_multiple%f(i) = lhs%f(i) * rhs
132      end do
133      call move_alloc(local_multiple,multiple)
134    end function
135
136    subroutine copy(lhs,rhs)
137      class(field) ,intent(in) :: rhs
138      class(periodic_2nd_order) ,intent(inout) :: lhs
139      select type(rhs)
140        class is (periodic_2nd_order)
141          lhs%f = rhs%f
142        class default
143          stop 'periodic_2nd_order%copy: unsupported copy class.'
144      end select
145    end subroutine
146
147    function df_dx(this)
148      class(periodic_2nd_order) ,intent(in) :: this
149      class(field) ,allocatable  :: df_dx
150      integer(ikind) :: i,nx, x_east, x_west
151      real(rkind) :: dx
152      class(periodic_2nd_order) ,allocatable :: local_df_dx
153
154      nx=size(x_node)
155      dx=2.*pi/real(nx,rkind)
156      allocate(local_df_dx)
157      allocate(local_df_dx%f(nx))
158      do i=1,nx
159        x_east = mod(i,nx)+1
160        x_west = nx-mod(nx+1-i,nx)
161        local_df_dx%f(i)=0.5*(this%f(x_east)-this%f(x_west))/dx
162      end do
163      call move_alloc(local_df_dx, df_dx)
164    end function
165
166    function d2f_dx2(this)
167      class(periodic_2nd_order)  ,intent(in)   :: this
168      class(field) ,allocatable :: d2f_dx2
```

```
169      integer(ikind) :: i,nx,x_east,x_west
170      real(rkind)                              :: dx
171      class(periodic_2nd_order) ,allocatable   :: local_d2f_dx2
172
173      nx=size(this%f)
174      dx=2.*pi/real(nx,rkind)
175      allocate(local_d2f_dx2)
176      allocate(local_d2f_dx2%f(nx))
177      do i=1, nx
178        x_east = mod(i,nx)+1
179        x_west = nx-mod(nx+1-i,nx)
180        local_d2f_dx2%f(i) = &
181        (this%f(x_east)-2.0*this%f(i)+this%f(x_west))/dx**2
182      end do
183      call move_alloc(local_d2f_dx2, d2f_dx2)
184    end function
185
186    subroutine output(this)
187      class(periodic_2nd_order) ,intent(in) :: this
188      integer(ikind) :: i
189      do i=1,size(x_node)
190        print *, x_node(i), this%f(i)
191      end do
192    end subroutine
193
194  end module
```

Figure 12.1. Periodic, 2nd-order central differences concrete field class.

Table 12.1. *Runtime profile for central difference Burgers solver procedures on an IBM smp system*

Procedure	Runtime Share
d2f_dx2	24.5%
df_dx	24.0%
multiple	18.7%
copy	15.1%
product	6.2%
total	5.8%
difference	5.4%
Total	99.7%

The evidence in Table 12.1 suggests the greatest impact of any optimization effort will come through reducing the time occupied by the seven listed procedures. These procedures collectively occupy more than 99% of the runtime. First efforts might best be aimed at the top four procedures, which collectively occupy a runtime share exceeding 80%, or even just the top two, which occupy a combined share near 50%.

Amdahl's Law quantifies the overall speed gain for a given gain in a subset of a process. In doing so, it also facilitates calculating the maximum gain in the limit as the given subset's share of the overall completion time vanishes. Imagine breaking some process into N distinct portions with the ith portion occupying P_i fraction of the overall completion time. Then imagine ordering the portions such that the Nth portion subsumes all parts of the overall process that have fixed costs and therefore cannot be sped up. Define the *speedup* of the ith portion as:

$$S_i \equiv \frac{t_{original}}{t_{optimized}} \tag{12.1}$$

where the numerator and denominator are the original and optimized completion times, respectively, and where $S_N \equiv 1$ by definition. Amdahl's Law states that the speedup of the overall process is:

$$S_{overall}(\vec{P},\vec{S}) = \frac{1}{P_N + \sum_{i=1}^{N-1} \frac{P_i}{S_i}} \tag{12.2}$$

where $\vec{P}$ and $\vec{S}$ are vectors with i^{th} elements P_i and S_i, respectively.

The previous paragraph makes no assumptions about whether the original completion time involves some optimization. Nor does it assume anything about the optimization process. It does not even stipulate that the process in question involves a computer! Amdahl's Law applies as well to one's daily commute, for example, as it does to scientific computing. A driver might optimize highway portions of the trip by taking on a passenger and driving in a carpool lane, whereas nonhighway portions are likely to have fixed costs. In scientific computing, the optimization might involve algorithmic redesign, compiler optimizations, hardware improvements, or parallelization. In the parallelization case, a speedup of 5 might involve going from serial code to code running on five execution units (e.g., cores) or it might involve scaling from 10,000 processors to 50,000.

In addition to calculating the actual speedup in various scenarios, Amdahl's Law places limits on the theoretical maximum speedup in various scenarios. These scenarios derive from taking the limit as various components of $\vec{S}$ grow without bound. Allowing all components to grow yields the upper bound on all scenarios. This ultimate bound is $S_{max} \equiv 1/P_N$.

S_{max} sets an important limit on the potential utility of parallel hardware. To expose this limitation, consider an optimization strategy limited to parallelization alone. Distributing the execution across M hardware units, the problem-independent maximum speedup such hardware can provide is M. With this observation, we define the parallel efficiency:

$$E \equiv \frac{S_{overall}}{M} \tag{12.3}$$

where E = 100% corresponds to the maximal use of the available hardware. When $S_{max} < M$, it is impossible take full advantage of all available execution units. Thus, efficient use of massively parallel systems requires very large S_{max} and correspondingly tiny P_N, implying that every dusty corner of a code must scale. Any portion that does not scale becomes the rate-limiting step.

After the procedures listed in Table 12.1, the next highest (unshown) as ranked by their runtime share are compiler-generated, memory-management system calls.

Each of these procedures occupies a share of 0.1% or lower, totaling 0.3%. Given that the time occupied by system calls lies beyond an application's control and given that reducing the number of calls would require significant algorithmic redesign, it seems reasonable to treat the system calls as a fixed cost and take $P_N \approx 0.003$ and $S_{max} \approx 333$. Hence, it would be impossible to take full advantage of more than $\tilde{3}33$ execution units in parallel. With the recent advent of quad-socket, quad-core nodes, the number of cores on even a 32-node cluster would exceed 333. At the time of this writing, a relatively modest $150,000 research equipment grant is likely to garner 512–1,024 cores, putting it out of reach for full utilization by a code with even a 0.3% share of nonscalable execution.

This discussion neglects the overhead costs associated with parallel execution. Such overhead might include, for example, the spawning and joining of threads or communication between processes. Also, we considered only the ideal case of $S_i \to \infty \forall i$. With finite speedup on portions 1 through $N-1$, the $S_{overall}$ might continue to improve with increasing numbers of execution units beyond the aforementioned limits. A code might then make effective use of greater than 250 execution units, although $S_{overall} \leq S_{max}$ would still hold.

Finally, the parallel efficiency definition in equation (12.3) applies to *strong scaling*, the case in which the problem size remains fixed as the number of execution units increases. The alternative, *weak scaling* describes the case where the problem size grows proportionately to the number of execution units. Which type of scaling matters more depends on the application domain.

"Vision without Execution Is Hallucination."
Thomas Edison

Regarding code efficiency, a viewpoint held without the backing of execution data represents an unfounded belief. Profiling procedures' runtime share guides developers toward the chief bottlenecks. Amdahl's law determines the speedup achievable by focusing on specific bottlenecks.

12.1.2 Automatic Parallelization: Compiler Technology

Chapter 1 argues that significantly speeding up overall solution times requires optimizing execution times in ways that do not significantly increase development times. Guided by that principle, Sections 12.1.2–12.1.5 emphasize parallelization strategies that minimally impact the programming process. The treatment progresses from a strategy with no impact on programming but concomitantly less benefit through strategies with varying degrees of programmer burden but with greater control and varying benefit.

There is no free lunch; someone has to do the hard work to get an application to scale, but much of that work can be shifted to library or compiler developers. Along the way, it becomes clear that ABSTRACT CALCULUS lends itself naturally to parallelization. With ABSTRACT CALCULUS, getting an application to scale reduces to the problem of getting each operator in an overloaded expression to scale. Furthermore, the purely functional (`intent(in)`) requirements of Fortran-defined operators help eliminate complex data dependencies that might otherwise inhibit parallelizing in those operations: Programmers and the compiler agree that no inadvertent updates on operands occur during operator evaluations (though operators might overlap by reading the same memory location). These operator evaluations can take place asynchronously, with synchronization occurring only at the end of the final assignment. ABSTRACT CALCULUS thus clears a straightforward path toward scalable execution.

The simplest strategy leverages the automatic parallelization of code by an optimizing compiler. Auto-parallelization generally comprises two categories of techniques:

1. Multithreading: splitting a loop into multiple threads of execution or
2. Auto-vectorization: replacing scalar operations (performed on one array element at a time) with vector operations (performed on multiple elements at a time) inside a loop.

Auto-vectorization requires supporting hardware. The growing of popularity of single instruction multiple data (SIMD) processing brings auto-vectorization downmarket from the high-end platforms to the mainstream ones. Optimizing compilers can now exploit vectorization capabilities on commodity hardware via libraries such as the Mathematical Acceleration Subsystem (MASS) libraries.[1] on the IBM PowerPC family of processors.[2]

SIMD hardware comprises multiple processing elements capable of executing the same operation on multiple data concurrently. Consider lines 130–132 in the user-defined multiplication operator implementation of Figure 12.1:

```
do i=1, size(lhs%f)
  local_multiple%f(i) = lhs%f(i) * rhs
end do
```

A vectorizing compiler might unroll this loop by a factor of n, where n is the number of floating point units on the target SIMD hardware. For instance a 16-byte-wide SIMD unit can process four single-precision, floating-point values simultaneously, corresponding to $n = 4$. For double-precision values, $n = 2$. In any case, the transformed loop would resemble:

```
do i=1, size(lhs%f), n
  local_multiple%f(i:i+n-1) = lhs%f(i:i+n-1) * rhs
end do
!..residual loop processing..
```

where a SIMD unit can carry out the load, store, and multiplication on the array sections `local_multiple%f(i:i+n-1)+` and `lhs%f(i:i+n-1)+` as if operating on a single pair of operand or pair of operands. The speedup attained via SIMD instructions depends on n, which depends on the available vector length and the floating-point unit length. On an IBM Power5+ machine, we have observed a speedup of 2 on the `multiple` function with auto-vectorization. Similar speedups obtain on all defined operators.

[1] http://www-01.ibm.com/software/awdtools/mass/
[2] http://tinyurl.com/2wmrjb7

Vectorization requires that the array sections occupy contiguous memory. The aforementioned example satisfies this condition because the `f` component, an instance of `periodic_2nd_order`, contains an allocatable array and is thereby guaranteed to be contiguous. By contrast, `pointer` arrays may pose challenges in this respect since they can be associated with noncontiguous array targets. In the case of SIMD architectures, alignments can also pose challenges to compilers applying vectorization techniques (Eichenberger et al. 2004).

The other popular auto-parallelization technique exploits symmetric multiprocessing (SMP) systems – that is systems in which each processor has an identical view of the memory. SMP computers (and SMP nodes on multinode, distributed clusters) can run multiple threads to parallelize a loop. The advent of multicore chips bolsters the popularity of such multithreading. The technique requires a mechanism for spawning and joining threads in shared memory as Section 12.1.3 explains. To maximize the speedup of a loop using all available cores, a compiler may automatically parallelize the loop, that is, divide it into multiple segments and bind each segment to an available core[3].

Considering the aforementioned loop from `multiple`, the concept of the auto-parallelization can be explained using the following pseudo-code:

```
  CHUNK_SIZE = size(lhs%f) / NUM_THREADS
  ! start parallel region
  call create_thread(run_by_one_thread, 1, CHUNK_SIZE)
  call create_thread(run_by_one_thread, CHUNK_SIZE+1, 2*CHUNK_SIZE)
  ...
  call create_thread(run_by_one_thread, &
                     (NUM_THREADS-1)*CHUNK_SIZE+1, size(lhs%f))
  ! join threads & end of parallel region
... subroutine run_by_one_thread (lbound, ubound)
  integer, intent( in ) :: lbound, ubound
  do i=lbound, ubound
    local_multiple%f(i) = lhs%f(i) * rhs
  end do
end subroutine
```

Nothing fundamental differentiates between the automated parallelized loops executed at runtime by compiler-automated means versus the directive-based technique we discuss next. For the simple loop used in `multiple`, the compiler can easily determine the loop is independent in execution order and therefore parallelizable. In practice, however, many loops are not so amenable to simple analysis, and compilers may have difficulty removing data dependencies between loop iterations. In those cases, the programmer usually has an edge in knowing the feasibility of parallelizing a loop.

[3] Although binding each thread to its own core avoids resource conflicts and is allowable for a compiler to do, the OpenMP multithreading technology discussed in Section 12.1.3 does not yet facilitate a standard, portable mechanism for such bindings.

"We Don't Believe That... Joe the Programmer Should Have to Deal with Parallelism in an Explicit Way."
Kunle Olukotun

Going parallel need not mean thinking parallel. Some parallelism can be achieved by optimizing compilers without direct programmer intervention, sometimes with results that outperform what a novice programmer can accomplish when given direct control. The next section demonstrates that greater programmer intervention does not necessarily lead to greater performance than a sophisticated compiler generates at least for modest numbers of cores.

12.1.3 Directive-Based Parallelization: OpenMP

The next step up in programmer involvement commonly involves multithreading via the OpenMP application programming interface (API). The OpenMP API consists of compiler directives, library routines, and environment variables (Chapman et al. 2007). Multithreading follows the fork-join paradigm in which a program forks into multiple threads of execution for some period of time before joining these threads back into one. The join operation awaits completion of each thread before continuing. In the most common approach, this happens multiple times during a run, as depicted in Figure 12.2.

OpenMP supports a shared-memory programming model. Each thread has access to the same data unless the programmer explicitly specifies that each thread have private copies of a given variable. The shared-memory viewpoint at the application programming level need not imply shared memory at the hardware level. The operating system on the Silicon Graphics, Inc. (sgi) Altix[4] supercomputer product line, for example, provides programs a shared view of memory despite the memory being distributed across multiple nodes. Even when the hardware and operating systems views of memory are distributed, it is possible to extend OpenMP to run on distributed-memory platforms as with the Intel Corp.'s Cluster OpenMP API[5].

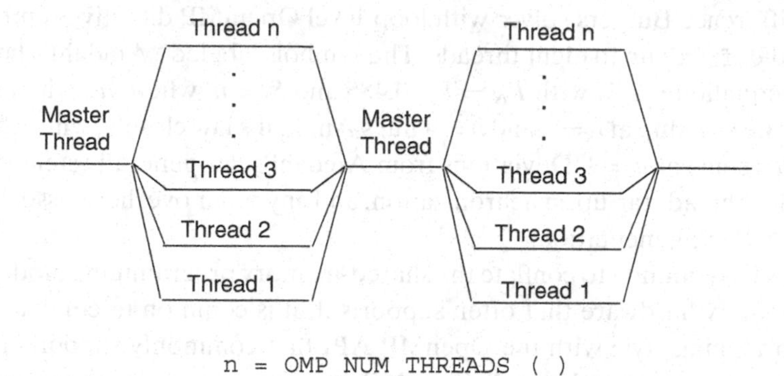

`n = OMP_NUM_THREADS ( )`

Figure 12.2. Fork-join programming model: A typical multithreaded program forks and joins multiple threads of execution many times throughout a given run.

[4] http://www.sgi.com/products/servers/
[5] http://software.intel.com/en-us/articles/cluster-openmp-for-intel-compilers/

Most current compilers support OpemMP. Table 12.1 shows the runtime profile for a serial, central-difference, Burgers-solver executable generated without automatic parallelization by the IBM XL Fortran compiler version 13.1 on an eight-core platform running the AIX operating system. Since d2f_dx2 and df_dx combine to occupy a 48.5% runtime share, optimizing these two procedures will have the greatest impact on speeding up the code. The simplest approach for doing so is with OpenMP bracing the loops in these procedures with OpenMP directives to the compiler. For df_dx, the result is:

```
  !$OMP parallel do  private (x_east, x_west)
  do i=1,nx
    x_east = mod(i,nx)+1
    x_west = nx-mod(nx+1-i,nx)
    local_df_dx%f(i)=0.5*(this%f(x_east)-this%f(x_west))/dx
  end do
 !$OMP end parallel do
```

where the !$OMP OpenMP sentinels prompt an OpenMP-enabled compiler to interpret the trailing directives. A compiler without OpenMP capabilities ignores the entire line, seeing the leading exclamation mark as the harbinger for a comment. The parallel do/end parallel do pair simultaneously fork threads and delimit a work-sharing region. At the beginning of this region, an OpenMP compiler forks a number of threads equal to the OMP_NUM_THREADS environment variable[6]. At the end, the compiler joins all the threads into one. The !$OMP end parallel do directive implicitly synchronizes the computation, acting as a barrier that forces the master thread to wait for all other threads to finish their work before closing the parallel region. Finally, the private clause gives each thread its own copies of the listed variables. These variables can be considered local to each thread.

When the number of threads exceeds the number of available execution units, performance generally degrades. Hence, the maximum number of threads tested on the aforementioned platform is eight. Figure 12.3 shows the performance of the central-difference Burgers solver with loop-level OpemMP directives embedded in df_dx and d2f_dx2 up to eight threads. The symbols labeled "Amdahl's law" correspond to equation (12.2) with $P_N = 1 - 0.485$ and $S_i = n$, where n is the number of threads. The speedup at $n = 2$ and $n = 4$ hugs Amdhal's law closely, while the performance slips some at $n = 8$. Deviations from Amdahl's law generally stem from costs incurred for thread startup, synchronization, and any other overhead associated with the OpenMP implementation.

Just as it is common to conflate the shared-memory programming model with the shared-memory hardware that often supports it, it is common to conflate the fork-join programming style with the OpenMP API that commonly supports it. In fact, one can employ other styles with OpenMP, including the more prevalent parallel programming style: Single Program Multiple Data (SPMD). With SPMD OpenMP,

[6] The setting of environment variables is platform-dependent. In a Unix bash shell, for example, typing "export OMP_NUM_THREADS=8" at the command line sets the variable to 8.

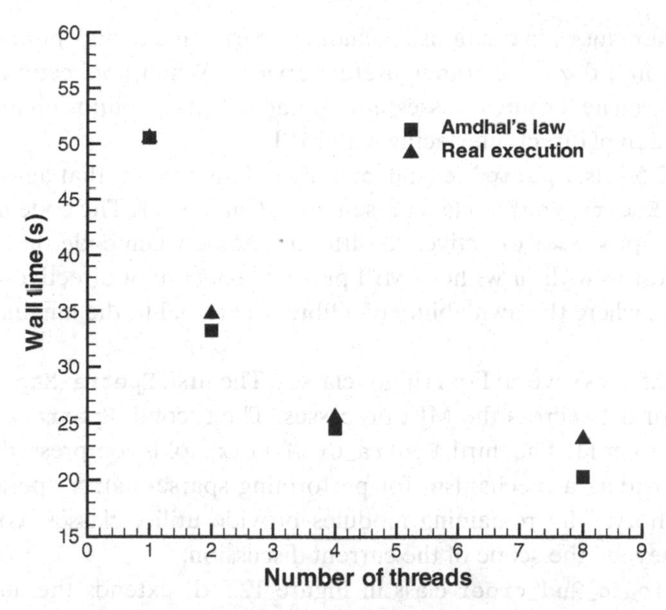

Figure 12.3. Scalability of a central difference Burgers solver using OpenMP directives.

one forks all threads at the beginning of program execution and joins them only at the end of execution (Wallcraft 2000; Wallcraft 2002; Krawezik et al. 2002). Multiple copies of the same program run simultaneously on separate execution units with the work organized via explicit references to a thread identification number rather than being shared out automatically by the compiler as with the aforementioned loop-level directives.

Although there has been some work on automating the transformation of more traditional OpenMP code to a SPMD style (Liu et al. 2003), manual SPMD OpenMP programming requires considerably greater effort and care than loop-level OpenMP. In the process, OpenMP loses some of its luster relative to the more common SPMD API: the Message Passing Interface (MPI). The knowledge and effort required to write SPMD OpenMP is comparable to that required for MPI. Furthermore, SPMD OpenMP and MPI share the same motivation: Both scale better to high numbers of execution units. Hence, when massive scalability drives the design, one more naturally turns to MPI. The next section discusses a MPI-based solver.

12.1.4 Library-Based Parallelization: ForTrilinos and MPI

Programmers commonly describe MPI as the assembly language of parallel programming. Characteristic of this description, moving to MPI would typically be the zenith of programmer involvement in the parallelization process. It would also be very difficult for raw MPI code to adhere to the style suggestions set forth in Part I of this book. Most MPI calls involve cryptic argument names and side effects, making it challenging to write code that comments itself or code that gives visual cues to its structure as recommended in Section 1.7.

Embedding MPI in an object-oriented library that supports domain-specific abstractions lowers the barrier to its use. The Trilinos project does this for various applied mathematics abstractions, including sparse linear and nonlinear solvers, eigensolvers, and dozens of other useful services. When applications developers extend or aggregate Trilinos classes in building solvers, Trilinos eliminates almost all of the burden of direct interfacing with MPI.

Figure 12.5 lists a `periodic_2nd_order` implementation that aggregates an instance of the `Epetra_Vector` class presented in Chapter 11. The code in Figure 12.5 employs C preprocessor directives to differentiate between code that will be compiled on platforms with or without MPI present. This use of directives typifies MPI programming, where the availability of a library external to the language cannot be guaranteed.

Figure 12.5 uses several ForTrilinos classes. The first, `Epetra_Map`, describes the distribution of data across the MPI processes. The second, `Epetra_Vector`, is described in Chapter 11. The third, `Epetra_CrsMatrix`, holds compressed-row-storage matrices, providing a mechanism for performing sparse-matrix operations in distributed memory. The remaining modules provide utility classes, constants, and procedures beyond the scope of the current discussion.

The `periodic_2nd_order` class in Figure 12.5(d) extends the abstract `field` class and aggregates an `Epetra_Vector` object. Because all `Epetra_Vector` objects in this case are constructed with the same `Epetra_Map`, Figure 12.5(d) stores the map in a single object with module scope. See Chapter 8 for additional discussion of the trade-offs associated with using module variables. The various operators in Figure 12.5 delegate their operations to `Epetra_Vector` by invoking the appropriate methods on their `Epetra_Vector` component.

Figure 12.5(e) lists a ForTrilinos-based Burgers-equation solver `main` program employing ABSTRACT CALCULUS. That main program illustrates a hallmark of Trilinos-based solvers: The only MPI calls are for process startup (`Mpi_Init`), shutdown (`Mpi_Finalize`), and timing (`MPI_Wtime`). Figure 12.4 shows the parallel efficiency of the ForTrilinos-based solver.

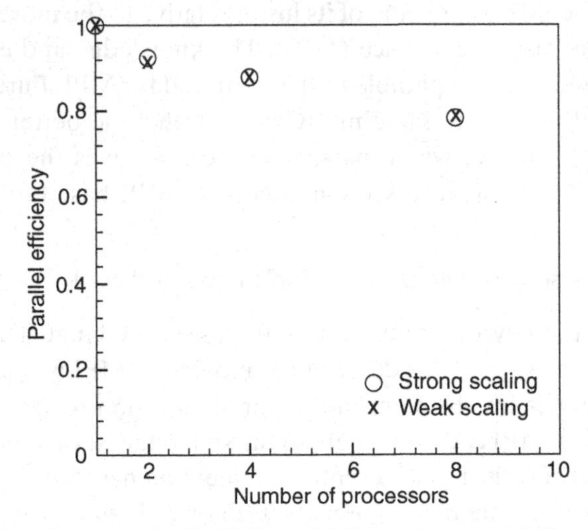

Figure 12.4. Parallel efficiency of a ForTrilinos/MPI-based central difference Burgers solver: strong scaling with 64^3 grid points and weak scaling with 64^3M points for M MPI processes. All simulations were run for 1,000 time steps.

Figure 12.5 (a)

```
module field_module
  use iso_c_binding, only: c_int, c_double
  use FEpetra_Comm, only:Epetra_Comm
  implicit none
  private
  public :: field, initial_field

  type ,abstract :: field
  contains
    procedure(field_op_field) ,deferred :: add
    procedure(field_eq_field) ,deferred :: assign
    procedure(field_op_field) ,deferred :: subtract
    procedure(field_op_field) ,deferred :: multiply_field
    procedure(field_op_real)  ,deferred :: multiply_real
    procedure(real_to_real)  ,deferred :: runge_kutta_2nd_step
    procedure(derivative) ,deferred :: x  ! 1st derivative
    procedure(derivative) ,deferred :: xx ! 2nd derivative
    procedure(output_interface) ,deferred  :: output
    generic :: operator(+)    => add
    generic :: operator(-)    => subtract
    generic :: operator(*)    => multiply_real,multiply_field
    generic :: assignment(=) => assign
    procedure(force_finalize_interface) ,deferred :: force_finalize
  end type

  abstract interface
    real(c_double) pure function initial_field(x)
      import :: c_double
      real(c_double) ,intent(in) :: x
    end function
    function field_op_field(lhs,rhs)
      import :: field
      class(field) ,intent(in) :: lhs,rhs
      class(field) ,allocatable :: field_op_field
    end function
    function field_op_real(lhs,rhs)
      import :: field,c_double
      class(field) ,intent(in)  :: lhs
      real(c_double) ,intent(in) :: rhs
      class(field) ,allocatable :: field_op_real
    end function
    real(c_double) function real_to_real(this,nu,grid_resolution)
      import :: field,c_double,c_int
      class(field) ,intent(in)  :: this
      real(c_double) ,intent(in) :: nu
      integer(c_int),intent(in) :: grid_resolution
    end function
    function derivative(this)
      import :: field
      class(field) ,intent(in)  :: this
      class(field) ,allocatable :: derivative
    end function
    subroutine field_eq_field(lhs,rhs)
      import :: field
```

```
55        class(field) ,intent(inout) :: lhs
56        class(field) ,intent(in) :: rhs
57      end subroutine
58      subroutine output_interface(this,comm)
59        import :: field,Epetra_Comm
60        class(field) ,intent(in) :: this
61        class(Epetra_Comm), intent(in) :: comm
62      end subroutine
63      subroutine force_finalize_interface(this)
64        import :: field
65        class(field), intent(inout)  :: this
66      end subroutine
67    end interface
68  end module
```

Figure 12.5 (b)

```
1   module field_factory_module
2     use FEpetra_Comm, only:Epetra_Comm
3     use iso_c_binding ,only :c_int
4     use field_module ,only : field,initial_field
5     implicit none
6     private
7     public :: field_factory
8
9     type, abstract :: field_factory
10    contains
11      procedure(create_interface), deferred :: create
12    end type
13
14    abstract interface
15      function create_interface(this,initial,grid_resolution,comm)
16        import :: field, field_factory ,initial_field,c_int,Epetra_Comm
17        class(Epetra_Comm),intent(in) :: comm
18        class(field_factory), intent(in) :: this
19        class(field) ,pointer :: create_interface
20        procedure(initial_field) ,pointer :: initial
21        integer(c_int) ,intent(in) :: grid_resolution
22      end function
23    end interface
24  end module
```

Figure 12.5 (c)

```
1   module periodic_2nd_order_factory_module
2     use FEpetra_Comm ,only:Epetra_Comm
3     use field_factory_module ,only : field_factory
4     use field_module ,only : field,initial_field
5     use periodic_2nd_order_module ,only : periodic_2nd_order
6     implicit none
7     private
8     public :: periodic_2nd_order_factory
9
10    type, extends(field_factory) :: periodic_2nd_order_factory
11    contains
12      procedure :: create=>new_periodic_2nd_order
```

```
13    end type
14
15  contains
16    function new_periodic_2nd_order(this,initial,grid_resolution,comm)
17      use iso_c_binding, only:c_int
18      class(periodic_2nd_order_factory), intent(in) :: this
19      class(field) ,pointer :: new_periodic_2nd_order
20      procedure(initial_field) ,pointer :: initial
21      integer(c_int) ,intent(in) :: grid_resolution
22      class(Epetra_Comm), intent(in) :: comm
23      new_periodic_2nd_order=> &
24        periodic_2nd_order(initial,grid_resolution,comm)
25    end function
26  end module
```

Figure 12.5 (d)

```
1   module periodic_2nd_order_module
2     use FEpetra_Comm, only:Epetra_Comm
3     use FEpetra_Map!, only: Epetra_Map
4     use FEpetra_Vector!, only:Epetra_Vector
5     use FEpetra_CrsMatrix, only:Epetra_CrsMatrix
6     use ForTrilinos_enum_wrappers
7     use ForTrilinos_error
8     use field_module ,only : field,initial_field
9     use iso_c_binding ,only : c_double,c_int
10    use ForTrilinos_assertion_utility ,only: &
11      error_message,assert,assert_identical
12    implicit none
13    private
14    public :: periodic_2nd_order
15
16    type ,extends(field) :: periodic_2nd_order
17      private
18      type(Epetra_Vector) :: f
19    contains
20      procedure :: add => total
21      procedure :: subtract => difference
22      procedure :: assign => copy
23      procedure :: multiply_field => product
24      procedure :: multiply_real => multiple
25      procedure :: runge_kutta_2nd_step => rk2_dt
26      procedure :: x  => df_dx     ! 1st derivative w.r.t. x
27      procedure :: xx => d2f_dx2   ! 2nd derivative w.r.t. x
28      procedure :: output
29      procedure :: force_finalize
30    end type
31
32    real(c_double) ,parameter                    :: pi=acos(-1._c_double)
33    real(c_double) ,dimension(:) ,allocatable :: x_node
34    type(Epetra_Map)             ,allocatable :: map
35
36    interface periodic_2nd_order
37      procedure constructor
38    end interface
```

```
39 contains
40   function constructor(initial,grid_resolution,comm) result(this)
41     type(periodic_2nd_order) ,pointer  :: this
42     procedure(initial_field) ,pointer :: initial
43     integer(c_int) ,intent(in) :: grid_resolution
44     integer(c_int) :: i,j
45     class(Epetra_Comm), intent(in) :: comm
46     integer(c_int) :: NumGlobalElements
47     integer(c_int),dimension(:),allocatable :: MyGlobalElements
48     integer(c_int)       :: NumMyElements,IndexBases=1,status
49     real(c_double) ,dimension(:) ,allocatable :: f_v
50     type(error) :: ierr
51     NumGlobalElements=grid_resolution
52     allocate(this)
53     if (.not. allocated(x_node)) x_node = grid()
54     if (.not. allocated(map)) then
55       allocate(map,stat=status)
56       ierr=error(status,'periodic_2nd_order: create map')
57       call ierr%check_allocation()
58       map = Epetra_Map(NumGlobalElements,IndexBases,comm)
59     end if
60     NumMyElements= map%NumMyElements()
61     allocate(MyGlobalElements(NumMyElements))
62     MyGlobalElements = map%MyGlobalElements()
63     allocate(f_v(NumMyElements))
64     forall(i=1:NumMyElements)f_v(i)=initial(x_node(MyGlobalElements(i)))
65     this%f=Epetra_Vector(map,zero_initial=.true.)
66     call this%f%ReplaceGlobalValues(NumMyElements,f_v,MyGlobalElements)
67   contains
68     pure function grid()
69       integer(c_int) :: i
70       real(c_double) ,dimension(:) ,allocatable :: grid
71       allocate(grid(grid_resolution))
72       forall(i=1:grid_resolution) &
73         grid(i)= 2.*pi*real(i-1,c_double)/real(grid_resolution,c_double)
74     end function
75   end function
76
77   subroutine copy(lhs,rhs)
78     class(field) ,intent(in) :: rhs
79     class(periodic_2nd_order) ,intent(inout) :: lhs
80     select type(rhs)
81       class is (periodic_2nd_order)
82         lhs%f = rhs%f
83       class default
84         stop 'periodic_2nd_order%copy: unsupported copy class.'
85     end select
86   end subroutine
87
88   real(c_double) function rk2_dt(this,nu, grid_resolution)
89     class(periodic_2nd_order) ,intent(in) :: this
90     real(c_double) ,intent(in) :: nu
91     integer(c_int) ,intent(in) :: grid_resolution
92     real(c_double)             :: dx, CFL, k_max
93     dx=2.0*pi/grid_resolution
```

```
94      k_max=grid_resolution/2.0_c_double
95      CFL=1.0/(1.0-cos(k_max*dx))
96      rk2_dt = CFL*dx**2/nu
97    end function
98
99    function total(lhs,rhs)
100     class(periodic_2nd_order) ,intent(in) :: lhs
101     class(field) ,intent(in) :: rhs
102     class(field) ,allocatable :: total
103     type(periodic_2nd_order) ,allocatable :: local_total
104     select type(rhs)
105       class is (periodic_2nd_order)
106         allocate(periodic_2nd_order::local_total)
107         local_total%f=Epetra_Vector(map,zero_initial=.true.)
108         call local_total%f%Update( &
109           1._c_double,lhs%f,1._c_double,rhs%f,0._c_double)
110         call move_alloc(local_total,total)
111       class default
112         stop 'periodic_2nd_order%total: unsupported rhs class.'
113     end select
114   end function
115
116   function difference(lhs,rhs)
117     class(periodic_2nd_order) ,intent(in) :: lhs
118     class(field) ,intent(in) :: rhs
119     class(field) ,allocatable :: difference
120    type(periodic_2nd_order) ,allocatable :: local_difference
121     select type(rhs)
122       class is (periodic_2nd_order)
123         allocate(periodic_2nd_order::local_difference)
124         local_difference%f=Epetra_Vector(map,zero_initial=.true.)
125         call local_difference%f%Update(&
126           1._c_double,lhs%f,-1._c_double,rhs%f,0._c_double)
127         call move_alloc(local_difference,difference)
128       class default
129         stop 'periodic_2nd_order%difference: unsupported rhs class.'
130     end select
131   end function
132
133  function product(lhs,rhs)
134    class(periodic_2nd_order) ,intent(in) :: lhs
135    class(field) ,intent(in) :: rhs
136    class(field),allocatable :: product
137    type(periodic_2nd_order) ,allocatable :: local_product
138    select type(rhs)
139     class is (periodic_2nd_order)
140       allocate(periodic_2nd_order::local_product)
141       local_product%f=Epetra_Vector(map,zero_initial=.true.)
142       call local_product%f%Multiply(1._c_double,lhs%f,rhs%f,0._c_double)
143       call move_alloc(local_product,product)
144      class default
145       stop 'periodic_2nd_order%product: unsupported rhs class.'
146    end select
147   end function
148
```

```
149    function multiple(lhs,rhs)
150      class(periodic_2nd_order) ,intent(in) :: lhs
151      real(c_double) ,intent(in)  :: rhs
152      class(field) ,allocatable :: multiple
153      type(periodic_2nd_order) ,allocatable :: local_multiple
154      allocate(periodic_2nd_order::local_multiple)
155      local_multiple%f=Epetra_Vector(map,zero_initial=.true.)
156      call local_multiple%f%Scale(rhs,lhs%f)
157     call move_alloc(local_multiple,multiple)
158    end function
159
160    function df_dx(this)
161      class(periodic_2nd_order) ,intent(in) :: this
162      class(field) ,allocatable  :: df_dx
163      type(Epetra_Vector) :: x
164      type(Epetra_CrsMatrix) :: A
165      type(error) :: err
166      real(c_double) ,dimension(:)   ,allocatable :: c
167      real(c_double) :: dx
168      integer(c_int) :: nx
169      type(periodic_2nd_order), allocatable :: df_dx_local
170      integer(c_int),dimension(:),allocatable :: MyGlobalElements
171      integer(c_int),dimension(:),allocatable :: MyGlobalElements_diagonal
172      integer(c_int),dimension(:),allocatable :: NumNz
173      integer(c_int) :: NumGlobalElements,NumMyElements,i
174      integer(c_int) :: indices(2), NumEntries
175      real(c_double) ::values(2)
176      real(c_double),parameter :: zero =0.0
177      integer(c_int),parameter :: diagonal=1
178
179      ! Executable code
180      nx=size(x_node)
181      dx=2.*pi/real(nx,c_double)
182      NumGlobalElements = nx
183
184  ! Get update list and number of local equations from given Map
185      NumMyElements = map%NumMyElements()
186      call assert_identical( [NumGlobalElements,map%NumGlobalElements()] )
187      allocate(MyGlobalElements(NumMyElements))
188      MyGlobalElements = map%MyGlobalElements()
189
190  ! Create an integer vector NumNz that is used to build the Epetra Matrix
191  ! NumNz(i) is the number of non-zero elements for the ith global eqn.
192  ! on this processor
193    allocate(NumNz(NumMyElements))
194
195  ! We are building a tridiagonal matrix where each row has (-1 0 1)
196  ! So we need 2 off-diagonal terms (except for the first and last eqn.)
197    NumNz = 3
198
199  ! Create a Epetra_Matrix
200    A = Epetra_CrsMatrix(FT_Epetra_DataAccess_E_Copy,map,NumNz)
201
202  ! Add rows one at a time
203  ! Need some vectors to help
```

```
204 ! off diagonal values will always be -1 and 1
205   values(1) = -1.0/(2.0*dx)
206   values(2) = 1.0/(2.0*dx)
207   do i=1,NumMyElements
208     if (MyGlobalElements(i)==1) then
209       indices(1) = NumGlobalElements
210       indices(2) = 2
211       NumEntries = 2
212     else if(MyGlobalElements(i)==NumGlobalElements) then
213       indices(1) = NumGlobalElements-1
214       indices(2) = 1
215       NumEntries = 2
216     else
217       indices(1) = MyGlobalElements(i)-1
218       indices(2) = MyGlobalElements(i)+1
219       NumEntries = 2
220     end if
221      call A%InsertGlobalValues(&
222        MyGlobalElements(i),NumEntries,values,indices,err)
223      call assert( [err%error_code()==0_c_int] , &
224        [error_message('A%InsertGlobalValues: failed')] )
225  !Put in the diagonal entry
226      MyGlobalElements_diagonal=MyGlobalElements+i-1
227      call A%InsertGlobalValues(MyGlobalElements(i), &
228        diagonal,[zero],MyGlobalElements_diagonal,err)
229      call assert( [err%error_code()==0_c_int] , &
230        [error_message('A%InsertGlobalValues: failed')] )
231   end do
232
233   !Finish up
234     call A%FillComplete()
235   !create vector x
236     x=Epetra_Vector(A%RowMap())
237     Call A%Multiply_Vector(.false.,this%f,x)
238     allocate(c(NumMyElements))
239     c=x%ExtractCopy()
240  !create vector of df_dx
241     allocate(periodic_2nd_order::df_dx_local)
242     df_dx_local%f=Epetra_Vector(map,zero_initial=.true.)
243     call df_dx_local%f%ReplaceGlobalValues(&
244       NumMyElements,c,MyGlobalElements)
245     call move_alloc(df_dx_local, df_dx)
246   end function
247
248   function d2f_dx2(this)
249     class(periodic_2nd_order) ,intent(in) :: this
250     class(field) ,allocatable  :: d2f_dx2
251     type(Epetra_Vector) :: x
252     type(Epetra_CrsMatrix) :: A
253     type(error) :: err
254     real(c_double) ,dimension(:)    ,allocatable :: c
255     real(c_double) :: dx
256     integer(c_int) :: nx
257     type(periodic_2nd_order) ,allocatable :: d2f_dx2_local
258     integer(c_int),dimension(:),allocatable :: MyGlobalElements,NumNz
```

```
259      integer(c_int),dimension(:),allocatable :: MyGlobalElements_diagonal
260      integer(c_int) :: NumGlobalElements,NumMyElements,i
261      integer(c_int) :: indices(2),NumEntries
262      real(c_double) :: values(2),two_dx2
263      integer(c_int),parameter :: diagonal=1
264
265    ! Executable code
266     nx=size(x_node)
267     dx=2.*pi/real(nx,c_double)
268      NumGlobalElements = nx
269
270   ! Get update list and number of local equations from given Map
271     NumMyElements = map%NumMyElements()
272     call assert_identical( [NumGlobalElements,map%NumGlobalElements()] )
273     allocate(MyGlobalElements(NumMyElements))
274     MyGlobalElements = map%MyGlobalElements()
275
276   ! Create an integer vector NumNz that is used to build the Epetra Matrix
277   ! NumNz(i) is the number of non-zero elements for the ith global eqn.
278   ! on this processor
279     allocate(NumNz(NumMyElements))
280
281   ! We are building a tridiagonal matrix where each row has (1 -2  1)
282   ! So we need 2 off-diagonal terms (except for the first and last eqn.)
283     NumNz = 3
284
285   ! Create a Epetra_Matrix
286     A = Epetra_CrsMatrix(FT_Epetra_DataAccess_E_Copy,map,NumNz)
287
288   ! Add rows one at a time
289   ! Need some vectors to help
290   ! off diagonal values will always be 1 and 1
291     values(1) = 1.0/(dx*dx)
292     values(2) = 1.0/(dx*dx)
293     two_dx2   =-2.0/(dx*dx)
294     do i=1,NumMyElements
295       if (MyGlobalElements(i)==1) then
296         indices(1) = NumGlobalElements
297         indices(2) = 2
298         NumEntries = 2
299       else if(MyGlobalElements(i)==NumGlobalElements) then
300         indices(1) = NumGlobalElements-1
301         indices(2) = 1
302         NumEntries = 2
303       else
304         indices(1) = MyGlobalElements(i)-1
305         indices(2) = MyGlobalElements(i)+1
306         NumEntries = 2
307       end if
308        call A%InsertGlobalValues( &
309          MyGlobalElements(i),NumEntries,values,indices,err)
310        call assert( [err%error_code()==0_c_int] , &
311          [error_message('A%InsertGlobalValues: failed')] )
312    !Put in the diagonal entry
313        MyGlobalElements_diagonal=MyGlobalElements+i-1
```

```
314        call A%InsertGlobalValues( MyGlobalElements(i) &
315          ,diagonal,[two_dx2],MyGlobalElements_diagonal,err)
316        call assert( [err%error_code()==0_c_int] , &
317          [error_message('A%InsertGlobalValues: failed')] )
318    end do
319
320    !Finish up
321      call A%FillComplete()
322    !create vector x
323      x=Epetra_Vector(A%RowMap())
324      Call A%Multiply_Vector(.false.,this%f,x)
325      allocate(c(NumMyElements))
326      c=x%ExtractCopy()
327    !create vector of df_dx
328      allocate(periodic_2nd_order::d2f_dx2_local)
329      d2f_dx2_local%f=Epetra_Vector(map,zero_initial=.true.)
330      call d2f_dx2_local%f%ReplaceGlobalValues( &
331        NumMyElements,c,MyGlobalElements)
332      call move_alloc(d2f_dx2_local, d2f_dx2)
333    end function
334
335    subroutine output(this,comm)
336      class(periodic_2nd_order) ,intent(in) :: this
337      class(Epetra_Comm),intent(in) ::comm
338      integer(c_int) :: i,NumMyElements,NumGlobalElements
339      integer(c_int), dimension(:), allocatable :: MyGlobalElements
340      real(c_double), dimension(:), allocatable :: f_v
341      real(c_double), dimension(:), allocatable :: f
342      NumGlobalElements=map%NumGlobalElements()
343      NumMyElements=map%NumMyElements()
344      allocate(MyGlobalElements(NumMyElements))
345      MyGlobalElements=map%MyGlobalElements()
346      allocate(f_v(NumMyElements))
347      f_v=this%f%ExtractCopy()
348      allocate(f(NumGlobalElements))
349      call comm%GatherAll(f_v,f,NumMyElements)
350      do i=1,NumGlobalElements
351        if (comm%MyPID()==0) write(20,'(2(E20.12,1x))') x_node(i),f(i)
352      enddo
353    end subroutine
354
355    subroutine force_finalize(this)
356      class(periodic_2nd_order), intent(inout) :: this
357      call this%f%force_finalize
358    end subroutine
359  end module
```

Figure 12.5 (e)

```
1  program main
2  #include "ForTrilinos_config.h"
3  #ifdef HAVE_MPI
4    use mpi
5    use FEpetra_MpiComm, only:Epetra_MpiComm
6  #else
```

```
7   use FEpetra_SerialComm, only:Epetra_SerialComm
8 #endif
9   use ForTrilinos_utils, only : valid_kind_parameters
10  use iso_c_binding, only : c_int,c_double
11  use field_module ,only : field,initial_field
12  use field_factory_module ,only : field_factory
13  use periodic_2nd_order_factory_module,only: periodic_2nd_order_factory
14  use initializer ,only : u_initial,zero
15  implicit none
16 #ifdef HAVE_MPI
17  type(Epetra_MpiComm) :: comm
18 #else
19  type(Epetra_SerialComm) :: comm
20 #endif
21  class(field), pointer :: u,half_uu,u_half
22  class(field_factory), allocatable :: field_creator
23  procedure(initial_field) ,pointer :: initial
24  real(c_double) :: dt,half=0.5,t=0.,t_final=0.6,nu=1.
25  real(c_double) :: t_start,t_end
26  integer(c_int) :: tstep
27  integer(c_int), parameter :: grid_resolution=(64**3)/4
28  integer(c_int)       :: MyPID, NumProc
29  logical              :: verbose
30  integer :: rc,ierr
31
32  if (.not. valid_kind_parameters()) &
33    stop 'C interoperability not supported on this platform.'
34 #ifdef HAVE_MPI
35  call MPI_INIT(ierr)
36  t_start=MPI_Wtime()
37  comm = Epetra_MpiComm(MPI_COMM_WORLD)
38 #else
39  call cpu_time(t_start)
40  comm = Epetra_SerialComm()
41 #endif
42  allocate(periodic_2nd_order_factory :: field_creator)
43  initial => u_initial
44  u => field_creator%create(initial,grid_resolution,comm)
45  initial => zero
46  half_uu => field_creator%create(initial,grid_resolution,comm)
47  u_half => field_creator%create(initial,grid_resolution,comm)
48  do tstep=1,20 !2nd-order Runge-Kutta:
49   dt = u%runge_kutta_2nd_step(nu ,grid_resolution)
50    half_uu = u*u*half
51    u_half = u + (u%xx()*nu - half_uu%x())*dt*half ! first substep
52    half_uu = u_half*u_half*half
53    u  = u + (u_half%xx()*nu - half_uu%x())*dt ! second substep
54    t = t + dt
55    if (comm%MyPID()==0) print *,'timestep=',tstep
56  end do
57  if (comm%MyPID()==0) write(10,*) 'u at t=',t
58 #ifdef HAVE_MPI
59  t_end= MPI_Wtime()
60 #else
61  call cpu_time(t_end)
```

```
62 #endif
63   if (comm%MyPID()==0) write(10,*) 'Elapsed CPU time=',t_end-t_start
64   call u%output(comm)
65   call half_uu%force_finalize
66   call u_half%force_finalize
67   call u%force_finalize
68   call comm%force_finalize
69 #ifdef HAVE_MPI
70   call MPI_FINALIZE(rc)
71 #endif
72 end program
```

Figure 12.5. ForTrilinos-based Burgers solver: (a) field, (b) field_factory, (c) periodic_2nd_factory, (d) periodic_2nd_order, and (3) main program.

12.1.5 Intrinsic Parallelization: Fortran 2008 Coarrays

Our journey toward scalable execution concludes with Fortran 2008 coarrays.[7] As compared to the ways in which we used automatic, multithreaded, and distributed parallel programming models in the preceding sections, coarrays beautifully blend simplicity, power, and portability. Originally designed as a language extension to Fortran 95 by Numrich and Reid (1998), and named by analogy with covariant indices in tensor calculus (Numrich 2005), coarrays aim to help Fortran programmers convert their serial applications to efficient parallel applications with ease.

Reid (2005) summarized the coarray language design as follows:

> The coarray programming model is designed to answer the question 'What is the smallest change required to convert Fortran into a robust and efficient parallel language?'. Our answer is a simple syntactic extension. It looks and feels like Fortran and requires Fortran programmers to learn only a few new rules. These rules are related to two fundamental issues that any parallel programming model must resolve, work distribution and data distribution.

Based on the SPMD programming model, a coarray application replicates itself across multiple execution units. The number of replications is set either at compile time or at program start up time. Each replication is called an **image** and has its own local data. All images execute asynchronously until they encounter a user-defined **synchronization** point.

With the exception of **coarrays**, all data objects declared within the program are local to each image and can be referenced only within that image. References to non-local data are represented syntactically by surrounding co-indices with square brackets as opposed to the usual parentheses that bracket array indices. This minimalistic manner of stipulating communication between images produces significantly simpler code than MPI and yet has the full potential to rival MPI's performance on large, parallel applications. One expects nearly identical MPI and coarray performance with compilers that support coarrays by using MPI under the hood as the open-source Gnu Fortran (gfortran) developers plan to do.[8] Coarray performance

[7] The Fortran 2008 standard was published in October 2010. The final committee draft can be found at ftp://ftp.nag.co.uk/sc22wg5/N1801-N1850/N1830.pdf.
[8] Cite Google archive of comp.lang.fortran

can even exceed that of MPI on platforms where a compiler can take advantage of special hardware support for the coarray programming model. Cray compilers and hardware exhibit this behavior (Numrich 2005).

Coarrays thus provide a powerful duo of simplicity and platform-agnosticism. Whereas syntax greatly enhances the scalability of development-time processes, operating at an abstraction level slightly above other parallel programming models enhances the scalability of runtime processes. The language standard body moved to this higher abstraction level by using the "image" terminology without tying the term to a specific technology. Hence, one compiler development team might implement its coarray support in MPI, whereas another development team might implement it in a hardware-specific instruction set. A third could use OpenMP. And with current trends toward increasing single-socket parallelism with multicore chips, the greatest benefit might ultimately derive from mapping coarray syntax to a mixed model that distributes work locally via multithreading (e.g., with OpenMP) and communicating nonlocally via message passing (e.g., with MPI).

Consider again the 2nd-order, central-difference Burgers equation solver. The following factors arise in attempting to get a supporting ABSTRACT CALCULUS to scale on parallel systems using coarrays:

- Data distribution.
- Work load distribution, or operational model.
- Necessary synchronization.

Next we address these factors in the order listed.

One naturally envisions evenly distributing the grid values and the operations on those values across all images. We therefore embed a coarray component in the `periodic_2nd_order` class. Figure 12.6 illustrates the design of that class. Figure 12.7 shows the implementation.[9] As with our other Burgers equation solvers, `field` remains abstract. A concrete `periodic_2nd_order` extends `field` and stores a reference to the grid values in its `global_f` coarray component. As shown at line 11 in Figure 12.7, the following syntax declares an allocatable coarray:

```
real(rkind) ,allocatable :: global_f(:)[:]
```

The same component could be represented with the alternate syntax:

```
real(rkind),dimension(:),codimension[:],allocatable :: global_f
```

Allocations of `global_f` take the form shown at line 55 in Figure 12.7 and repeated here:

```
allocate (this%global_f(local_grid_size)[*])
```

[9] At the time of this writing, no fully compliant Fortran 2008 compilers have been released. Most compilers with partial compliance have not reached a sufficient maturity in terms of features and robustness to fully test the coarray code in this section. The GNU Fortran compiler is transitioning towards supporting the coarray syntax for single-image runs. All code in this section has been checked for syntactical correctness by compiling it with the development version of GNU Fortran 4.6.0 20100508 (experimental) installed via macports on a Mac OS X 10.6 platform. Where the compiler appears to be in error, we have consulted the draft Fortran 2008 standard for guidance.

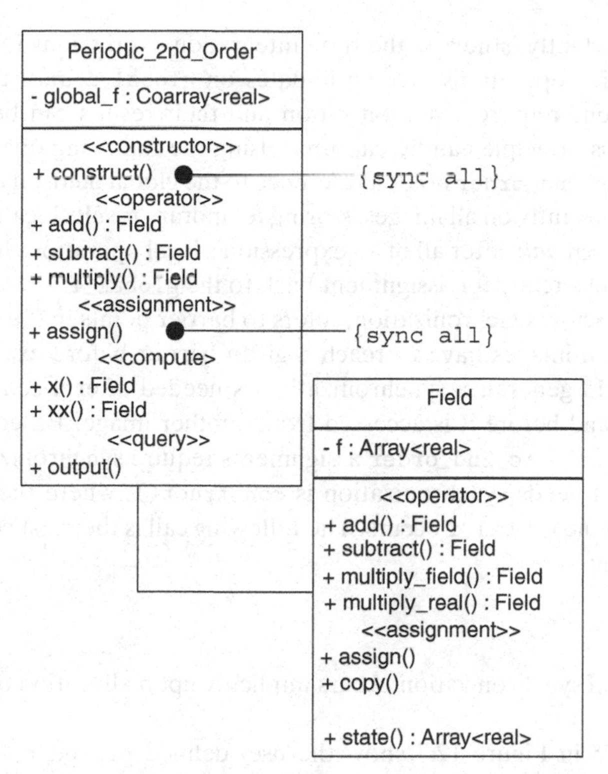

Figure 12.6. Class diagram of periodic_2nd_order class using coarray grid field. Methods constuct() and assign() require synchronization as noted by constraints sync.

where the asterisk represents the number of images that will be set at compile-time or runtime. The language standard precludes any more explicit specification of the number of images to be allocated.

When allocated, `global_f` behaves as though it consists of many rank-one arrays, each affiliated with an image. An image can access its "own" `global_f` array using regular array syntax. It can also access the `global_f` array on another image using an image index syntax that encloses the index in square brackets. Computing the spatial derivatives `x()` and `xx()`, for example, requires cross-image data access at lines 141, 144, and 148 in Figure 12.7.

Considering the second of the aforementioned factors, the work load distribution highlights a critically significant distinction between the `field` interface and its implementation in `periodic_2nd_order` in the context of a coarray ABSTRACT CALCULUS: Each `periodic_2nd_order` type-bound function, such as the operators + and * and the differentiation methods `x()` and `xx()`, returns a `field` instance rather than a `periodic_2nd_order` one. (Figures 12.8 defines the `field` class.) This circumvents Fortran's prohibition against functions returning objects that contain coarray components.[10]

[10] Fortran's prohibition against allocatable coarray components inside function addresses performance concerns. The standard mandates automatic deallocation of allocatable components of a function result after the result's use. Deallocating allocatable coarrays requires synchronizations across all images, which leads to segment ordering difficulties. In our 2nd-order field case, it would also be superfluous to reallocate and copy the global mesh since it doesn't change at all.

More importantly, studying the time integration expressions reveals that none of the supporting operations involve updates on `global_f` until the final assignment. Thus, none require synchronization and their results can be stored locally in a `field`. This principle can be captured using the following operational model: While `periodic_2nd_order` holds a reference to the global field, all its operators are evaluated concurrently on all images, storing temporary results locally. Nonlocal updates need happen *only* after all of an expression's local operations have completed, making the results ready for assignment back to the `global_f`.

The third factor, synchronization, refers to barrier points in the executing code. By definition, all images have to reach a given barrier before executing any subsequent code. In general, a synchronization is needed after a coarray is updated on one image and before it is accessed from another image. Based on our operational model, `periodic_2nd_order` assignments require synchronization. The only other place that needs synchronization is `construct()`, where the allocation and initialization of the `global_f` occurs. The following call is the most common form of synchronization:

```
sync all
```

Another form of synchronization occurs implicitly upon allocation of an allocatable coarray.

Lines 45–75 in Figure 12.7 show the user-defined `periodic_2nd_order` constructor. This procedure could be written as a subroutine due to the aforementioned restriction on functions returning objects containing a coarray component. The `num_images()` intrinsic function employed in the constructor queries the total number of images in the run. The constructor uses that value to compute the local grid size assuming the field is evenly distributed on all images. Function `this_image()` queries the image index of the executing image and uses that value to calculate the initial field values.

Figure 12.7

```
1
2  module periodic_2nd_order_module
3    use kind_parameters ,only : rkind, ikind
4    use field_module ,only : field
5
6    implicit none
7    private
8    public :: periodic_2nd_order, initial_field
9    type periodic_2nd_order
10     private
11     real(rkind), allocatable :: global_f(:)[:]
12   contains
13     procedure :: construct
14     procedure :: assign          => copy
15     procedure :: add             => add_field
16     procedure :: multiply        => multiply_field
17     procedure :: x               => df_dx
18     procedure :: xx              => d2f_dx2
19     procedure :: output
```

```
20      generic   :: assignment(=) => assign
21      generic   :: operator(+)   => add
22      generic   :: operator(*)   => multiply
23    end type
24
25    real(rkind) ,parameter :: pi=acos(-1._rkind)
26
27    abstract interface
28      real(rkind) pure function initial_field(x)
29        import :: rkind
30        real(rkind) ,intent(in) :: x
31      end function
32    end interface
33
34    contains
35
36    subroutine output (this)
37      class(periodic_2nd_order), intent(in) :: this
38      integer(ikind) :: i
39      do i = 1, size(this%global_f)
40          print *, (this_image()-1)*size(this%global_f) + i, &
41                   this%global_f(i)
42      end do
43    end subroutine
44
45    subroutine construct (this, initial, num_grid_pts)
46      class(periodic_2nd_order), intent(inout) :: this
47      procedure(initial_field) ,pointer :: initial
48      integer(ikind) ,intent(in) :: num_grid_pts
49      integer :: i, local_grid_size
50
51      !<-- assume mod(num_grif_pts, num_images()) == 0
52      local_grid_size = num_grid_pts / num_images()
53
54      ! set up the global grid points
55      allocate (this%global_f(local_grid_size)[*])
56
57      this%global_f(:) = grid()
58
59      do i = 1, local_grid_size
60          this%global_f(i) = initial(this%global_f(i))
61      end do
62
63      sync all
64
65    contains
66      pure function grid()
67        integer(ikind) :: i
68        real(rkind) ,dimension(:) ,allocatable :: grid
69        allocate(grid(local_grid_size))
70        do i=1,local_grid_size
71          grid(i)  = 2.*pi*(local_grid_size*(this_image()-1)+i-1) &
72                     /real(num_grid_pts,rkind)
73        end do
74      end function
```

```
75    end subroutine
76
77    real(rkind) function rk2_dt(this,nu, num_grid_pts)
78      class(periodic_2nd_order) ,intent(in) :: this
79      real(rkind) ,intent(in) :: nu
80      integer(ikind) ,intent(in) :: num_grid_pts
81      real(rkind)              :: dx, CFL, k_max
82      dx=2.0*pi/num_grid_pts
83      k_max=num_grid_pts/2.0_rkind
84      CFL=1.0/(1.0-cos(k_max*dx))
85      rk2_dt = CFL*dx**2/nu
86    end function
87
88    ! this is the assignment
89    subroutine copy(lhs,rhs)
90      class(periodic_2nd_order) ,intent(inout) :: lhs
91      class(field) ,intent(in) :: rhs
92
93      ! update global field
94      lhs%global_f(:) = rhs%state()
95      sync all
96    end subroutine
97
98    function add_field (this, rhs)
99      class(periodic_2nd_order), intent(in) :: this
100     class(field), intent(in) :: rhs
101     class(field), allocatable :: add_field
102
103     allocate (add_field)
104     add_field = rhs%state()+this%global_f(:)
105   end function
106
107   function multiply_field (this, rhs)
108     class(periodic_2nd_order), intent(in) :: this, rhs
109     class(field), allocatable :: multiply_field
110
111     allocate (multiply_field)
112     multiply_field = this%global_f(:)*rhs%global_f(:)
113   end function
114
115   function df_dx(this)
116     class(periodic_2nd_order), intent(in) :: this
117     class(field) ,allocatable  :: df_dx
118     integer(ikind) :: i,nx, me, east, west
119     real(rkind) :: dx
120     real(rkind), allocatable :: tmp_field_array(:)
121
122     nx=size(this%global_f)
123     dx=2.*pi/(real(nx,rkind)*num_images())
124
125     allocate(df_dx,tmp_field_array(nx))
126
127     me = this_image()
128
129     if (me == 1) then
```

```
130          west = num_images()
131          east = 2
132       else if (me == num_images()) then
133          west = me - 1
134          east = 1
135       else
136          west = me - 1
137          east = me + 1
138       end if
139
140       tmp_field_array(1) = &
141         0.5*(this%global_f(2)-this%global_f(nx)[west])/dx
142
143       tmp_field_array(nx) = &
144         0.5*(this%global_f(1)[east]-this%global_f(nx-1))/dx
145
146       do i=2,nx-1
147         tmp_field_array(i)=&
148           0.5*(this%global_f(i+1)-this%global_f(i-1))/dx
149       end do
150
151       df_dx = tmp_field_array
152     end function
153
154     function d2f_dx2(this)
155       class(periodic_2nd_order), intent(in) :: this
156       class(field) ,allocatable   :: d2f_dx2
157       integer(ikind) :: i,nx, me, east, west
158       real(rkind) :: dx
159       real(rkind), allocatable :: tmp_field_array(:)
160
161       nx=size(this%global_f)
162       dx=2.*pi/(real(nx,rkind)*num_images())
163
164       allocate(d2f_dx2,tmp_field_array(nx))
165
166       me = this_image()
167
168       if (me == 1) then
169          west = num_images()
170          east = 2
171       else if (me == num_images()) then
172          west = me - 1
173          east = 1
174       else
175          west = me - 1
176          east = me + 1
177       end if
178
179       tmp_field_array(1) = &
180         (this%global_f(2)-2.0*this%global_f(1)+this%global_f(nx)[west])&
181         /dx**2
182
183       tmp_field_array(nx) =&
184         (this%global_f(1)[east]-2.0*this%global_f(nx)+this%global_f(nx-1))&
```

```
185        /dx**2
186
187      do i=2,nx-1
188        tmp_field_array(i)=&
189          (this%global_f(i+1)-2.0*this%global_f(i)+this%global_f(i-1))&
190          /dx**2
191      end do
192
193      d2f_dx2 = tmp_field_array
194    end function
195  end module
```

Figure 12.7. Definition of periodic 2nd-order central difference field class contains a coarray global field. Operation results are stored locally using field type.

Figure 12.8

```
1   module field_module
2     use kind_parameters ,only : rkind, ikind
3     implicit none
4     private
5     public :: field
6     type :: field
7       real(rkind), allocatable :: f(:)
8     contains
9       procedure :: add             => total
10      procedure :: subtract        => difference
11      procedure :: multiply_field => product
12      procedure :: multiply_real  => multiple
13      procedure :: assign          => assign_field_f
14      procedure :: copy            => copy_filed
15      procedure :: state           => field_values
16      generic   :: operator(+)     => add
17      generic   :: operator(-)     => subtract
18      generic   :: operator(*)     => multiply_real,multiply_field
19      generic   :: assignment(=)   => assign, copy
20    end type
21
22  contains
23
24    function field_values (this)
25      class(field), intent(in) :: this
26      real(rkind), allocatable :: field_values(:)
27      field_values = this%f
28    end function
29
30    subroutine assign_field_f (lhs, rhs)
31      class(field), intent(inout) :: lhs
32      real(rkind), intent(in) :: rhs(:)
33      lhs%f = rhs
34    end subroutine
35
36    subroutine copy_filed (lhs, rhs)
37      class(field), intent(inout) :: lhs
38      class(field), intent(in) :: rhs
39      lhs%f = rhs%f
```

```
40    end subroutine
41
42    function total(lhs,rhs)
43      class(field) ,intent(in) :: lhs
44      class(field) ,intent(in) :: rhs
45      class(field) ,allocatable :: total
46      allocate (total)
47      total%f = lhs%f + rhs%f
48    end function
49
50    function difference(lhs,rhs)
51      class(field) ,intent(in) :: lhs
52      class(field) ,intent(in)  :: rhs
53      class(field) ,allocatable :: difference
54      allocate (difference)
55      difference%f = lhs%f - rhs%f
56    end function
57
58    function product(lhs,rhs)
59      class(field) ,intent(in) :: lhs
60      class(field) ,intent(in)  :: rhs
61      class(field) ,allocatable :: product
62      allocate(product)
63      product%f = lhs%f * rhs%f
64    end function
65
66    function multiple(lhs,rhs)
67      class(field) ,intent(in) :: lhs
68      real(rkind) ,intent(in)  :: rhs
69      class(field) ,allocatable :: multiple
70      allocate(multiple)
71      multiple%f = lhs%f * rhs
72    end function
73  end module
```

Figure 12.8. Definition of field type that stores operation results from periodic_2nd_order class.

12.2 Case Studies: Multiphysics Modeling

This section presents several of the case studies from which we culled the patterns that appear in this book. In addition to inspiring some of the patterns, these cases demonstrate the broad spectrum of physics subdisciplines to which the design principles in this book apply. Fluid dynamics lies at the core of each problem. Characteristic of multiphysics modeling, we couple numerous other phenomena to the flows under study. These phenomena include quantum vortices, magnetic induction and diffusion, electromagnetic wave propagation, and solid particle transport. We also vary the representation of the fluid flow itself from the most common, a deterministic continuum mechanics approximation (the Navier-Stokes and the magnetic induction equations), to the most exotic, a discrete vortex filament approximation (the superfluid equation and Biot-Savart law), and finally an inherently probabilistic particle representation (the lattice Boltzmann equation).

The following sections briefly describe the physics, the software architecture, and the simulation results. We completed most of the related work before the availability of Fortran 2003 compilers. As such, some of the symbols in the associated UML diagrams represent constructs or relationships we emulated without direct language support – namely the interface implementation (inheritance) relationship symbol and the abstract class notation. The emulation techniques build upon the programming styles outlined by Decyk et al. (1997b, 1997a, 1998) and Akin (2003) and supported by the memory management infrastructure described by Rouson et al. (2006).

12.2.1 Quantum Turbulence in Superfluid Liquid Helium

Section 4.3.2 provides an overview of the superfluid phenomenology. At sufficiently high driving velocities, the quantum vortices described in that section become a complicated tangle often referred to in the literature as *quantum turbulence* (Vinen and Niemela 2002). Despite the microscopic difference between this turbulent state and that of a classical fluid, numerous laboratory experiments have demonstrated that the macroscopic statistics closely resemble classical turbulence. The aim of the simulations was to investigate a widely hypothesized explanation for this similarity: vortex locking, the bundling of quantum vortices colocated and aligned with their normal-fluid counterparts so that the overall two-fluid mixture moves in tandem (Morris et al. 2008). Figure 12.9 demonstrates this bundling, colocation, and alignment.

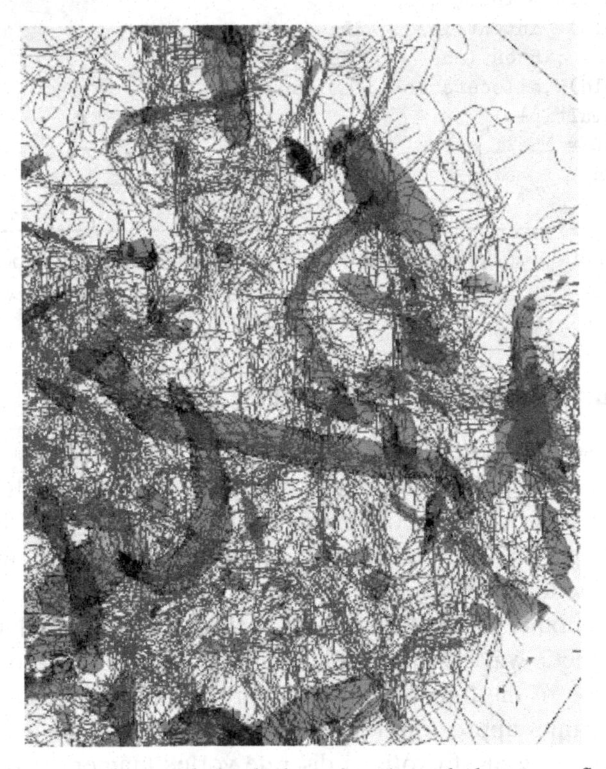

Figure 12.9. Quantum vortices and classical vortices in turbulent superfluid liquid helium. Image courtesy of Morris (2008). For a color version of this image, please visit http://www.cambridge.org/Rouson.

The Navier-Stokes equations govern the normal fluid velocity $\mathbf{u}$:

$$\mathbf{u}_t = -\mathbf{u}\cdot\nabla\mathbf{u} - \nabla p + \nu\nabla^2\mathbf{u} + \mathbf{f} \tag{12.4}$$

$$\nabla\cdot\mathbf{u} = 0 \tag{12.5}$$

where p is the hydrostatic pressure, ν is the kinematic viscosity, and $\mathbf{f}$ is the mutual friction force between the superfluid and normal fluid. The total material velocity is $\mathbf{u}+\mathbf{v}$. Morris et al. (2008) solved a transformed system of two equations: one corresponding to the Laplacian of one component of equation (12.4) and a second corresponding to the parallel component of the curl of equation (12.4). Choosing the second component in the Cartesian system $\{x_1,x_2,x_3\}$ and defining the vorticity $\omega \equiv \nabla\times\mathbf{u}$ and the nonlinear product $\mathbf{H} \equiv \mathbf{u}\times\omega$ yields:

$$\frac{\partial}{\partial t}\nabla^2 u_2 = -\frac{\partial}{\partial x_2}(\nabla\cdot\mathbf{H}) + \nabla^2 H_2 + \nu\nabla^2(\nabla^2 u_2) \tag{12.6}$$

$$\frac{\partial}{\partial t}\omega_2 = (\nabla\times\mathbf{H})_2 + \nu\nabla^2\omega_2 \tag{12.7}$$

where the mutual friction force was dropped under the assumption that it influences the normal fluid motion much less than it does the massless superfluid vortex lines. The advantage of equations (12.6)–(12.7) over (12.4) lies in the global conservation of energy (when $\nu = 0$) as well as in the reduction of the number of dependent variables from four to two.[11]

Section 4.3.2 presented the superfluid evolution equations (4.7)–(4.8), which we repeat here for convenience:

$$\mathbf{v} = \frac{\kappa}{4\pi}\int\frac{(\mathbf{S}-\mathbf{r})\times d\mathbf{S}}{\|\mathbf{S}-\mathbf{r}\|^3} \tag{12.8}$$

$$\frac{d\mathbf{S}}{dt} = \mathbf{v} - \alpha\mathbf{S}'\times(\mathbf{u}-\mathbf{v}) - \alpha'\mathbf{S}'\times[\mathbf{S}'\times(\mathbf{u}-\mathbf{v})] \tag{12.9}$$

Section 4.3.2 defines each of the quantities in the latter two equations. Morris et al. (2008) solved equations (12.6)–(12.7) on the semiopen, cubical domain $x_j \in [0,2\pi), j = 1,2,3$ with periodic boundary conditions in each direction.

Figure 12.10 shows the class model emulated by the solver of Morris et al. (2008). At the heart of this class diagram are two physics abstractions: a Fluid class that encapsulates equations (12.6)–(12.7) and a Tangle class that encapsulates equations (4.7)–(4.8). The depicted Integrand abstract class specifies the interface physics abstractions must satisfy in order to be integrated over time by the marching algorithm of Spalart et al. (1991). Appendix A describes that scheme. An exercise at the end of the current chapter asks the reader to write ABSTRACT CALCULUS versions of the governing equations and to enhance the class diagram to show the public methods and the likely private data that would support these expressions.

Two patterns appear in Figure 12.10. The Superfluid class embodies the PUPPETEER pattern, whereas the Integrand class embodies an ABSTRACT CALCULUS for

[11] The differentiations required to go from equation (12.4) to equation (12.7) destroy information that can be recovered by solving the original equations averaged over the two directions of differentiation: x_1 and x_3. See Rouson et al. (2008b) for more detail.

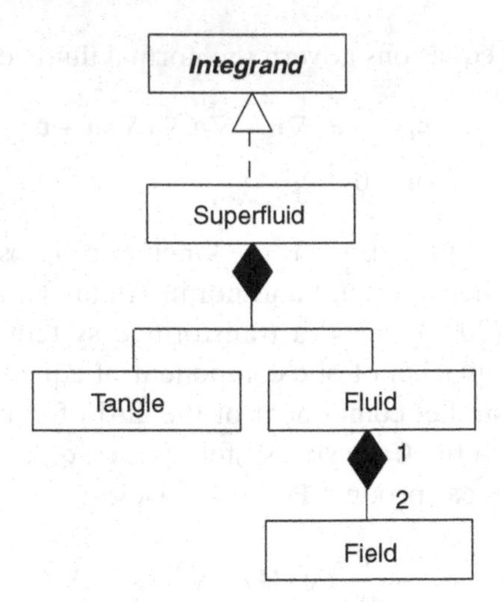

Figure 12.10. Quantum turbulence solver architecture.

time integration. That figure also presents several additional opportunities for using patterns. Whereas the Field class embodies an ADT calculus for spatial differentiation, use of the ABSTRACT CALCULUS pattern would increase the flexibility of the design by replacing the concrete Field with an abstract class. To facilitate this, one might also add an ABSTRACT FACTORY and a FACTORY METHOD to the design. Furthermore, the time-integration process could be made more flexible via a STRATEGY and a supporting SURROGATE. Another exercise at the end of the chapter asks the reader to add these elements to the design.

In addition to the time advancement role the Superfluid PUPPETEER plays, it also shuttles information between the Fluid and the quantum vortex Tangle. Specifically, the Superfluid queries the Tangle for the locations of each vortex filament grid point. The Superfluid queries the Fluid for the local velocities at each location vortex point and passes these velocities to the Tangle for use in equation (4.8). Likewise, in a fully consistent simulation that accounts for the influence of the quantum vortices on the normal fluid motion, the Superfluid would query the Tangle for the mutual friction vectors at each vortex point and pass these to the Fluid for incorporation into the Navier-Stokes equation (12.4).

In many applications, the Superfluid might also referee the competing stability and accuracy requirements of its puppets. A likely algorithm for doing so would involve polling the puppets for their suggested time steps based on their private evolution equations and state vectors at a given point in time. The Superfluid would then choose the smallest answer so as to satisfy the accuracy and stability requirements of each puppet. It might be useful to design an abstract POLLSTER to conduct such polls when passed a pointer array of poll respondents. To facilitate this, each physics abstraction could extend a Respondent type, and a polymorphic `poll()` procedure could be written to collect the responses and set the time step.

12.2.2 Lattice-Boltzman Biofluid Dynamics

This section briefly introduces the lattice Boltzmann equation (LBE) method and an applies it to a biofluid engineering problem: blood flow around a stent. The

Boltzmann equation originates from the kinetic theory of gases with the two assumptions: binary collisions and a lack of correlation between a molecule's velocity and its position (Huang 1963). The Boltzmann equation follows:

$$\frac{\partial f}{\partial t} + \mathbf{e}\Delta f = \Omega(f) \tag{12.10}$$

with the primary variables $f(\mathbf{x},\mathbf{e},t)$ representing the particle probability distribution functions, $\mathbf{x}$ describing the particles' locations, $\mathbf{e}$ denoting the their velocities, and Ω containing collisional information. For incompressible flow, the collision term can be simplified using Bhatnagar-Gross-Krook (BGK) model (Bhatnagar et al. 1954) as $-(f - f^{eq})/\tau$, where f^{eq} is the particle distribution function at the state of thermal equilibrium and τ is the relaxation time. We restrict ourselves to the BGK model in this section.

The LBE method borrows ideas from cellular automata fluids (Frisch et al. 1986). A regular lattice covers the physical space domain. Particles are distributed in each lattice with probability distribution function f. This approach can also be viewed as a finite-difference approximation to the Boltzmann equation (He and Luo 1997). Particles travel to neighbor lattices along different directions and collide with each other. Figure 12.11 shows the lattice geometry and velocity vectors in the two-dimensional nine-speed *D2Q9* model(He and Luo 1997). A first-order-accurate discretization in time leads to:

$$f_i(\mathbf{x}+e_i\Delta t, t+\Delta t) - f_i(\mathbf{x},t) = -\frac{1}{\tau}(f_i(\mathbf{x},t) - f_i^{eq}(\mathbf{x},t)) \quad (i = 0,1,\cdots,8) \tag{12.11}$$

where the particle velocities e_i is given by:

$$e_i = \begin{cases} (0,0) & i=0 \\ c(cos(i-1)\pi/2, sin(i-1)\pi/2) & i=1,2,3,4 \\ \sqrt{2}c(cos(2i-9)\pi/2, sin(2i-9)\pi/2) & i=5,6,7,8 \end{cases} \tag{12.12}$$

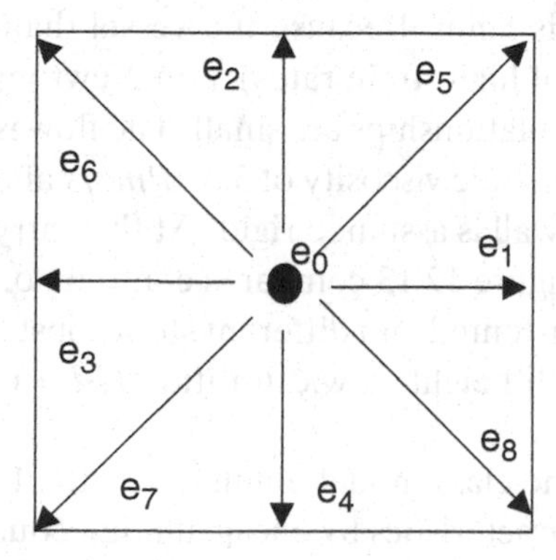

Figure 12.11. The lattice of D2Q9 model.

where here $c = \Delta x/\Delta t$ is the lattice velocity. The equilibrium form of the distribution function is approximated as:

$$f_i^{eq} = \rho\omega_i \left[1 + \frac{1}{c_s^2} e_i \cdot \mathbf{u} + \frac{1}{2c_s^4}(e_i \cdot \mathbf{u})^2 - \frac{1}{2c_s^2}\mathbf{u} \cdot \mathbf{u} \right] \tag{12.13}$$

where $c_s = c/\sqrt{3}$ is the "LBE sound speed" (He and Luo 1997), the weighting functions $\omega_0 = 4/9$, $\omega_i = 1/9$ for $i = 1,2,3,4$, and $\omega_i = 1/36$ for $i = 5,6,7,8$. The fluid density ρ and velocity $\mathbf{u}$ is obtained by:

$$\rho = \sum_{i=0}^{8} f_i, \qquad \rho\mathbf{u} = \sum_{i=0}^{8} e_i f_i \tag{12.14}$$

This completes the mathematical model.

The LBE method has been applied to many engineering and science problems including turbulence (Cosgrove et al. 2003), porous media flow (Manz et al. 1999), multiphase flows (Boyd et al. 2004) and physiological flows (Artoli et al. 2006). One of the advantages of using the LBE method is that data communications between lattices are always local, which makes the method extremly efficient for large-scale parallel computations. Since LBE method is derived based on kinetic equations in mesoscopic scales, it is also very useful for as a scale-bridge for multiscale simulations.

The motivation for the work described here stems from the problems in the use of stents in aneurysms. Despite stents' ability to reduce blood flow into an an aneurysm sac, complications occur that require a deeper understanding of the fluid dynamics. Xu and Lee (2008) used D2Q9 LBE method to perform a simulation of flows in aneurysms with and without a transverse stent obstacle on the upper wall to analyze the flow inside the artery. Won et al. (2006) investigated problems encountered during and after stentgraft treatment, including stentgraft migration, stentgraft folding, cerebral ischemia and mechanical failure. The porosity and size of the stent and the difficulty in implementation to complex vessel geometry are notable problems. The modification of the flow field also causes coagulation in the aneurysm leading to its permanent occlusion after treatment(Lieber et al. 1997). Therefore, minimal flow changes by the disease treatment and the ease of implementation to the patient-specific geometry are very important.

As shown in Figure 12.12, different shapes of obstacles situated before an aneurysm were used. The parent vessel is 4mm in diameter, and the diameter of the sac of the aneurysm is 5mm. Because the vessel diameter is larger than 1mm (i.e., it falls in the region of high strain rates), non-Newtonian effect associated with complicated shear-strain relationships are small. The flow is therefore assumed to be Newtonian fluid with a dynamic viscosity of $3.510^6 m^2/s$ and a density of $1060 kg/m^3$. For simplicity, the vessel wall is assumed rigid. At the entry of the artery, a pulsatile inlet velocity was used. Figure 12.13 compares contours of the vorticity (the curl of the velocity) between nonstented and different stent (obstacle) shapes. It shows that the rectangular stents with height-to-width ratio $H/d = 0.2$ resulted in the largest vorticity reduction,

Figure 12.14 shows the class model emulated in the LBE solver. A Fluid class implements the blood characteristics by encapsulating equation 12.11. The "stream-and-collide" time marching algorithm is implemented in the Integrand abstract class.

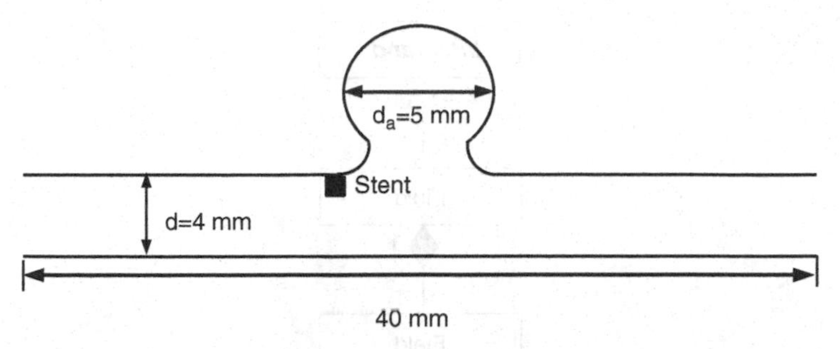

Figure 12.12. The computational domain of the aneurysm model.

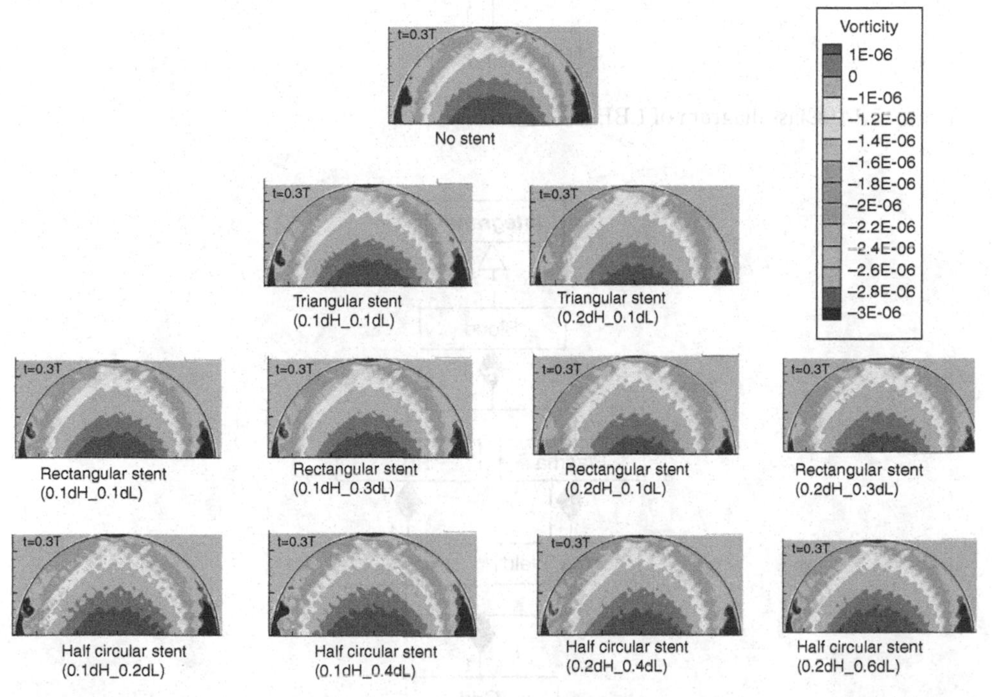

Figure 12.13. Vorticity contours inside the sac of aneurysm under the effects of different shapes of stent. For a color version of this image, please visit http://www.cambridge.org/Rouson.

Using the basic two-step "stream-and-collide" algorithm, equation 12.11 can be discretized into two equations as:

$$Collisionstep : \tilde{f}_i(\mathbf{x},t) = f_i(\mathbf{x},t) - \frac{1}{\tau}(f_i(\mathbf{x},t) - f_i^{eq}(\mathbf{x},t)) \tag{12.15}$$

$$StreamingStep : f_i(\mathbf{x} + e_i\Delta t, t + \Delta t) = \tilde{f}_i(\mathbf{x},t) \tag{12.16}$$

The lattice information is stored in the Grid class, including the additional boundary lattice information used to treat the complicated geometries. The information in the Grid class is hidden from Fluid class and can only be accessed from the Field class, which represents the particle distribution functions in nine directions.

With this code structure, the solver can be extended to three-dimensional code by simply modifying the Field and Grid classes to 3D domains while keeping the

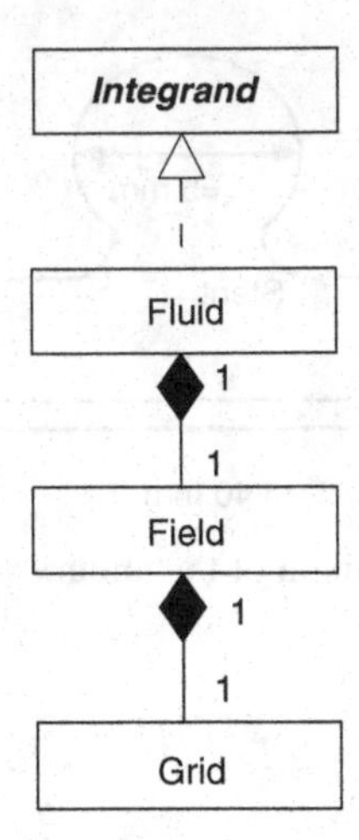

Figure 12.14. Class diagram of LBE solver.

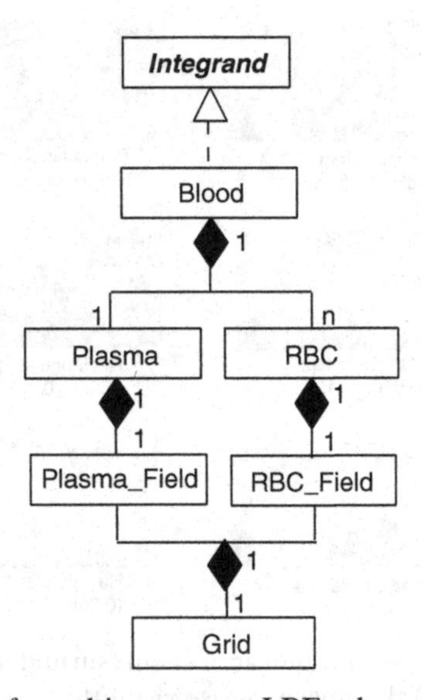

Figure 12.15. Class diagram of a multicomponent LBE solver.

Fluid and Integrand classes the same. One could also develop a general solver for both 2D and 3D flows by replacing the Field and Grid classes with ABSTRACT FACTORY class and adding FACTORY METHOD to 2D or 3D application. For the bioengineering problems we discussed in this section, the blood is treated as a Newtonian fluid due to the large diameter of the artery is. When people are more interested in better understanding the blood characteristics, detailed numerical simulations will be needed to study the interactions among the blood plasma, blood cells, and elastic vessels. Due to the different properties of blood plasma and blood cells, a multicomponent solver would be needed for such a study. Figure 12.15 shows a class diagram that extends the single-component LBE solver to a multicomponent solver. The RBC class represents the red blood cells in the blood plasma. The PUPPETEER pattern implemented in the Blood class facilitated the extension.

12.2.3 Particle Dispersion in Magnetohydrodynamics

Magnetic fields often provide the only practical means for pumping, stirring, and otherwise controlling the motions of molten metals (Mulcahy 1991; Lu et al. 2002). In one important class of industrial applications, semisolid metals processing, the medium so manipulated consists of solid nucleation sites dispersed throughout a liquid metal casting. The mechanical properties of semisolids render them highly manipulable, offering the prospect of manufacturing so-called near-net-shape products that require little filing or other materially wasteful final preparations. Semisolids might well be found also in natural settings near solid/liquid phase boundaries such as the inner edge of the earth's liquid mantle.

Particle dispersions in purely hydrodynamical flows – that is those without magnetic field effects – often exhibit preferential concentration, wherein particles concentrate in certain flow regions. Less is known about the particle dispersion in magnetohydrodynamic turbulent flow. One can investigate this behavior in semisolid metals by solving the magetohydrodynamics (MHD) equations coupled to a set of Lagrangian equations that describe the motion of immersed particles. MHD combines the Navier-Stokes equations with Maxwell's equations in the form of the magnetic induction equation (Davidson 2001). The incompressible, homogeneous MHD flow can be described by the incompressible MHD equations:

$$u_{i,i} = 0 \tag{12.17}$$

$$b_{i,i} = 0 \tag{12.18}$$

$$\partial_t u_i = -P_{,i} + H_i + \nu u_{i,jj} + b_j b_{i,j} + B_j^{ext} b_{i,j} + B_{i,j}^{ext} b_j \tag{12.19}$$

$$\partial_t b_i = -u_j b_{i,j} + b_j u_{j,i} - u_j B_{i,j}^{ext} + B_j^{ext} u_{i,j} + \eta b_{ii} \tag{12.20}$$

where u_i and b_i are the components of velocity and magnetic fluctuation fields, ν is the kinematic viscosity, η is the magnetic diffusivity, B_i^{ext} is the magnitude of externally applied magnetic field, $P = p/\rho + 1/2u_i u_i$ is the modified pressure, ρ is the fluid density, p is the fluctuating pressure, $H_i = \epsilon_{i,j,k} u_j \omega_k$, $\epsilon_{i,j,k}$ is the alternating unit symbol, and $\omega_i = \epsilon_{i,j,k} u_{k,j}$ is the vorticity. Assuming particles follow the drag law of Stokes (Stokes 1851), the equations of motion for each particle are:

$$\frac{d\mathbf{r}}{dt} = \mathbf{v} \tag{12.21}$$

$$\frac{d\mathbf{v}}{dt} = \frac{1}{\tau_p}[u(\mathbf{r},\mathbf{t}) - \mathbf{v}] \tag{12.22}$$

$$\tau_p \equiv \frac{2\rho_p a^2}{9\rho\nu} \tag{12.23}$$

where $\mathbf{r}$ and $\mathbf{v}$ are the particle position and velocity, receptively; $\mathbf{u}(\mathbf{r},\mathbf{t})$ is the undisturbed fluid velocity in the neighborhood of the particle; and τ_p, a and ρ_p are the particle's hydrodynamic response time, radius, and material density, respectively. Equation (12.22) balances particle inertia (as characterized by its mass $4\pi\rho_p a^3/3$) against the molecular momentum exchange integrated over the particle surface (as characterized by the transport coefficient ν). The undisturbed fluid velocity, $\mathbf{u}(\mathbf{r},\mathbf{t})$, interpolated to particle location, $\mathbf{r}$, estimates the appropriate freestream velocity in the

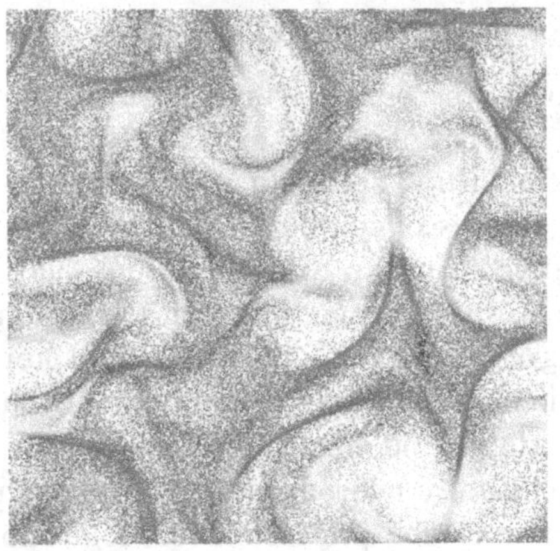

Figure 12.16. Planar projection of inertial particles in liquid-metal MHD turbulence under the influence of an externally applied magnetic field orthogonal to the viewing plane: fastest (red), slowest (blue). For a color version of this image, please visit http://www.cambridge.org/Rouson.

neighborhood of the particle. The dependence on ρ_p indicates that denser particles respond more slowly to changes in the neighboring fluid velocity. The dependence of the dynamic viscosity, $\mu \equiv \rho\nu$, and particle radius a indicates that smaller particles in more viscous fluids more rapidly track changes in the local fluid velocity. The particles can be treated as passive scalar when the particle volume fraction is low.

Rouson et al. (2008a) simulated the particle dispersion in the initially isotropic MHD turbulence under the influence of a spatially constant external magnetic field at low magnetic Reynolds number. In MHD turbulence, the particles tend to cluster near the large-strain regions where both magnetic filed and the fluid velocity field appear strong fluctuations. Figure 12.16 shows the particle positions projected onto the plane which is perpendicular to the direction of magnetic force. The apparent voids driven by the large strain represent approximately cylindrical evacuated regions extending the entire length of the problem domain, thin sheets of particle clusters surrounding the voids are observed.

The class model shown in Figure 12.17 emulates the solver coupling MHD and Lagrangian particle equations. The Magnetofluid class represents the MHD physical abstraction and encapsulates equations (12.17)–(12.20). The Cloud class is the abstraction of inertial particles and encapsulates equations (12.21)–(12.23). The time marching algorithm of Spalart et al. (1991) is implied in the Integrand abstract class. The PUPPETEER pattern is implemented in two classes: Semisolid class and Magnetofluid class, whereas the ABSTRACT CALCULUS pattern appears in the Integrand class.

12.2.4 Radar Scattering in the Atmospheric Boundary Layer

Remote sensing observations via lidar and millimeter-wavelength radar are revealing the fine-scale features of turbulence in the earth's atmosphere with unprecedented detail. The length scales at which researchers have recently observed scattered return signals in clear atmospheres, however, pose challenges for both the theoretical

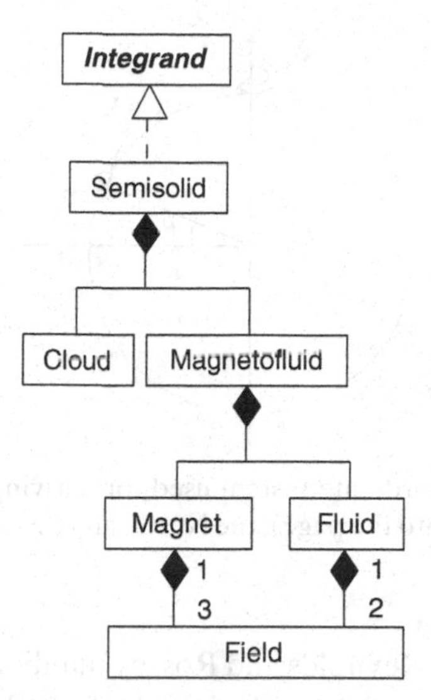

Figure 12.17. Semisolid MHD solver architecture.

understanding and computational exploration (Linde and Ngo 2008). Theoretically, the standard statistical treatments of turbulence posit a spectrum of eddy sizes that cuts off above the lengths required to scatter radar at the observed frequencies (Tatarskii 1961; Atlas et al. 1966; Doviak and Zrnic 1984). Computationally, the broadness of the spectrum of eddy sizes makes direct numerical simulation (DNS) of all dynamically relevant scales infeasible on even the largest and fastest computers.

To gain insights while rendering the problem tractable, Morris et al. (2010) recently performed simulations of the turbulent Ekman layer, a model problem relevant to the dynamics of the atmosphere boundary layer (ABL) near the earth's surface. In the atmosphere, the Ekman layer is the vertical region over which the wind transitions from its prevailing "geostrophic" direction imposed by atmospheric pressure imbalances to its near-surface direction strongly influenced by the earth's rotation and friction at the ground. Figure 12.18 shows the geometry and coordinate-system configuration imagined in an Ekman-layer simulation. In the case Morris and colleagues studied, $\theta = \pi/2$; hence, the angular velocity vector Ω points along the x_3 direction. The simulation domain is a small rectangle with one boundary containing the x_1-x_2 ground plane in Figure 12.18.

Morris et al. (2010) solved a dimensionless form of equations (12.4)–(12.5):

$$\mathbf{u}_t = -\mathbf{u}\cdot\nabla\mathbf{u} - \nabla\mathbf{p} - \frac{1}{Ro}\mathbf{e}_\Omega \times (\mathbf{u} - \mathbf{e_G}) + \frac{1}{R}\nabla^2\mathbf{u} \qquad (12.24)$$

$$\nabla\cdot\mathbf{u} = 0 \qquad (12.25)$$

where the velocity $\mathbf{u}$ has components u_1, u_2, and u_3 in the x_1, x_2, and x_3 coordinate directions, which Morris et al. refer to as the streamwise, wall-normal, and spanwise directions, respectively. In equations (12.4)–(12.5), all variables are rendered nondimensional by normalizing them by the geostrophic wind speed G, the computational domain height δ, and the fluid density ρ. The dimensionless parameters $R \equiv G\delta/\nu$

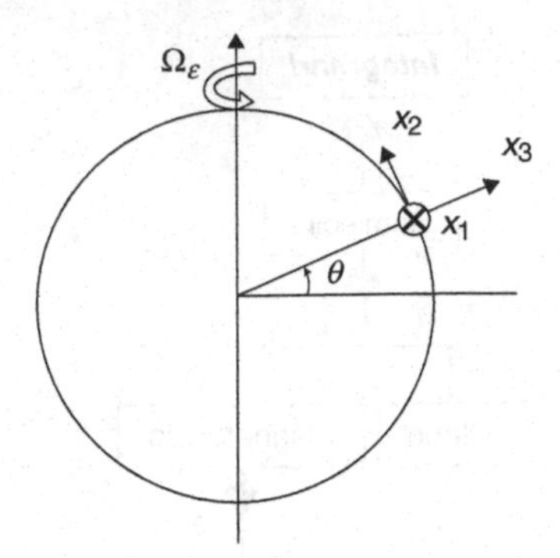

Figure 12.18. Geophysical coordinate system used for deriving the governing equations: x_1, x_2, and x_3 point to the East (into the page), the North, and the vertical directions, respectively.

and $Ro \equiv G/(2\Omega\delta)$ are the Reynolds and Rossby numbers; ν is the kinematic viscosity; Ω is the angular velocity of the system; and p is the hydrodynamic pressure plus appropriate potentials.

The unit vectors associated with the system rotation and the geostrophic velocity are $\mathbf{e}_\Omega$ and $\mathbf{e}_G$, respectively. In equation (12.24), $\mathbf{e}_G = \mathbf{e}_{x_1}$, and $\mathbf{e}_\Omega = \mathbf{e}_{x_2}$. These choices generate Coriolis forces solely in the horizontal (x_1-x_3) plane and an effective driving uniform pressure gradient in the negative x_3 direction.

Figure 12.19 depicts the physics of interest in simulating the ABL for purposes of understanding radar return signals. Figure 12.19(a) demonstrates streamwise velocity component contours characteristic of vertical plumes that spread horizontally away from the ground. Figure 12.19(b) demonstrates the high mechanical energy dissipation rates that surround the depicted plumes. Dissipative eddies make up the flow structures with the shortest length and time scales and therefore represent the most likely candidates for scattering short-wavelength electromagnetic waves. Morris et al. demonstrated highly non-Gaussian statistics of the dissipation and the corresponding much greater likelihood of the presence of the smallest-scale features than would otherwise be estimated based on average dissipation rates.

A useful next step in understanding the anomalous return signals would be to predict variation of quantities known to cause scattering. In the scattering mechanism identified by Tatarskii (1961), this principally includes the relative humidity. Additionally, the relevant plume dynamics would likely be influenced by other scalar thermodynamic variables such as the absolute temperature. A complete study would then involve predicting the return signals of the simulated scattering fields.

Figure 12.20 diagrams a solver architecture for simulating scalar transport in the ABL. The depicted Atmosphere PUPPETEER aggregates zero or more scalar quantities that would presumably be simulated via a scalar advection/diffusion equation of the form:

$$\phi_t = -\mathbf{u} \cdot \nabla\phi + \frac{1}{Sc}\nabla^2\phi \tag{12.26}$$

where ϕ is the transported scalar and $Sc \equiv G\delta/\Gamma$ is the dimensionless parameter characterizing the relative importance of bulk transport and molecular diffusion with a diffusion coefficient Γ. In cases where these scalars actively influence the air flow – for example, temperature variations that impart momentum to the air via buoyancy

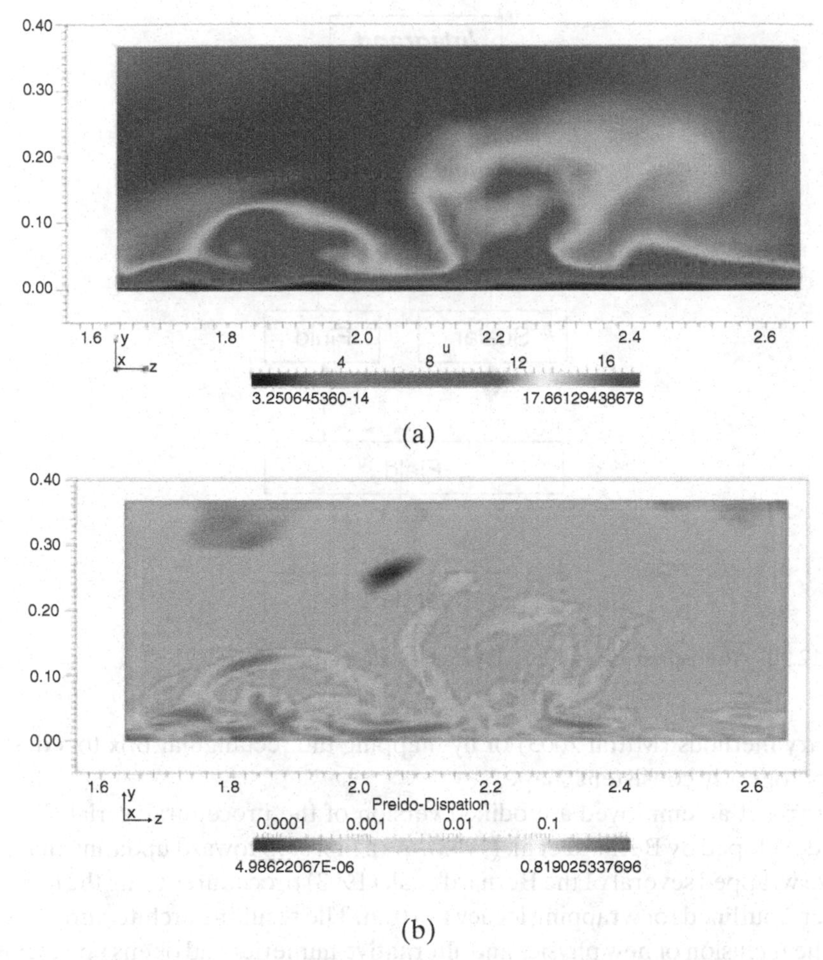

Figure 12.19. Instantaneous contours in a vertical slice of an Ekman layer: (a) wind velocity component, and (b) pseudo-dissipation. For a color version of this image, please visit http://www.cambridge.org/Rouson.

forces – the Atmosphere abstraction would mediate the communication of coupling terms in the manner described in Chapter 8.

The multiphysics solver discussions in Sections 12.2.1–12.2.3 neglected the impact of boundary condition specification on the solver architectures. The simplicity of the geometry and boundary conditions in those cases facilitated the implicit inclusion of boundary constraints on a Field into the the Field ABSTRACT CALCULUS abstraction itself. Ultimately, realistic simulation of the ABL problem is likely to require studying a sufficiently broad a range of boundary condition types, values and geometries as to greatly increase the complexity of the Field class should it subsume all of the related possibilities. In such cases, Shalloway and Trott (2002) recommend, "Find what varies and encapsulated it." The Boundary class in Figure 12.20 thus encapsulates the boundary condition values, types, and geometry to insulate the Field class from the associated complexities. In the aforementioned rectangular box geometry, a Field would aggregate six boundaries, each corresponding to a face of the box. More complicated ground topographies might include additional boundaries by immersed

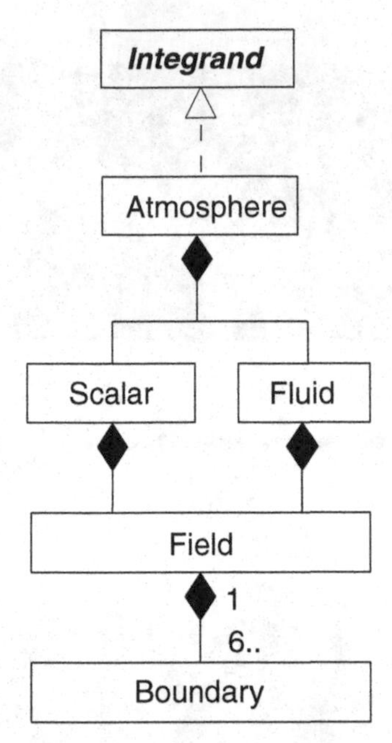

Figure 12.20. Atmospheric boundary layer class diagram.

boundary methods (Mittal 2005) or by mapping the rectangular box to a deformed but topologically consistent shape.

Morris et al. employed a modified version of the procedural, serial Fortran 77 solver developed by Bernard et al. (1993). As a first step toward updating that solver, we have wrapped several of the Bernard et al. (1993) procedures using the techniques Chapter 2 outlined for wrapping legacy Fortran. The resulting architectural flexibility eases the inclusion of new physics and alternative numerics and opens up the possibility of progressive replacement of procedures with scalable numerical algorithms from Trilinos, leading to faster simulations with increased resolution of the flow features of interest. The completion of this new architecture remains an ongoing effort.

12.3 The Morfeus Framework

Morfeus is the Multiphysics Object-oriented Reconfigurable Fluid Environment for Unified Simulations. The acronym is one transliteration for the name of the ancient Greek god more commonly translated to English as "Morpheus." Morpheus had the ability to appear in people's dreams while taking on many different forms. The Morfeus project aims to bring this sort of flexibility to scientific software design via patterns. Morfeus also aims to be the first such pattern-based framework implemented in object-oriented Fortran 2003. It thereby provides a straightforward migration path to OOD patterns and OOP for Fortran programmers, for most of whom both OOP and patterns present significant new conceptual leaps.

Morfeus consists of an open-source repository[12] of interfaces (abstract classes) and example concrete implementations of those interfaces in the context of

[12] As of this writing, the Trilinos project leadership has agreed to host Morfeus in a repository for terminal packages, that is, applications that build atop Trilinos packages but on which Trilinos does

multiphysics applications. A suggested Morfeus use case involves end-users substituting their own concrete classes for the Morfeus examples and writing applications that access those concrete classes only through the interfaces Morfeus publishes. The determination of what concrete implementation gets invoked would then happen at runtime via dynamic polymorphism. Morfeus provides only the structure through which user-defined methods get invoked on user-defined objects. Morfeus thus exemplifies a framework in the manner Gamma et al. (1995) contrast with applications and libraries: An application is any code that has a `main` program, whereas a library is code that user code calls, and a framework is code that calls user code.

Early candidates for inclusion in Morfeus include a tensor field ABSTRACT CALCULUS, including scalar, vector, and dyadic tensor field interfaces. Each of these will publish generic operators and corresponding deferred bindings for arithmetic and spatial differentiation, integration, and arithmetic. These will be based on the coordinate-free programming concepts of Grant, Haveraaen, and Webster (2000), and will be developed in collaboration with the second of those three authors. In Fortran syntax, the coordinate-free programming paradigm involves writing differential operations on tensors in forms such as `.grad.`, `.div.`, and `.curl.` without specifying a coordinate system, deferring to a lower-level abstraction the resolution of these forms into coordinate-specific calculations based on the selection of a manifold in which the tensors are embedded. That manifold could provide a metric tensor in the case of general curvilinear coordinates or scale factors in the special case of orthogonal coordinates (Aris 1962). Figure 12.21 depicts several tensor field interfaces. That figure specifies no relationships between the abstract classes that describe the interfaces. An exercise at the end of the chapter asks the reader to discuss the pros and cons of various approaches to relating those interfaces based on the design principles discussed in this book.

The problem decomposition described in Figure 12.21 carries with it at least two scalability benefits. First, it naturally lends itself to a purely functional, side-effect-free programming style that supports asynchronous expression evaluation as noted in Section 12.1.2. This becomes increasingly important as massively parallel hardware platforms evolve toward communication dominating computation in the run-time budget. Synchronization implies communication. A design patterns that lights a path to highly asynchronous computation thus lights a path to massively parallel computation. In this sense, scalable design leads directly to scalable execution.

Second, Numrich (2005) demonstrated a beautiful, formal analogy between array indices and contravariant tensor indices, and likewise between coarray indices and covariant tensor indices. This analogy naturally generates a powerful notation for expressing parallel algorithms for matrix and vector algebra. Moreover, the analogy supports very direct mapping of objects onto distributed memory via co- and contravariant partitioning operators. All of this enables an exceptionally direct translation of parallel mathematical algorithms into compilable, efficient code that closely mirrors the original mathematical syntax. This makes coarray Fortran 2008 an obvious candidate for inclusion into an ABSTRACT CALCULUS. We envision a scenario in which continuum tensor calculus expressions written at the highest abstraction levels

not, in turn, depend. The first release of the Morfeus source is expected to occur around the time of publication of this book.

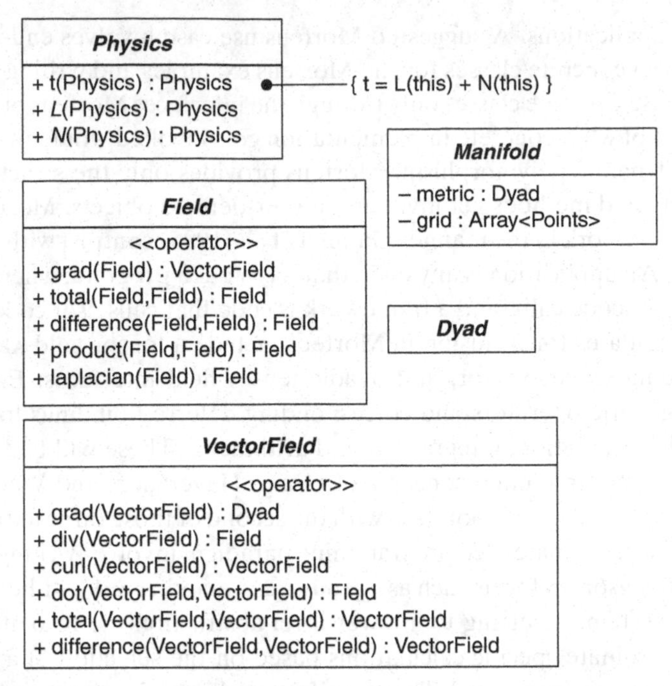

Figure 12.21. Abstract classes in the Morfeus framework.

get resolved into discrete, tensor algebra expressions at the lowest abstraction levels, and that these tensor algebra calculations get translated very directly into coarray syntax via the analogy with covariant tensor forms.

Another early candidate for inclusion in Morfeus will be a Physics abstraction that publishes an abstract interface for an evolution equation by defining a public time derivative method, `t()`, as in Table 3.1. For Runge-Kutta schemes, `t()` represents the RHS function evaluation in equations of the form 4.1. For semi-implicit time advancement, wherein it proves useful to advance implicit terms differently from explicit ones, the default implementation for `t()` might simply sum the linear and nonlinear differential operators `L` and `N`, respectively, as indicated in Figure 12.21. Appendix A.5.2.4 provides a description of the semi-implicit algorithm used in several of the case studies described in Section 12.2.1.

Another candidate would be a Puppeteer abstraction that defines several default methods for polling puppets for information that requires coordination such as time step selection, cross-coupling terms and Jacobian construction. Additional design patterns described in Part II of the current text also provide likely candidates either on their own or incorporated into the other patterns. For example, the STRATEGY and SURROGATE patterns could be incorporate into an ABSTRACT CALCULUS to provide flexibility in the decision which discrete approximations will be employed to approximate the continuous forms.

Between design and execution lies development. In this context, scalable development refers to practices for managing the complexities of programming writ large: multideveloper, multiplatform, multilanguage, multipackage, multiuser software construction. Scalable development methodologies must address build automation, documentation, repository management, issue tracking, user feedback, and testing.

Hosting Morfeus on the Trilinos terminal application repository brings with it professional software engineering solutions to each of these problems. These solutions include:

- Nightly automated multi-platform builds via CMake,[13]
- Nightly automated regression testing via the CMake tool CTest,
- Automated e-mail notification of test failures with links to an online test matrix dashboard constructed by the CMake tool CDash,
- State-of-the-art repository management and version control via Git,[14]
- User, developer, and leadership mailing lists managed by Mailman,[15]
- Web-based issue tracking via Bugzilla,[16]
- Automated hypertext markup of documentation by Doxygen.[17]

We expect that each of these technologies will contribute to an end-to-end, ubiquitous scalability – that is, scalable design leading to scalable development leading to scalable execution with an inclusive vision of domain-specific abstractions that disruptively reduce design and development times and thereby bring more scientific software developers into the once-protected domain of massively parallel execution.

"[T]he Most Dangerous Enemy of a Better Solution Is an Existing Codebase That Is Just Good Enough."
Eric S. Raymond

The Morfeus framework aims to increase the flexibility of multiphysics software architectures by encouraging the use of patterns. In particular, it aims to demonstrate a scalable ABSTRACT CALCULUS that enables application developers to operate at an abstraction level characteristic of their application domain: tensor calculus. Moreover, the promise of coarray Fortran 2008 suggests the possibility of writing tensor-algebraic approximations to the tensor calculus in ways that translate directly into coarray syntax via analogies with covariant tensors. Such an approach could represent a disruptive technology if it meets the challenge of overcoming attachments to working codes that get the job done for the small subset of researchers with the skills and access to use them.

EXERCISES

1. The switch from 6th-order Padé differences to 2nd-order central differences can be accomplished with minimal code revision by designing a new concrete factory `periodic_2nd_factory` to replace `periodic_6th_factory` in Figure 9.1(e). Write this new factory and the revise the `main` program of Figure 9.1(a) to use your new factory instead of `periodic_6th_factory`.
2. Calculate the cumulative, overall theoretical maximum speedups that could result from progressive optimization of the procedures in Table 12.1.

[13] http://www.cmake.org
[14] http://git-scm.com/
[15] http://www.gnu.org/software/mailman/
[16] http://www.bugzila.org
[17] http://www.doxygen.org

3. Progressively augment the type-bound procedures in `periodic_2nd_order` module in Figure 12.1 with loop-level (`parallel do`) directives. Use the Fortran `cpu_time()` and `system_time()` procedures to measure cumulative speedup on 1, 2, and 4 cores for progressive parallelization of the first four procedures listed in Table 12.1.
4. Write ABSTRACT CALCULUS versions of the governing equations (12.6)–(12.9) as they would appear inside a time derivative method on the Integrand class of Figure 12.10. Enhance the class diagram in that figure to include the public methods that appear in your expressions. Further enhance that figure with private data in each class that could support the calculations you specified.
5. Replace the Field class in Figure 12.10 with an ABSTRACT FACTORY and a FACTORY METHOD capable of constructing a `periodic_6th_order` concrete product. Also add a STRATEGY to the Integrand class diagram to facilitate swapping between 2nd- and 4th-order accurate Runge-Kutta algorigthms.
6. Discuss the pros and cons of choosing aggregation or composition versus inheritance to relate the tensor field abstractions in Figure 12.21.

APPENDIX A

Mathematical Background

This chapter provides the mathematical background necessary for the material in this book. The level of the treatment targets seniors and beginning graduate students in engineering and the physical sciences.

A.1 Interpolation

A.1.1 Lagrange Interpolation

Solving the quantum vortex problem described in Chapters 4 and 5 requires estimating the 3D fluid velocity vector field $\mathbf{u}(\mathbf{x},\mathbf{t})$ at each quantum vortex mesh point $\mathbf{S}$. Without loss of generality, we consider here interpolating a scalar field $u(\mathbf{x})$ that can be interpreted as a single component of $\mathbf{u}$ at a given instant of time. We estimate the desired values with 3D linear Lagrange polynomial interpolation. Given a collection of data points $u(x_0,y_0,z_0),u(x_1,y_1,z_1),\ldots,u(x_k,y_k,z_k)$ across a 3D space, each Lagrange polynomial in a series expansion of such polynomials evaluates to unity at the location of one datum and zero at the locations of all other data. In 1D, this yields:

$$\ell_j(x) = \prod_{i=0,i\neq j}^{k} \frac{x-x_i}{x_j-x_i} = \frac{(x-x_0)\cdots(x-x_{j-1})(x-x_{j+1})\cdots(x-x_k)}{(x_j-x_0)\cdots(x_j-x_{j-1})(x_j-x_{j+1})\cdots(x_j-x_k)}. \tag{A.1}$$

When the problem is well resolved in space, it can be preferable for both accuracy and efficiency reasons to use only the nearest-neighbor end points of the smallest interval (in 1D) or corners of the smallest cube (in 3D) that brackets the point of interest. Each polynomial for the 1D case then reduces to a single factor of the form:

$$\ell_j(x) = \frac{(x-x_i)}{(x_j-x_i)}. \tag{A.2}$$

where x_i is an endpoint of the interval between x_i and x_j. This facilitates writing the series:

$$u(x) = \sum_{j=0}^{1} u_j \ell_j(x) \tag{A.3}$$

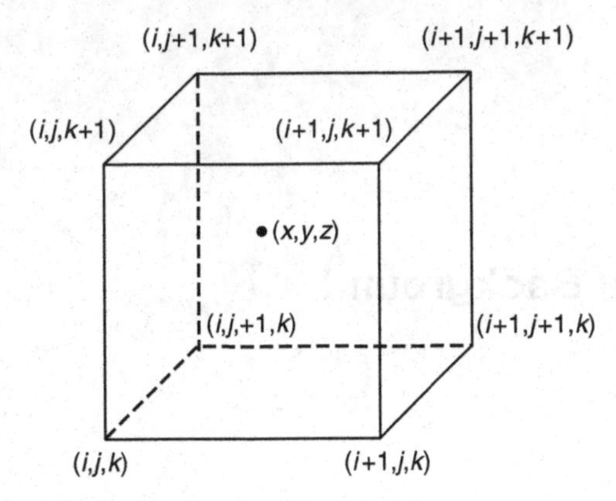

Figure A.1. 3D Lagrange interpolation in uniform grid.

In 3D, the $\ell_j(x,y,z)$ can be obtained by multiplying 1D Lagrange polynomials:

$$\ell_j(x,y,z) = \ell_j(x)\ell_j(y)\ell_j(z) \tag{A.4}$$

In the book, we only use the nearest-neighbor points. Thus:

$$u(x,y,z) = \sum_{j=0}^{8} u_j \ell_j(x,y,z) \tag{A.5}$$

where:

$$\ell_j(x,y,z) = \frac{(x-x_i)}{(x_j-x_i)} \frac{(y-y_i)}{(y_j-y_i)} \frac{(z-z_i)}{(z_j-z_i)} \tag{A.6}$$

For the uniform grid as shown in Fig. A.1, the distances between neighbor points in x, y, and z direction are all equal to δ, and we have:

$$\begin{aligned}
u(x,y,z) = {} & \frac{(x-x_{i,j,k})(y-y_{i,j,k})(z-z_{i,j,k})}{\delta^3} u_{i,j,k} \\
& + \frac{(x-x_{i+1,j,k})(y-y_{i+1,j,k})(z-z_{i+1,j,k})}{\delta^3} u_{i+1,j,k} \\
& + \frac{(x-x_{i,j+1,k})(y-y_{i,j+1,k})(z-z_{i,j+1,k})}{\delta^3} u_{i,j+1,k} \\
& + \frac{(x-x_{i+1,j+1,k})(y-y_{i+1,j+1,k})(z-z_{i+1,j+1,k})}{\delta^3} u_{i+1,j+1,k} \\
& + \frac{(x-x_{i,j,k+1})(y-y_{i,j,k+1})(z-z_{i,j,k+1})}{\delta^3} u_{i,j,k+1} \\
& + \frac{(x-x_{i+1,j,k+1})(y-y_{i+1,j,k+1})(z-z_{i+1,j,k+1})}{\delta^3} u_{i+1,j,k+1} \\
& + \frac{(x-x_{i,j+1,k+1})(y-y_{i,j+1,k+1})(z-z_{i,j+1,k+1})}{\delta^3} u_{i,j+1,k+1} \\
& + \frac{(x-x_{i+1,j+1,k+1})(y-y_{i+1,j+1,k+1})(z-z_{i+1,j+1,k+1})}{\delta^3} \\
& \times u_{i+1,j+1,k+1}
\end{aligned} \tag{A.7}$$

which provides a good approximation given well-resolved data. Higher-order Lagrange polynomials offer formally better estimates but suffer from the possibility of oscillations related to the lack of control of the polynomial derivatives. These oscillations can be particularly accurate near the edges of the domain over which the polynomial is defined. All polynomials ultimately diverge at a rate determined by the order of the polynomial, and such divergence renders higher-order polynomials inaccurate near domain boundaries.

A.2 Linear Solvers

A.2.1 Gaussian Elimination

Gaussian elimination solves a system of linear algebraic equations in the form:

$$\mathbf{Ax} = \mathbf{b} \tag{A.8}$$

when the determinant of matrix $\mathbf{A}$ does not vanish. When $\det A = 0$, the system is ill-posed and the equations have no unique solution. The basic Gaussian elimination procedure includes two steps. The first step, forward elimination, transfers the system of equations into an upper triangular form. The second step solves the triangular system using back substitution.

Consider the following equations:

$$\begin{bmatrix} a_{11} & a_{12} & \cdots & a_{1n} \\ a_{21} & a_{22} & \cdots & a_{2n} \\ \vdots & \vdots & \ddots & \vdots \\ a_{n1} & a_{n2} & \cdots & a_{nn} \end{bmatrix} \begin{bmatrix} x_1 \\ x_2 \\ \vdots \\ x_n \end{bmatrix} = \begin{bmatrix} b_1 \\ b_2 \\ \vdots \\ b_n \end{bmatrix} \tag{A.9}$$

The augmented matrix of the above linear system is the matrix $[\mathbf{A}|\,\mathbf{b}]$:

$$\left[\begin{array}{cccc|c} a_{11} & a_{12} & \cdots & a_{1n} & b_1 \\ a_{21} & a_{22} & \cdots & a_{2n} & b_2 \\ \vdots & \vdots & \ddots & \vdots & \vdots \\ a_{n1} & a_{n2} & \cdots & a_{nn} & b_n \end{array}\right] \tag{A.10}$$

This system retains the same solution under certain elementary operations, including row interchanges, multiplication of a row by a nonzero number, and replacement of one row with the sum of two rows. Gaussian elimination uses these operations to transform the equation system into upper triangular form. In the first step, one chooses an equation with a nonzero x_1 coefficient as the "pivot" equation and swaps this equation's coefficients and RHS with those in the first row of the augmented matrix. Upon switching, one zeroes out the x_1 coefficients listed below the pivot equation by multiplying the first row of the augmented matrix by a_{i1}/a_{11} and subtracting the result from the i_{th} row for all $i = 2, 3, \cdots, n$.

A similar process based on choosing an equation with a nonzero x_2 coefficient as the pivot and swapping that equation with the second equation zeros ultimately produces zeros below the diagonal in the second column. This process continues until it produces an equivalent upper triangular system:

$$\left[\begin{array}{cccc|c} a_{11} & a_{12} & \cdots & a_{1n} & b_1 \\ 0 & a'_{22} & \cdots & a'_{2n} & b'_2 \\ \vdots & \vdots & \ddots & \vdots & \vdots \\ 0 & 0 & \cdots & a'_{nn} & b'_n \end{array}\right] \tag{A.11}$$

where primes denote values that have been modified by the Gaussian elimination process. Upon obtaining the upper triangular form, the back substitution process proceeds as follows: the bottom equation immediately yields:

$$x_n = b'_n / a'_{nn}. \tag{A.12}$$

The penultimate equation then yields:

$$x_{n-1} = (b'_{n-1} - a'_{nn} x_n)/a'_{n-1} \tag{A.13}$$

and so forth, working up the equation listing until the first equation has been solved.

The accuracy of Gaussian elimination can deteriorate by round-off errors accumulated through the many algebraic operations. Approximately $n^3/3$ multiplications and $n^3/3$ additions are required to solve n equations using Gaussian elimination. The accuracy also depends on the arrangement of the equation system. Rearranging the equation in order to put the coefficients with largest magnitude on the main diagonal tends to improve accuracy. This rearrangement process is called partial pivoting. Figure A.2 shows corresponding pseudocode.

Figure A.2

```
1   input: A(n,n), b(n)
2   output: x(n)
3
4   !transform matrix A to upper diagonal form
5   for i=1 to n-1 do
6     !find the pivot row with the maximum value in main diagonal
7     for j=i+1 to n do
8       if absolute(A(j,i))>absolute(A(i,i)) then
9         swap row i and row j of matrix A and vector b
10      end if
11    end for
12
13    !eliminate all terms in the ith colume of the rows below the pivot row
14    for j=i+1 to n do
15      for k=i to n do
16        factor=A(j,i)/A(i,i)
17        A(j,k)=A(j,k)-factor*A(i,k)
18      end for
19    end for
20  end for
21
22  !back substitute
23  x(n) = b(n)/A(n,n)
24  for i=n-1 to 1 do
```

```
25     known=0.0
26     for j=i+1 to n do
27        known=known+A(i,j)*x(j)
28     end for
29     x(i) = (b(i) - known)/A(i,i)
30  end for
```

Figure A.2. Pseudocode of Gaussian elimination with partial pivoting.

Chapter 3 demonstrates the application of Gaussian elimination to a linear system that arises in solving a fin heat conduction problem, wherein one obtains a vector of temperature values T^{n+1} at time t_{n+1} from the temperature vector T^n at a previous time step $t_n = t_{n+1} - \Delta t$ for some time step Δt according to:

$$T^{n+1} = \left(I - \frac{\Delta t}{\alpha}\nabla^2\right)^{-1} T^n \tag{A.14}$$

where I is the identity matrix with unit values along the diagonal and zeros elsewhere. Given a uniformly spaced grid overlaid on the computational domain, applying the central difference formula of Section A.5.1 to the RHS of equation (A.14) leads to the linear algebraic equations in the form of (A.8), where:

$$\mathbf{A} = \begin{bmatrix} 1+\frac{2\Delta t}{\alpha(\Delta x)^2} & -\frac{\Delta t}{\alpha(\Delta x)^2} & 0 & \cdots & 0 & 0 \\ -\frac{\Delta t}{\alpha(\Delta x)^2} & 1+\frac{2\Delta t}{\alpha(\Delta x)^2} & -\frac{\Delta t}{\alpha(\Delta x)^2} & \cdots & 0 & 0 \\ \vdots & \vdots & \vdots & \ddots & \vdots & \vdots \\ 0 & 0 & 0 & \cdots & 1+\frac{2\Delta t}{\alpha(\Delta x)^2} & -\frac{\Delta t}{\alpha(\Delta x)^2} \\ 0 & 0 & 0 & \cdots & -\frac{\Delta t}{\alpha(\Delta x)^2} & 1+\frac{2\Delta t}{\alpha(\Delta x)^2} \end{bmatrix} \tag{A.15}$$

$$\mathbf{x} = \begin{bmatrix} T_1^{n+1} \\ T_2^{n+1} \\ \vdots \\ T_{m-1}^{n+1} \\ T_m^{n+1} \end{bmatrix} \tag{A.16}$$

$$\mathbf{b} = \begin{bmatrix} T_1^n + \frac{\Delta t}{\alpha(\Delta x)^2} T_{chip} \\ T_2^n \\ \vdots \\ T_{m-1}^n \\ T_m^n + \frac{\Delta t}{\alpha(\Delta x)^2} T_{air} \end{bmatrix} \tag{A.17}$$

where m is the number of discrete node points at which one seeks fin temperatures and Δx is the uniform space between each node. Matrix $\mathbf{A}$ is *diagonally dominant*, in the sense that the magnitude of each main diagonal element exceeds that of the other elements in any given row. Diagonal dominance eliminates the need for partial pivoting. Figure (A.3) provides a Fortran subroutine implementation of Gaussian elimination without partial pivoting useful for solving the above equations. Figure (A.4) presents an equivalent C++ code.

When **A** is tridiagonal as in equation (A.15), one can simplify the Gaussian elimination procedures and reduce the required storage by exploiting the zero coefficients. Thomas (1949) suggested this method. Using the Thomas algorithm, one transforms the tridiagonal matrix in equation (A.15) to an upper triangular form by replacing the diagonal coefficients with:

$$d_i = \left(1 + \frac{2\Delta t}{\alpha\,(\Delta x)^2}\right) - \frac{\left(\frac{\Delta t}{\alpha(\Delta x)^2}\right)^2}{1 + \frac{2\Delta t}{\alpha(\Delta x)^2}} \qquad i = 2, 3, \cdots, m \tag{A.18}$$

and replacing the vector **b** by:

$$T_2' = T_2^n + \frac{\frac{\Delta t}{\alpha(\Delta x)^2}}{1 + \frac{2\Delta t}{\alpha(\Delta x)^2}} \left(T_1^n + \frac{\Delta t}{\alpha\,(\Delta x)^2} T_{chip}\right) \tag{A.19}$$

$$T_m' = T_m^n + \frac{\frac{\Delta t}{\alpha(\Delta x)^2}}{1 + \frac{2\Delta t}{\alpha(\Delta x)^2}} \left(T_m^n + \frac{\Delta t}{\alpha\,(\Delta x)^2} T_{air}\right) \tag{A.20}$$

Figure A.3

```
1  module linear_solve_module
2    use kind_parameters ,only : rkind,ikind
3    implicit none
4  contains
5    function gaussian_elimination(lhs,rhs) result(x)
6      real(rkind) ,dimension(:,:) ,allocatable ,intent(in) :: lhs
7      real(rkind) ,dimension(:) ,allocatable ,intent(in) :: rhs
8      real(rkind) ,dimension(:)   ,allocatable :: x,b ! Linear system:
9      real(rkind) ,dimension(:,:) ,allocatable :: A   ! Ax = b
10     real(rkind)              :: factor
11     real(rkind) ,parameter :: pivot_tolerance=1.0E-02
12     integer(ikind)           :: row,col,n,p ! p=pivot row/col
13     n=size(lhs,1)
14     b = rhs ! Copy rhs side to preserve required intent
15     if ( n /= size(lhs,2) .or. n /= size(b)) &
16       stop 'gaussian_elimination: ill-posed system'
17     allocate(x(n))
18     A = lhs                  ! Copy lhs side to preserve required intent
19     !______ Gaussian elimination ________________________________________
20     do p=1,n-1             ! Forward elimination
21       if (abs(A(p,p))<pivot_tolerance) &
22         stop 'gaussian_elimination: use pivoting'
23       do row=p+1,n
24         factor=A(row,p)/A(p,p)
25         forall(col=p:n) A(row,col) = A(row,col) - A(p,col)*factor
26         b(row) = b(row) - b(p)*factor
27       end do
28     end do
29     x(n) = b(n)/A(n,n) ! Back substitution
30     do row=n-1,1,-1
31       x(row) = (b(row) - sum(A(row,row+1:n)*x(row+1:n)))/A(row,row)
```

```
32      end do
33    end function
34  end module
```

Figure A.3. Gaussian elimination procedure.

$$T_i' = T_i^n + \frac{\frac{\Delta t}{\alpha(\Delta x)^2}}{1+\frac{2\Delta t}{\alpha(\Delta x)^2}} T_{i-1}^n \qquad i = 3,4,\cdots,m-1 \tag{A.21}$$

where $d_1 = 1 + \frac{2\Delta t}{\alpha(\Delta x)^2}$. The unknowns can be solved from back substitution according to:

$$T_m^{n+1} = \frac{T_m'}{d_m} \tag{A.22}$$

Figure A.4(a)

```
1   #ifndef GAUSSIAN_ELIMINATION_H_
2   #define GAUSSIAN_ELIMINATION_H_ 1
3
4   #include <exception>
5   #include "mat.h" // 2-D array
6
7   struct gaussian_elimination_error : public std::exception {
8       virtual ~gaussian_elimination_error() throw () {}
9   };
10
11  crd_t gaussian_elimination (const mat_t & lhs, const crd_t & rhs);
12
13  #endif
```

Figure A.4(b)

```
1   #include "gaussian_elimination.h"
2   #include <cmath>
3
4   crd_t gaussian_elimination (const mat_t& lhs, const crd_t& rhs)
5   {
6       const real_t pivot_tolerance=0.01;
7       const int n = lhs.rows();
8
9       if ( n != lhs.cols() || n != rhs.size()) {
10          std::cerr << "gaussian_elimination: ill-posed system"
11              << std::endl;
12          throw gaussian_elimination_error();
13      }
14
15      // copy parameters to preserve lhs and rhs
16      crd_t b = rhs;
17      mat_t A = lhs;
18
19      //-----  Gaussian elimination -----
20      // forward elimination
21      for (int p = 0; p < n-1; ++p) {
```

```
22            if (fabs(A(p,p)) < pivot_tolerance) {
23                std::cerr << "gaussian_elimination: use pivoting"
24                    << std::endl;
25                throw gaussian_elimination_error();
26            }
27            for (int row = p+1; row < n; ++row) {
28                real_t factor = A(row,p) / A(p,p);
29
30                for (int col = p; col < n; ++col) {
31                    A(row,col) -= A(p,col)*factor;
32                }
33
34                b.at(row) -= b.at(p)*factor;
35            }
36        }
37
38        // Back substitution
39        crd_t x(n);
40
41        x.at(n-1) = b.at(n-1) / A(n-1, n-1);
42
43        for (int row = n-2; row >=0; --row) {
44            real_t sum = 0.0;
45
46            for (int col = row+1; col < n; ++ col) {
47                sum += A(row,col) * x.at(col);
48            }
49            x.at(row) = (b.at(row) - sum) / A(row, row);
50        }
51        return x;
52    }
53
```

Figure A.4. Gaussian elimination in C++. (a) Header file. (b) C++ code.

$$T_i^{n+1} = \frac{T'_i + \frac{\Delta t}{\alpha(\Delta x)^2} T_{i+1}^{n+1}}{d_i} \qquad i = m-1, m-2, \cdots, 1 \tag{A.23}$$

Figure (A.5) presents a Fortran implementation of the Thomas algorithm.

A.2.2 LU Decomposition

LU decomposition functions improves upon Gaussian elimination process by factoring the coefficient matrix into a form that can be reused with multiple RHS **b** vectors. The operation count for solving the resulting factored system is lower than the count for Gaussian elimination. This approach proves useful when a linear system must be solved at each time step of a differential equation solver and the coefficient matrix stays constant over time. Such is the case for equations (A.57) and (A.58).

LU decomposition factors any nonsingular coefficient matrix **A** (that is one with nonzero determinant) into the product of a lower triangular matrix L and an upper triangular one U. When no pivoting is required, the factorization takes the simplest form: $\mathbf{A} = \mathbf{LU}$. With pivoting, LU decomposition takes the form $\mathbf{PA} = \mathbf{LU}$, where $\mathbf{P}$ is a *permutation matrix*. Permutation matrices have one unit entry in each row and each column with zeros elsewhere. The multiplication of $\mathbf{PA}$ produces a new matrix by interchanging the rows in matrix $\mathbf{A}$ so that $\mathbf{PA}$ has an LU decomposition.

Figure A.5

```
1  function thomasTimes(lhs,rhs)
2    real(rkind) ,dimension(:,:) ,allocatable ,intent(in) :: lhs
3    class(field)                             ,intent(in) :: rhs
4    type(field)                 ,allocatable :: tomasTimes
5    real(rkind) ,dimension(:)   ,allocatable :: x,b ! Linear system:
6    real(rkind) ,dimension(:,:) ,allocatable :: A   ! Ax = b
7    real(rkind)              :: factor
8    integer(ikind)           :: row,n
9    n=size(lhs,1)
10   b = rhs%node     ! Copy rhs side to preserve required intent
11   if ( n /= size(lhs,2) .or. n /= size(b)) &
12     stop 'thomasTimes: ill-posed system'
13   allocate(x(n))
14   A = lhs              ! Copy lhs side to preserve required intent
15   !______ Establish upper triangular matrix _______________________
16   do row=2,n
17     factor=A(row,row-1)/A(row-1,row-1)
18     A(row,row) = A(row,row) - factor*A(row-1,row)
19     b(row) = b(row) - factor*b(row-1)
20   end do
21   !______ Back supstitution  _______________________________________
22   x(n) = b(n)/A(n,n) ! Back substitution
23   do row=n-1,1,-1
24     x(row) = (b(row) - A(row,row+1)*x(row+1))/A(row,row)
25   end do
26   allocate(thomasTimes)
27   thomasTimes%node = x
28 end function
```

Figure A.5. Thomas algorithm procedure.

The following steps summarize the LU decomposition method with pivoting:

1. Obtain the permutation matrix $\mathbf{P}$, lower triangular matrix $\mathbf{L}$, and upper triangular matrix $\mathbf{U}$ such that $\mathbf{A} = \mathbf{PLU}$.
2. Calculate $\mathbf{Pb}$, which interchanges the rows of vector $\mathbf{b}$ and replaces the RHS of (A.8) so $\mathbf{PAx} = \mathbf{Pb}$.
3. Solve $\mathbf{Ly} = \mathbf{Pb}$ for $\mathbf{y}$ using forward substitution where $\mathbf{y} = \mathbf{Ux}$.
4. Solve $\mathbf{Ux} = \mathbf{y}$ for solution $\mathbf{x}$ using back substitution.

A.3 Nonlinear Solvers

A.3.1 Newton's Method in 1D

The Newton's method, also known as the Newton-Raphson method, approximates the root of $f(x) = 0$. Expanding a 1D, real function $f(x)$, in a Taylor series about a

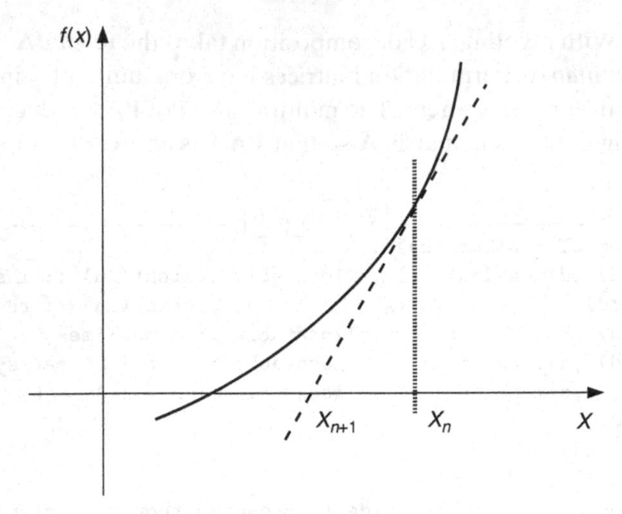

Figure A.6. One iteration of Newton's method (dashed line is the tangent line of $f[x]$ at x_n)

reference point x_n:

$$f(x_n + \Delta x) = f(x_n) + f'(x_n)\Delta x + \cdots \quad \text{(A.24)}$$

where $\Delta x = x_{n+1} - x_n$ and f' is the derivative of f with respect to x, Newton's method proceeds from one approximation, x_n, to the next, x_{n+1}, by truncating the series at the second term and setting $f(x_{n+1}) = f(x_n + \Delta x) = 0$, where:

$$\Delta x = x_{n+1} - x_n = \frac{f(x_n)}{f'(x_n)}. \quad \text{(A.25)}$$

Figure A.6 illustrates this process. One finds the root of $f(x)$ by repeating this process until $|\Delta x|$ is sufficiently small. Starting sufficiently close to the desired root of a smooth function, Newton's method converges faster than some competing methods and it can diverge for poor initial guesses or ill-behaved functions. One pathological case occurs when the initial guess lies at a stationary point – that is, a point with a horizontal tangent line, in which case the iteration step (A.25) requires division by 0.

A.3.2 Newton's Method in Multiple Dimensions

One can readily extend the 1D Newton's method to multiple dimensions. Given m nonlinear equations $\mathbf{f}(\mathbf{x}) = \mathbf{0}$, where $\mathbf{f}$ and $\mathbf{x}$ have components f_i and x_i for $i = 1,2,\cdots,m$, one represents successive approximations to the roots as:

$$\mathbf{x}^{\mathbf{n+1}} = \mathbf{x}^{\mathbf{n}} + \mathbf{J}^{-1}(\mathbf{x}^{\mathbf{n}})\mathbf{f}(\mathbf{x}^{\mathbf{n}}) \quad \text{(A.26)}$$

where the Jacobian matrix $\mathbf{J}$ generalizes the 1D derivative f'. The ij^{th} element of $\mathbf{J}$ is:

$$J_{ij} = \frac{\partial f_i(\mathbf{x})}{\partial x_j} \quad i = 1,2,\cdots,m \;\; j = 1,2,\cdots,m. \quad \text{(A.27)}$$

An approximate solution to the nonlinear equations obtains from iterating on equation (A.26), starting from an appropriate initial guess. Much like the pathological stationary point in the 1D case, the multidimensional Newton method diverges with singular or nearly singular Jacobian matrices.

A.4 Partial Differential Equations

A.4.1 The Heat Equation

Throughout Part I of this text, we discussed heat conduction in a solid within which energy conservation arguments applied throughout a simply connected volume V bounded by a closed surface S, neglecting other energy forms, leads to:

$$\frac{\partial}{\partial t}\int_V \rho e dV = -\int_S \mathbf{q}\cdot d\mathbf{S} \tag{A.28}$$

where ρ, e, and $\mathbf{q}$ are the material density, thermal energy per unit mass, and surface heat flux, respectively, and where the LHS and RHS integrals are volume and surface integrals, respectively. The LHS of equation (A.28) represents the total thermal energy in the volume, whereas the RHS represents the net heat transfer into the volume at the boundary.

One can combine the two sides of equation (A.28) into a single term by applying Gauss's divergence theorem to convert the RHS of equation (A.28) to a volume integral so that:

$$\int_V (\frac{\partial}{\partial t}[\rho e) + \nabla\cdot q]dV = 0 \tag{A.29}$$

Because there is nothing special about the volume V, equation (A.29) holds for any arbitrary volume, including one shrunk to within an infinitesimally small neighborhood of a point. This forces the integrand to vanish at all points:

$$\frac{\partial}{\partial t}(\rho e) + \nabla\cdot q = 0 \tag{A.30}$$

Writing the energy, $e \equiv cT$, in terms of the specific heat c and temperature T, and writing Fourier's law of heat conduction $\mathbf{q} \equiv -\mathbf{k}\nabla\mathbf{T}$, where k is the thermal conductivity, yields:

$$\frac{\partial T}{\partial t} = \alpha\nabla^2 T \tag{A.31}$$

where $\alpha \equiv k/(\rho c)$ is the thermal diffusivity. Equation (A.31) is the heat equation.

We now present the analytical solutions for the 1D heat equation problem solved numerically throughout Part I of this text. In equation 1.1, the heat equation is subjected to the initial condition $T(x,0) = T_{air}$ and the boundary condition $T(0,t) = T_{chip}$ and $T(L_{fin},t) = T_{air}$. If $T(0,t) = T(L_{fin},t) = 0$, the heat equation can be solved by the method of separation variables and yields the general solution:

$$T(x,t) = \sum_{n=0}^{n=\infty} B_n sin\frac{n\pi}{L_{fin}}xe^{-n^2\pi^2\alpha t/L_{fin}^2} \tag{A.32}$$

where:

$$B_n = \frac{2}{L_{fin}}\int_0^{L_{fin}} T_{air} sin\frac{n\pi}{L_{fin}}xdx \tag{A.33}$$

In order to solve equation 1.1 subjected to nonzero boundary condition, we assume $T = T_1 + T_2$, where T_1 satisfies the final steady state, that is, the point in time after which the temperature becomes independent of time. Therefore, one can determine T_1 from:

$$\frac{\partial^2 T_1}{\partial x^2} = 0 \tag{A.34}$$

subject to the boundary conditions:

$$T_1(0) = T_{chip} \quad T_1(L_{fin}) = T_{air} \tag{A.35}$$

The solution is $T_1 = A + Bx$ where $A = T_{chip}$ and $B = (T_{air} - T_{chip})/L_{fin}$. The solution T_1 does not satisfy the initial condition. In order to have $T(x,0) = T_1(x) + T_2(x,0) = T_{air}$, T_2 must satisfy the initial condition:

$$T_2(x,0) = T_{air} - T_{chip} - \frac{T_{air} - T_{chip}}{L_{fin}} x \tag{A.36}$$

T_2 also must satisfy the heat equation and the boundary conditions:

$$T_2(0,t) = 0, \; T_2(L_{fin},t) = 0 \tag{A.37}$$

According to equation A.32, the solution of T_2 will be:

$$T_2(x,t) = \sum_{n=0}^{n=\infty} B_n sin \frac{n\pi}{L_{fin}} x e^{-n^2\pi^2 \alpha t / L_{fin}^2} \tag{A.38}$$

where:

$$B_n = \frac{2}{L_{fin}} \int_0^{L_{fin}} \left(T_{air} - T_{chip} - \frac{T_{air} - T_{chip}}{L_{fin}} x \right) sin \frac{n\pi}{L_{fin}} x dx \tag{A.39}$$

Thus, the analytical solution of the temperature in equation 1.1 is $T(x,t) = T_1(x) + T_2(x,t)$.

A.4.2 The Burgers Equation

Of much younger vintage, the Burgers equation at first appears more daunting due to its nonlinearity:

$$u_t + u u_x = \nu u_{xx}. \tag{A.40}$$

Nonetheless, as mentioned in Section 4.3.3, the Burgers equation can be transformed into the heat equation and solved exactly for the case of periodic boundary conditions employed in the text. Section 4.3.3 provides the resulting analytical solution. Here we briefly sketch the analytical solution process. The initial condition is $u(x,0) = A\sin(x)$ and the boundary condition is $u(x+n\pi,t) = u(x,t) = 0$ for all integers n. As described in Section 4.3.3, the transformed Burgers equation is:

$$\phi_t = \nu \phi_{xx}. \tag{A.41}$$

where $u = -2\nu\phi_x/\phi$. Integrating the initial condition $\phi(x,0)$ yields:

$$\phi(x,0) = e^{-\frac{1}{2\nu}\int A\sin(x)dx} = e^{\frac{1}{2\nu}A\cos(x)} \tag{A.42}$$

Using the method of separation of variables, we obtain the general solution of Eq A.41:

$$\phi = \sum_{n=-\infty}^{\infty} c_n e^{-n^2 t} \cos(nx) \tag{A.43}$$

where:

$$\begin{aligned} c_n &= \frac{1}{2\pi} \int_{-\pi}^{\pi} \phi(x,0)\cos(nx)dx \\ &= \frac{1}{2\pi} \int_{-\pi}^{\pi} e^{\frac{1}{2\nu}A\cos x} \cos(nx)dx \end{aligned} \tag{A.44}$$

The above integration can be expressed by Bessel function I_n as (Benton and Platzman 1972):

$$c_n = (-1)^n I_n(-\frac{A}{2\nu}) \tag{A.45}$$

Hence, the solution is:

$$u = \frac{4\nu \sum_{n=1}^{n=\infty} n c_n e^{-n^2 t} \sin(nx)}{c_0 + 2\sum_{n=1}^{n=\infty} c_n e^{-n^2 t} \cos(nx)} \tag{A.46}$$

A.5 Numerical Analysis

A.5.1 Finite Differences

All of the numerical PDE solvers in this book employ the semidiscrete approach, wherein one first discretizes the spatial domain by replacing the the spatial derivatives with difference approximations. This results in a set of ODEs with continuous dependence on time. One then discretizes the time domain by choosing a numerical quadrature that marches the equations forward in time over a sequence of finite steps.

In order to represent the spatial derivatives using a finite difference approximation, one overlays a grid, or mesh, on the continuous problem domain. Figure (A.7) shows a typical finite-difference stencil: The layout of the grid points employed in approximating x- and y-direction derivatives at a point (x_0,y_0). One can define a finite-difference approximation to an x-direction derivative of a function $f(x,y)$ at a point (x_0,y_0) as:

$$\frac{\delta f}{\delta x} \equiv \frac{f(x_0+\Delta x, y_0) - f(x_0,y_0)}{\Delta x} \tag{A.47}$$

When Δx is sufficiently small, (A.47) approximates $\partial f/\partial x$ closely. A Taylor-series expansion of f about (x_0,y_0) motivates equation (A.47):

$$f(x_0+\Delta x, y_0) = f(x_0,y_0) + \left.\frac{\partial f}{\partial x}\right|_{(x_0,y_0)} \Delta x + \left.\frac{\partial^2 f}{\partial x^2}\right|_{(x_0,y_0)} \frac{\Delta x^2}{2} + \cdots \tag{A.48}$$

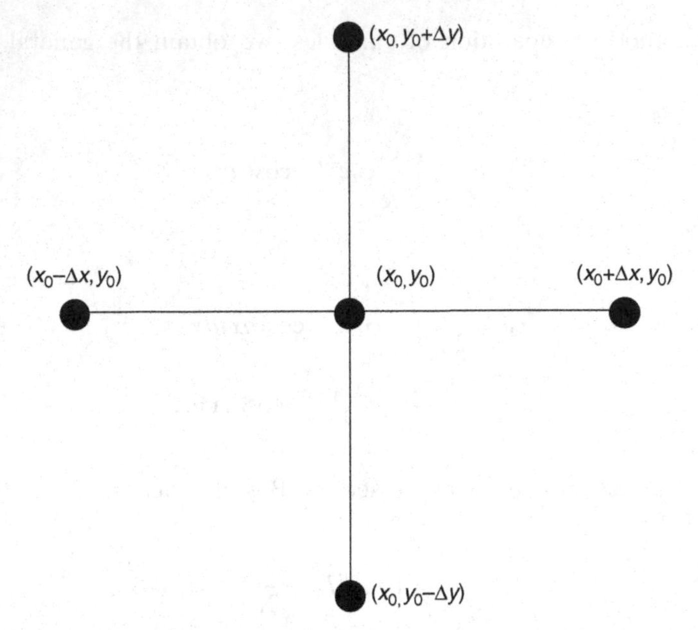

Figure A.7. A finite-difference stencil at the point (x_0, y_0).

The finite-difference approximation to $\partial f/\partial x$ derives from rearranging equation (A.48) and combining it with equation (A.47) so that:

$$\frac{\delta f}{\delta x} = \left.\frac{\partial f}{\partial x}\right|_{(x_0,y_0)} + \left.\frac{\partial^2 f}{\partial x^2}\right|_{(x_0,y_0)} \frac{\Delta x}{2} + \cdots \tag{A.49}$$

where the second and subsequent RHS terms represent the truncation error, which is the difference between the $\partial f/\partial x$ and $\delta f/\delta x$. The truncation error behaves asymptotically like its leading term, which is usually specified with the "order of" (O) notation, so the truncation error behaves like:

$$e \equiv \left.\frac{\partial f}{\partial x}\right|_{(x_0,y_0)} - \frac{\delta f}{\delta x} = O(\Delta x) \tag{A.50}$$

Refining the mesh by shrinking Δx decreases the truncation error proportionately with the first power Δx. Equation (A.47) is therefore referred to as a first-order, forward difference.

There are an infinite number of finite-difference approximations to $\partial f/\partial x$ with different orders of truncation error. Jumping backward to $(x_0 - \Delta x, y_0)$ and forming the Taylor series expansion about (x_0, y_0) yields:

$$f(x_0 - \Delta x, y_0) = f(x_0, y_0) - \left.\frac{\partial f}{\partial x}\right|_{(x_0,y_0)} \Delta x + \left.\frac{\partial^2 f}{\partial x^2}\right|_{(x_0,y_0)} \frac{\Delta x^2}{2} + \cdots \tag{A.51}$$

from which we can extract the first-order, backward difference formula:

$$\left.\frac{\delta f}{\delta x}\right|_{(x_0,y_0)} \equiv \frac{f(x_0,y_0)-f(x_0-\Delta x,y_0)}{\Delta x} \tag{A.52}$$

$$= \left.\frac{\partial f}{\partial x}\right|_{(x_0,y_0)} + O(\Delta x) \tag{A.53}$$

The leading-order truncation error for this approximation equals that of the forward difference formula.

Higher-order finite-difference approximations result from combining Taylor series expansions of f at different points. For example, one derives the second-order, central difference formula by subtracting equation (A.51) from equation (A.48) and rearranging so that:

$$\left.\frac{\delta f}{\delta x}\right|_{(x_0,y_0)} \equiv \frac{f(x_0+\Delta x,y_0)-f(x_0-\Delta x,y_0)}{\Delta x} \tag{A.54}$$

which has a leading-order error of order $O[(\Delta x)^2]$. We leave the derivation of the precise form of this error as an exercise for the reader. Other combinations of Taylor series produce approximations to higher-order derivatives. Adding equation (A.51) to equation (A.48) gives a second-order-accurate approximation to the second partial derivative of f with respect to x:

$$\left.\frac{\delta^2 f}{\delta x^2}\right|_{(x_0,y_0)} \equiv \frac{f(x_0+\Delta x,y_0)-2f(x_0,y_0)+f(x_0-\Delta x,y_0)}{(\Delta x)^2} \tag{A.55}$$

which has a leading-order error of order $O[(\Delta x)^2]$.

One arrives at corresponding finite-difference estimates of derivatives with respect to other variables, such as y, similarly. For example, the first-order, forward finite-difference approximation to $\partial f/\partial y$ is:

$$\left.\frac{\delta f}{\delta y}\right|_{(x_0,y_0)} \equiv \frac{f(x_0,y_0+\Delta y)-f(x_0,y_0)}{\Delta y} \tag{A.56}$$

which has a leading-order error of order $O(\Delta y)$, and likewise in the z direction.

Higher-order finite difference methods result from canceling additional error terms. To avoid broadening the stencil, compact schemes can be constructed with special care. These generally express the derivative approximation implicitly, necessitating the solution of a linear system of equations. The solution to this system yields the derivative approximation at all points in the domain simultaneously. The classical Padé schemes take this form (Moin 2001) for the fourth-order first derivative:

$$\left.\frac{\partial f}{\partial x}\right|_{(x_i+\Delta x)} + \left.\frac{\partial f}{\partial x}\right|_{(x_i-\Delta x)} + 4\left.\frac{\partial f}{\partial x}\right|_{x_i} = \frac{3}{\Delta x}(f_{x_i+\Delta x}-f_{x_i-\Delta x}) \tag{A.57}$$

and fourth-order second derivative:

$$\left.\frac{1}{12}\frac{\partial^2 f}{\partial x^2}\right|_{(x_i-\Delta x)} + \left.\frac{10}{12}\frac{\partial^2 f}{\partial x^2}\right|_{x_i} + \left.\frac{1}{12}\frac{\partial^2 f}{\partial x^2}\right|_{(x_i+\Delta x)} = \frac{f_{x_i+\Delta x}-2f_{x_i}+f_{x_i-\Delta x}}{\Delta x^2} \tag{A.58}$$

both of which yield linear systems with tridiagonal coefficient matrices.

A.5.2 Numerical Methods for Differential Equations

As indicated in Chapter 4, the governing equations in most multiphysics simulations can be described as:

$$\frac{\partial}{\partial t}\vec{U}(\vec{x},t) = \vec{\Im}(\vec{U}(\vec{x},t)), \ \vec{x} \in \Omega, \ t \in (0,T] \tag{A.59}$$

with boundary and initial conditions:

$$\vec{B}(\vec{U}) = \vec{C}(\vec{x},t), \ \vec{x} \in \Gamma \tag{A.60}$$

$$\vec{U}(\vec{x},0) = \vec{U}_0(\vec{x}), \ \vec{x} \in \Omega \tag{A.61}$$

where $\vec{U} \equiv \{U_1, U_2, ..., U_n\}^T$ is the problem state vector; $\vec{x}$ and t are coordinates in the space-time domain $\Omega \times (0,T]$; and $\vec{\Im} \equiv \{\Im_1, \Im_2, ..., \Im_n\}^T$ is a vector-valued operator that couples the state vector components via a set of governing ordinary-, partial-, or integro-differential equations. The space domain Ω is bounded by Γ, $\vec{B}(\vec{U})$ typically represents linear or nonlinear combinations of $\vec{U}$ and its derivatives, and $\vec{C}$ specifies the values of those combinations on Γ.

Numerical solution of (A.59) on a digital computer requires discretizing space and time. Adopting the aforementioned semidiscrete approach and applying the spatial finite difference methods of the previous section results in a system of ODEs of the form:

$$\frac{d}{dt}\vec{V} = \vec{R}(\vec{V}) \tag{A.62}$$

where $\vec{V}$ contains the discretized values of $\vec{U}$ on the grid and the vector function $\vec{R}$ represents the discrete approximation to the RHS of (A.59) derived via finite differences or another suitable approximation formalism.

Considering the following single, first-order ODE simplifies the presentation of approaches to solving (A.62):

$$\frac{dy}{dt} = f(y) \tag{A.63}$$

The numerical methods discussed next estimate the solution y_{n+1} at time $t_{n+1} = t_n + \Delta t$ based on a know solution at t_n.

A.5.2.1 Euler Method

One approach to deriving numerical time advancement schemes involves expanding the solution y in a Taylor-series forward in time from t_n to t_{n+1}:

$$y_{n+1} = y_n + \left.\frac{dy}{dt}\right|_{t=t_n} \Delta t + \left.\frac{d^2y}{dt^2}\right|_{t=t_n} \frac{\Delta t^2}{2} + \left.\frac{d^3y}{dt^3}\right|_{t=t_n} \frac{\Delta t^3}{6} + \cdots \tag{A.64}$$

According to equation (A.63), $dy/dt|_{t_n} = f(y_n)$. Using this information and truncating equation (A.64) after two RHS terms produces the *forward Euler* method:

$$y_{n+1} = y_n + \Delta t f(y_n), \tag{A.65}$$

which is also known as the *explicit Euler* method because the unknown quantity y_{n+1} appears only on the LHS.

Whereas studying the truncation error addresses the asymptotic behavior of a numerical method for infinitesimal Δt, ensuring accurate solutions in practice requires also considering the method's behavior at finite Δt. In the most pathological cases, errors that stay under control at sufficiently small Δt might explode at larger Δt. Ideally, numerical time-integration algorithms would remain stable whenever the exact solution is stable, and unstable only when the exact solution is unstable.

One gains some understanding of a time advancement scheme's stability properties by studying its behavior for a simple, model problem:

$$\frac{dy}{dt} = \lambda y, \tag{A.66}$$

where $\lambda \equiv \lambda_R + i\lambda_I$ and where λ_R and λ_I are the real and imaginary parts of λ. Since equation (A.66) admits an exponential solution, a sufficient condition for a stable exact solution is $\lambda_R \leq 0$. This bound thus holds also for the problems of interest in studying the stability of time advancement schemes.

Applying explicit Euler method (A.65) to the model problem (A.66) yields:

$$y_{n+1} = y_n + \lambda \Delta t y_n \tag{A.67}$$

Thus, the solution at time t_n can be written as:

$$y_n = y_0(1+\lambda\Delta t)^n = y_0(1+\lambda_R\Delta t + i\lambda_I\Delta t)^n = y_0\sigma^n \tag{A.68}$$

where $\sigma = (1+\lambda_R\Delta t + i\lambda_I\Delta t)$ is called the amplification factor. The Euler explicit scheme will be stable if the solution y_n is bounded when n becomes large. The necessary condition for Euler explicit method to be stable is $|\sigma| \leq 1$, so that:

$$|1+\lambda_R\Delta t + i\lambda_I\Delta t| = (1+\lambda_R\Delta t)^2 + \lambda_I^2(\Delta t)^2 \leq 1 \tag{A.69}$$

Therefore, the Euler explicit method is conditionally stable in the sense that the time step must be small enough to satisfy (A.69).

The expression just to the left of the inequality sign in equation (A.69) is a polynomial in Δt. All explicit time-advancement schemes generate stability criteria that place polynomial bounds on the time step. As such, explicit schemes suffer from the same problem that afflicts high-order polynomial interpolation schemes as mentioned in Section A.1.1: All polynomials diverge at infinity. This precludes unconditional stability because any polynomial will violate the inequality in (A.69) outside some finite region of the complex-λ plane. As we demonstrate next, implicit methods circumvent this restriction by generating stability bounds based on rational polynomials – that is, ratios of polynomials. Some implicit algorithms even offer unconditional stability: They are stable for all problems for which the exact solution is stable.

The implicit Euler scheme derives from expanding y in a Taylor series about the time t_{n+1}:

$$y_{n+1} = y_n - \left.\frac{dy}{dt}\right|_{t=t_{n+1}} \Delta t + \left.\frac{d^2y}{dt^2}\right|_{t=t_{n+1}} \frac{\Delta t^2}{2} - \left.\frac{d^3y}{dt^3}\right|_{t=t_{n+1}} \frac{\Delta t^3}{6} + \cdots \tag{A.70}$$

Implicit Euler truncates the series after two terms:

$$y_{n+1} = y_n + \Delta t f(y_{n+1}), \tag{A.71}$$

which is also called the backward Euler method. The additional stability afforded by implicit schemes comes at the cost of requiring iterative solutions when $f(y_{n+1})$ is nonlinear. This additional complexity generally manifests in increased solution costs relative to explicit methods.

Applying the implicit Euler scheme to the model problem (A.66) leads to:

$$y_{n+1} = y_n + \lambda \Delta t y_{n+1}, \tag{A.72}$$

which admits a solution at time t_n given by:

$$y_n = \sigma^n y_0, \tag{A.73}$$

where the amplification factor is the rational polynomial $\sigma = 1/(1-\lambda\Delta t)$. The method is stable if, and only if, $|\sigma| \leq 1$, which always holds when the exact solution is stable ($\lambda_R \leq 0$). Therefore, the implicit Euler method is unconditionally stable.

A.5.2.2 Trapezoidal Method

An alternative strategy for developing time-advancement algorithms integrates equation (A.63) over the interval $[y_n, y_{n+1}]$ to obtain:

$$y_{n+1} = y_n + \int_{t_n}^{t_{n+1}} f(y)\,dt \tag{A.74}$$

A large family of schemes derive from applying various Lagrange polynomial approximations to $f(y)$ in A.74 using the f values at the $m+2$ time steps $\{t_{n-m}, t_{n-m+1}, \ldots, t_n, t_{n+1}\}$ as the nodal data:

$$f[y(t)] \approx \sum_{i=n-m}^{n+1} f_i \ell_i(t), \tag{A.75}$$

where the Lagrange functions, ℓ_i, are as defined in Section A.1.1. Substituting equation (A.75) into the integral in equation (A.74) yields:

$$y_{n+1} = y_n + \sum_{i=n-m}^{n+1} c_i f_i, \tag{A.76}$$

$$c_i \equiv \int_{t_n}^{t_{n+1}} \ell_i(t)\,dt. \tag{A.77}$$

In the degenerate case of a single quadrature point, the only available interpolating polynomial is a constant. Choosing the point t_n, for example, generates the polynomial $\ell_n = 1$, which produces the explicit Euler formula (A.65). Choosing instead t_{n+1} generates the polynomial $\ell_{n+1} = 1$ and the implicit Euler formula (A.65). This general machinery outputs a time-integration algorithm in the form of coefficient values, $\{c_i | i = n-m, ..., n+1\}$, for each choice of quadrature points and resulting interpolating polynomials.

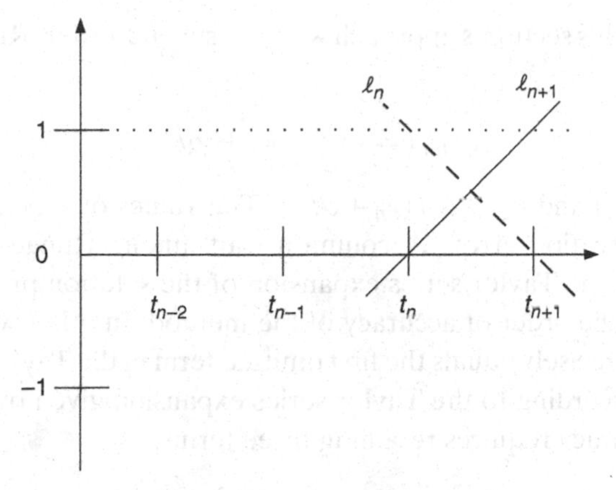

Figure A.8. Linear Lagrange polynomials for the case in which the f values at t_n and t_{n+1} are used to derive a numerical quadrature scheme.

Figure A.8 displays the case in which the two quadrature points are the previous time step, t_n, and the next one, t_{n+1}. With these points, the interpolating polynomials are the two lines shown: $\ell_n(t)$ and $\ell_{n+1}(t)$. In this simple case, one can evaluate the integral (A.77) by graphical inspection:

$$y_{n+1} = y_n + \frac{\Delta t}{2}[f(y_n) + f(y_{n+1})], \tag{A.78}$$

which is called the *trapezoidal rule.* The trapezoidal rule provides the time-integration mechanism in the *Crank-Nicolson method*, which couples trapezoidal time integration with central difference spatial approximation for the heat equation (1.1). In the 1D case, the Crank-Nicolson method is thus:

$$T_j^{n+1} = T_j^n + \frac{\alpha \Delta t}{2}\left[\frac{T_{j+1}^{n+1} - 2T_j^{n+1} + T_{j-1}^{n+1}}{(\Delta x)^2} + \frac{T_{j+1}^n - 2T_j^n + T_{j-1}^n}{(\Delta x)^2}\right] \tag{A.79}$$

where subscripts indicate location on a uniformly distributed 1D spatial grid and superscripts indicate time step. The Crank-Nicolson method is an unconditionally stable, implicit method.

Additional choices of quadrature points lead to multistep methods. When the point at t_{n+1} is included, equation (A.75) generates implicit schemes called *Adams-Moulton* methods. When the point at t_{n+1} is dropped, equation (A.75) generates explicit schemes called *Adams-Bashforth* methods.

A.5.2.3 Runge-Kutta Method

A third approach to deriving time-integration algorithms employs intermediate evaluations of f at points between t_n and t_{n+1}. These schemes are generally called *Runge-Kutta methods*. The approximate solution at t_n can then be calculated at once as a weighted sum of these f values. Because an equivalent estimate can also be constructed via intermediate predictions for y between t_n and t_{n+1} and successive corrections to ultimately arrive at the approximation at t_{n+1}, these schemes are also called *predictor-corrector methods*.

We illustrate this section's approach with the second-order, Runge-Kutta (RK2) method:

$$y_{n+1} = y_n + \gamma_1 k_1 + \gamma_2 k_2 \tag{A.80}$$

where $k_1 = \Delta t f(y_n)$ and $k_2 = \Delta t f(y_n + \alpha k_1)$. The values of γ_1, γ_2, and α determine the method's truncation error. A common trait among Runge-Kutta methods is that they truncate the Taylor series expansion of the solution precisely at the term corresponding to the order of accuracy of the method. In other words, the leading-order error term precisely equals the first omitted term in the Taylor expansion of the exact solution. According to the Taylor series expansion given by equation (A.64), second-order accuracy requires retaining three terms:

$$y_{n+1} = y_n + \Delta t f(y_n) + \frac{(\Delta t)^2}{2} \left.\frac{df}{dy}\right|_{t=t_n} f(y_n) \tag{A.81}$$

Comparing the estimated solution in the RK2 formula (A.80) to the truncated Taylor expansion (A.81) determines the the RK2 constants. To facilitate a direct comparison, one substitutes the Taylor series expansion of k_2 about y_n equation (A.80). This leads to:

$$y_{n+1} = y_n + (\gamma_1 + \gamma_2)\Delta t f(y_n) + \gamma_2 \alpha (\Delta t)^2 \left.\frac{df}{dy}\right|_{t=t_n} f(y_n) \tag{A.82}$$

Comparing equation (A.82) to equation (A.81) and matching the coefficients of the similar terms gives:

$$\begin{aligned} \gamma_1 + \gamma_2 &= 1 \\ \gamma_2 \alpha &= \frac{1}{2} \end{aligned} \tag{A.83}$$

Therefore, $\gamma_1 = 1 - 1/(2\alpha)$, and we have arrived at a one-parameter family of RK2 methods, each member of which corresponds to a different value for α. Thus:

$$k_1 = \Delta t f(y_n), \tag{A.84}$$

$$k_2 = \Delta t f(y_n + \alpha k_1), \tag{A.85}$$

$$y_{n+1} = y_n + \left(1 - \frac{1}{2\alpha}\right) k_1 + \frac{1}{2\alpha} k_2. \tag{A.86}$$

A predictor-corrector form of RK2 reads:

$$y^*_{n+1/2} = y_n + \frac{\Delta t}{2} f(y_n), \tag{A.87}$$

$$y_{n+1} = y_n + \Delta t f(y^*_{n+1/2}), \tag{A.88}$$

corresponding to $\alpha = 0.5$.

A.5.2.4 Semi-implicit Methods

To take advantage of the stabilizing effect of implicit schemes in the presence of diffusive terms that impose severe time step restrictions while avoiding iterative solution of nonlinear equations, it is attractive to advance linear and nonlinear terms in different manners. Spalart et al. (1991) published a semi-implicit algorithm that exemplifies this approach. Their method advances linear terms implicitly and nonlinear and inhomogeneous terms explicitly. For a solution vector **U** and differential equation system:

$$\mathbf{U_t} = \mathbf{L}(\mathbf{U}) + \mathbf{N}(\mathbf{U}) \equiv \mathbf{R}(\mathbf{u}) \tag{A.89}$$

where **L** contains all linear terms and **N** contains all nonlinear and inhomogeneous terms, the Spalart et al. algorithm takes the form:

$$\mathbf{U}' = \mathbf{U_n} + \Delta\mathbf{t_{n+1}}\{\mathbf{L}[\alpha_1\mathbf{U}' + \beta_1\mathbf{U_n}] + \gamma_1\mathbf{N}(\mathbf{U_n})\} \tag{A.90}$$

$$\mathbf{U}'' = \mathbf{U}' + \Delta\mathbf{t_{n+1}}\{\mathbf{L}[\alpha_2\mathbf{U}'' + \beta_2\mathbf{U}'] + \gamma_2\mathbf{N}(\mathbf{U_n}) + \zeta_1\mathbf{N}(\mathbf{U}')\} \tag{A.91}$$

$$\mathbf{U_{n+1}} = \mathbf{U}'' + \Delta\mathbf{t_{n+1}}\{\mathbf{L}[\alpha_3\mathbf{U_{n+1}} + \beta_3\mathbf{U}''] + \gamma_3\mathbf{N}(\mathbf{U}') + \zeta_2\mathbf{N}(\mathbf{U}'')\} \tag{A.92}$$

where primes denote substep values between the nth and $(n+1)^{th}$ time steps, and $\Delta t_{n+1} = t_{n+1} - t_n$. For **N**, this approach resembles an explicit Euler substep followed by two second-order Adams-Bashforth substeps **N**[1]. For **L**, this approach resembles trapezoidal integration.

Third-order accuracy can be achieved by matching the Taylor series:

$$U_i(t+\Delta t) = U_i + \Delta t R_i(\mathbf{U}) + \frac{\Delta t^2}{2}\sum_{j=1}^{3}\frac{\partial R_i}{\partial U_j}\Delta U_j \tag{A.93}$$

$$= + \frac{\Delta t^3}{6}\sum_{j=1}^{3}\sum_{k=1} 3\frac{\partial^2 R_i}{\partial U_k \partial U_j}\Delta U_j \Delta U_k + O(\Delta t^4) \tag{A.94}$$

This condition generates 11 unknown coefficients that must satisfy 17 equations. By further requiring that the length of the substeps be the same for **L** as for **N**, Spalart, Moser, and Rogers (1991) reduce this system to eight equations in eight unknowns. Unfortunately, they also reported that the system appears to be insoluble. They therefore sacrificed one constraint, ultimately achieving third-order accuracy for **N** and second-order accuracy for **L**. In subsequent work, they have typically used the values:

$$\{\alpha_1, \alpha_2, \alpha_3\} \equiv \{4/15, 1/15, 1/6\} \tag{A.95}$$

$$\{\beta_1, \beta_2, \beta_3\} \equiv \{4/15, 1/15, 1/6\} \tag{A.96}$$

$$\{\gamma_1, \gamma_2, \gamma_3\} \equiv \{8/15, 5/12, 3/4\} \tag{A.97}$$

$$\{\zeta_1, \zeta_2\} \equiv \{-17/60, -5/12\} \tag{A.98}$$

When we use their scheme in this book, we use these values.

A.5.2.5 Stability of PDE solvers

Multiphysics simulations typically involve coupled, multidimensional, nonlinear PDEs in complicated geometries with complex boundary conditions. Proving the

[1] Adams-Bashforth schemes are multistep schemes in that they require time values from multiple previous steps in order to move forward. See Moin (2001).

stability of numerical approximations in such contexts requires a great degree of mathematical sophistication and is often best approached by starting from a discretization strategy that supplies a rich mathematical formalism. Most such discretizations fall under the heading of weighted residual methods (WRM), including spectral and finite element methods and hybridizations thereof (Canuto, Hussaini, Quarteroni, and Zang 2006; 2007; Karniadakis and Sherwin 2005). In the interest of simplicity of presentation, we do not pursue such strategies in this text.

Both the finite difference approximations employed in this book and the WRM involve laying a grid over the spatial domain and marching the solution through time in discrete steps. The grid spacing, Δx, characterizes the spatial discretization, whereas the step size, Δt, characterizes the temporal discretization. In multiphysics modeling – specifically applications that couple fluid dynamics to other physical phenomena – stability arguments can often be couched in terms of limitations on the speed with which information propagates. Information propagation occurs either by organized motion of bulk matter (flow) or by random, disorganized motion of individual molecules (diffusion). The flow speed u characterizes the flow. A diffusion coefficient – say, the thermal diffusivity, α, in the case of thermal energy transport – typically characterizes diffusion.

Combining Δx, Δt, u, and α leads to two dimensionless parameters that generally determine the stability of PDE discretizations:

$$CFL \equiv u\Delta t/\Delta x \tag{A.99}$$

$$\beta \equiv \alpha\Delta t/(\Delta x)^2 \tag{A.100}$$

The first of these is the Courant-Freidrichs-Lewy number and characterizes the stability of explicit time advancement of flow equations. *The second parameter characterizes the stability of explicit time advancement of diffusion equations*. The stability criteria take the form of inequalities that restrict the values of these parameters to remain below values of order unity. Although the derivation of specific stability bounds is beyond the scope of this discussion, such bounds are enforced in most of the solvers throughout this book.

Since β varies inversely with the square of the grid spacing, the stability bounds on β tend to be more restrictive than the limits on the CFL number. Implicit time advancement circumvents such restrictions. Because most multiphysics simulations involve nonlinear processes, however, fully implicit time advancement requires costly iterations. It is common therefore to use semi-implicit time advancement, marching the nonlinear terms forward explicitly to avoid iteration, while simultaneously marching the implicit terms forward implicitly to relax the time step restrictions. The multiphysics simulations discussed in Sectioin 12.2.1 all employ the semi-implicit time advancement algorithm of Spalart, Moser, and Rogers (1991).

APPENDIX B

Unified Modeling Language Elements

This appendix summarizes the Unified Modeling Language (UML) diagrammatic notation employed throughout this book along with the associated terminology and brief definitions of each term. We consider the elements that appear in the five types of UML diagrams used in the body of the current text: use case, class, object, package, and sequence diagrams. At the end of the appendix, we give a brief discussion of Object Constraint Language (OCL), a declarative language for describing rules for UML models.

B.1 Use Case Diagrams

A use case is a description of a system's behavior as it responds to an outside request or input. It captures at a high level who does what for the system being modeled. Use cases describe behavior, focusing on the *roles* of each element in the system rather than on *how* each element does its works.

A use case diagram models relationships between use cases and external requests, thus rendering a visual overview of system functionality. Figure B.1 reexamines the fin heat conductor analyzer diagram from Figure 2.6, adding notations to identify the elements of the use case diagram.

Use case diagrams commonly contain the following elements:

1. **Actors:** people or external systems that interact with the system being modeled. Actors live outside the system and are the users of the system. Typically actors interact with the system through use cases. In UML, actors are drawn as stick figures. In the fin analyzer system example, system architect, thermal analyst, and numerical analyst are actors.
2. **Use cases:** abstractions that represent sets of sequences of actions that the system performs to produce measurable results to actors. A use case describes the actions the system takes in response to a request from an actor. It captures the essence of interactions between an actor and the system. In UML, a use case is drawn as an ellipse, with its name printed in the center of the ellipse. The fin analyzer example includes three use cases: specify problem, design numerical method, and predict heat conduction. All three use cases have actors associated with them.

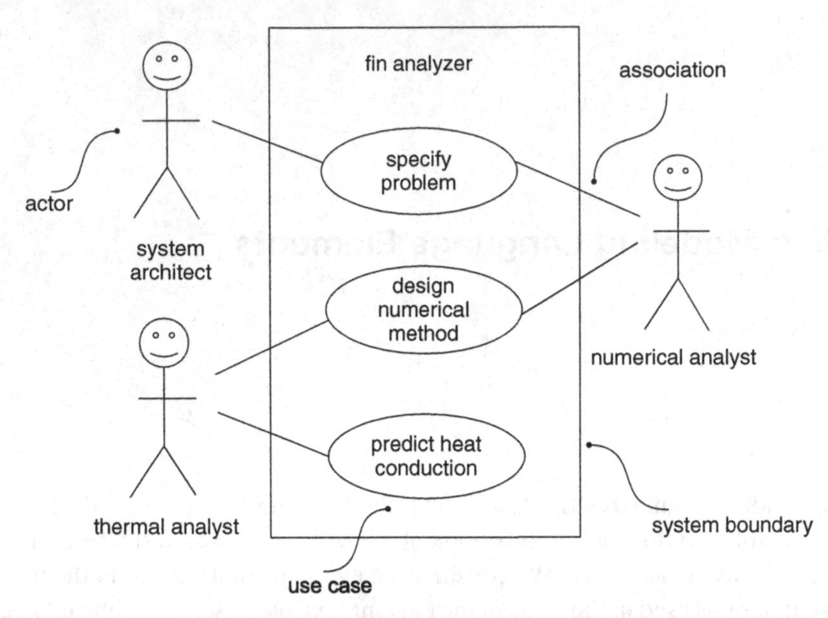

Figure B.1. Use case diagram for a fin heat conduction analysis system.

3. **Associations:** connections between use cases and actors. An association is established when an actor interacts with the system through a use case that describes the interaction. An association is drawn using a solid line connecting an actor and a use case. Optionally an arrow can be added to indicate the direction of interaction, that is, who initiates the interaction. In the fin analyzer system, five associations are identified with the following pairs of actor and use case: system architect—specify problem, numerical analyst—specify problem, numerical analyst—design numerical method, thermal analyst—design numerical method, and thermal analyst—predict thermal conduction.
4. **The system boundary:** shown as a rectangular box to indicate the scope of the system. All the use cases are inside the system boundary, and all actors are outside the system boundary.
5. **Other relationships:** relationships between actors and between use cases. Actor generalization and use case relationships, such as "include," "extend," and "generalization," are also common in practice (Booch et al. 1999).

A use case diagram can also include *packages* to group different modeling elements into chunks. The UML package diagram provides a convenient way to depict very general collections of model elements. The elements need not be object-oriented. For example, Figure 6.1 depicts a package containing a class and a stand-alone procedure that accepts an instance of that class as an argument. Because package diagrams see minimal use in the current text, appearing only in Figure 6.1, we go into no greater depth here.

In our fin analyzer example, the interaction drawn between the system architect and the fin analyzer system depicts use of the system to specify the heat conduction problem. This usage is depicted by specify problem use case. The output from this problem specification – including the production of a set of physical parameters such as tolerances, boundary conditions, initial heat conditions, and so on – becomes the input for another actor, the numerical analyst. Note the numerical analyst has two use case associations with the fin analyer. First, she receives data from the fin analyzer through the specify problem use case. This is depicted by the numerical analyst—specify problem association. Then she develops a numeric model based on these data from the fin analyzer. The latter interaction is depicted by the numerical analyst—design numerical method association. Similarly, the thermal analyst receives the numeric model produced by the numerical analyst and uses this model to predict the fin heat transfer performance.

Use cases diagrams produce a static view of a system. Applications of use case diagrams in modeling typically take two forms (Booch et al. 1999):

1. **Actor-centric:** model the context of a system. This modeling method places an emphasis on actors and their roles, focusing on the environment surrounding the system. It starts with identifying actor and organizing actors according to their roles, followed by pinpointing use cases and their associations with actors based on the actors' behaviors and requirements on the system. Once the use case diagram is completed, all use cases are associated with one or more actors. Our fin analyzer example follows this approach.
2. **System behavior-centric:** model the requirements of a system. Much like specifying a contract on the system functionalities, this approach focuses on the system behaviors. It begins with identifying actors and gathering their requirements to the system. Use cases are developed based on these requirements. Additional use cases may be developed based on additional system requirements, variants to existing use cases, and so on. (Booch et al. 1999) has provided an example of validation system for credit card transactions to illustrate this approach. A detect card fraud use case is an addition to the use case diagram due to security requirements. Similarly, they introduced a manage network outage use case to the diagram due to a requirement that the system provide reliable and continuous operations.

In each case, use case diagrams aid in visualizing and documenting the system behavior. They can also aid in documenting the business requirements. Ideally, these diagrams enhance designers, developers, and domain experts' understanding of the system before and during construction.

B.2 Class Diagrams

Class definitions comprise the most fundamental building blocks in any object-oriented application. One can model a system's classes using class diagrams, the most common UML diagram. In particular, class diagrams can convey class attributes and operations as well as the interactions and relationships among different classes. Class diagrams provide a static, structural overview of a modeled system. The remainder of this section covers the two key elements for class diagrams: representations of classes and the relationships among classes.

B.2.1 Classes

A class is an abstraction for a set of objects that share the same attributes, operations, and relationships to others. UML represents a class graphically using a

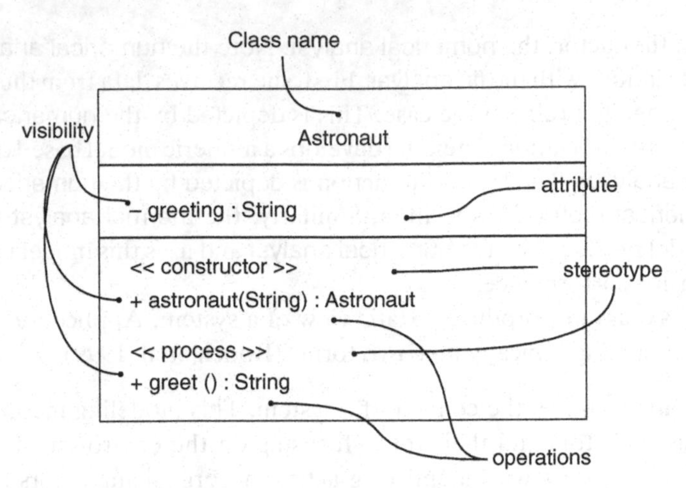

Figure B.2. Class diagram for astronaut class.

three-compartment box. In Figure B.2, we redraw the Astronaut class from Figure 2.2 to illustrate the following elements of a class:

- **Name:** The top compartment holds the class name, Astronaut (Figure 2.2). The name is the only mandatory part of a class diagram. In diagrams that contain large collections of classes (hundreds or thousands in some cases), it is common to omit details such as attributes and operations so readers can focus on the relationships between classes.
- **Attributes:** As shown in Figure B.2, the middle compartment of a class is used to show the class attributes. These are the Fortran-derived type components or the C++ data members. The attributes are normally drawn in one of two ways:
  1. By showing attribute names only.
  2. By specifying attribute names followed by their data types, separating the two by a colon. Additionally the diagram can provide default attribute values.

  Our example in Figure B.2 uses the second method. The Astronaut class contains one attribute, greeting, of type String.
- **Operations:** An operation is a service a class provides. Common Fortran operations include type-bound procedures, defined operators, and defined assignments. Common C++ operations include virtual member functions and overloaded operators. As shown in Figure B.2, the bottom compartment of a class box displays the operations. Similar to attributes, operations can be specified with one of two methods:
  1. By showing operation names only.
  2. By specifying operation names accompanied with their function signatures, including the procedure name and argument types followed by a colon and the return data type, if any.

  Our Astronaut example employs the latter method to reveal more information about the operations. Two operations are shown in the Astronaut class: an overridden constructor astronaut(), which takes a string and returns a constructed Astronaut object; and the procedure greet(), which produces String.

As with attributes, the list of operations of a class can be omitted from the diagram if the emphasis of the design is placed on other aspects of the system such as the interactions between classes.
- **Responsibilities:** Optionally, one can detail a class's responsibilities in an additional compartment drawn below that of the operations. The responsibility part of a class describes the contracts or obligations of the class (Booch et al. 1999).

In addition to these basic elements, there are some further notations commonly employed to specify additional information on classes:

- **Visibility:** Visibility refers to the accessibility of attributes or operations by other classes. It falls into one of the following three states:
  1. *Public:* Denoted by a plus sign (+), this state implies that code outside the class can access the attribute or operation. In our Astronaut class, the operations astronaut() and greet() are publicly accessible.
  2. *Private:* Denoted by a minus sign (–), this state indicates the attribute or operation is not visible to code outside the class that contains the attribute or operation. The attribute greeting in Astronaut class is private, preventing direct access from outside the Astronaut class.
  3. *Protected:* Denoted by a pound, or number, sign (#), an attribute or class declared as protected is visible only to the class and its direct subclass. C++ supports the declaration of protected attributes and operations. Fortran does not.[1]
- **Stereotypes:** Graphically represented using names enclosed by guillemets («»), a stereotype in UML is one of extensibility mechanisms that allow derivation of new and domain-specific modeling elements from existing ones. Our Astronaut class exhibits one common usage of stereotypes in a class diagram: organizing operations. «constructor» is routinely used in UML to denote that operations immediately following the stereotype are overloaded constructors, and similarly «process» marks the roles of the subsequent operations.
- **Abstract classes:** UML distinguishes an abstract class from a concrete one by putting the abstract class name in a ***bold italic*** font. The ***Integrand*** class in our ABSTRACT CALCULUS pattern is an example of an abstract class (see Figure 6.1).
- **Interfaces:** UML distinguishes an interface from a class by enclosing the interface in a circle instead of a box. Interface diagrams are more relevant to languages that have an explicit interface construct, for example, Java. In Fortran and C++, one typically models interfaces as abstract classes. For clarity, we use the abstract class notation in all class diagrams in this book.
- **Abstract operations:** An important functionality of an abstract class is to specify the services its subclasses must provide. Each specification usually takes the form of an abstract operation – that is, an operation signature defined without concrete implementations. These are Fortran *deferred* procedures bindings or C++ pure virtual member functions. An abstract operation is indicated by *italic* font. In the abstract class ***Integrand***, shown in Figure 6.1(b), operations *add*, *multiply* and *t* are all abstract.

[1] The "protected" keyword in Fortran denotes a module variable that is visible but not directly modifiable from outside a module.

- **Template classes:** A template class is a parameterized class. C++ template classes and Fortran parameterized derived types are both template classes. In UML, a template class is drawn as an ordinary class with an additional dashed box in the upper-right corner of the class. This additional box shows the template parameters used to instantiate an object of this class. As shown in Figure 2.3(a), the Array class in Chapter 2 is a template class, with Element being the sole template parameter.
- **Datatypes:** As one of the classifiers available in UML, a "datatype" represents a set of data values, including intrinsic data types and enumerations. The graphical representation of a datatype is similar to a stereotype. Figure 2.3(b) shows a common usage of datatypes in UML modeling: defining data types used in template classes. In the example, data types of Integer, Character, Real, and Boolean can be used as for the template parameter Element.

B.2.2 Relationships

In UML, a relationship is abstracted as a connection between two elements. In class diagrams, all relationships are represented lines connecting classes. The following three kinds of relationships are identified as the most important relationships in OOD: dependency, generalization, and association.

B.2.2.1 Dependency

A dependency is a relationship between two classes wherein one class depends on the behaviors of the other. Most often this indicates a use relationship between two classes: We say class A depends on class B when B is used either as a passed parameter or as a local variable for the operations of A. One draws dependencies in UML using a dashed line connecting two classes with an arrow pointing to the class on which the other class depends.

Derived from Figure 3.4, Figure B.3 emphasizes the class dependencies of the heat conduction system given in Section 3.1. We only draw the dependencies used in

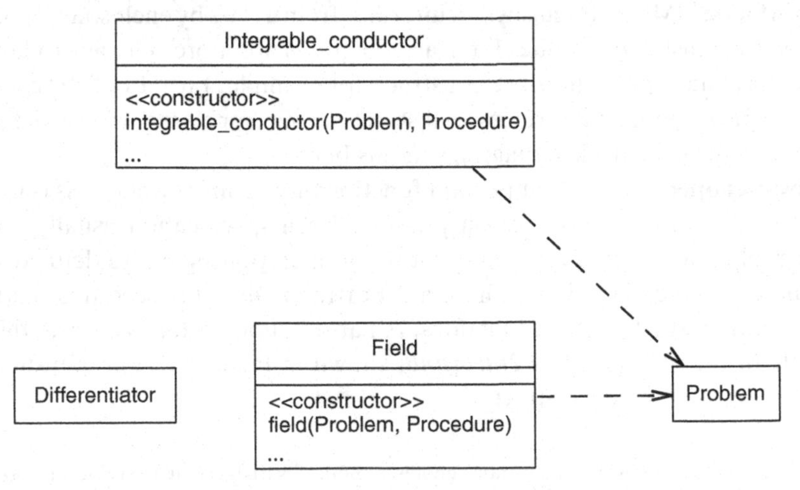

Figure B.3. Class dependencies in heat conduction class diagram.

the heat conduction system, omitting other relationships between classes. In the system, both the Integrable_conductor class and Field class have a dependency on the Problem class. The class definitions for Integrable_conductor and Field also highlight the cause of their dependencies on Problem: The constructions of Integrable_conductor and Field require the passing of a Problem object. From the corresponding code in Figure 3.2 and Figure 3.3, we can further see that the Field's construction is directly impacted by the implementation (or implementation changes) of Problem because the Field derives the nodal values, the stencil, and the system boundary conditions from the Problem attributes, whereas the dependency on Problem for constructing an Integrable_conductor object is pertinent to that of its `T_field` component. In this sense, we can say Field has a stronger dependency on Problem.

B.2.2.2 Generalization and Realization

In a class diagram, a generalization refers to a relationship between a base class (the parent) and one that extends it (the child). The generalization relationship is also known as inheritance or an "is-a" relationship. In a generalization, the parent represents the general properties of a class (i.e., attributes and operations), whereas the child represents a specialized extension of the parent. In application code, a child type is always substitutable for its parent type in procedure calls, but not the reverse. In UML, a generalization is graphically rendered as a solid line connecting the parent and the child, with an open arrowhead pointing to the parent.

Closely related to a generalization, a realization in a class diagram is a relationship between an interface and a class (an implementor) that realizes the operations specified by the interface. Since abstract classes in C++ and Fortran model interfaces, a realization relationship is established when a class extends an abstract class. Consequently, a realization relationship automatically replaces a generalization relationship when the base class is abstract. The interface consistency in a realization is reaffirmed by the fact that both C++ and Fortran mandate the implementor to keep the same procedure interfaces when implementing the inherited operations. In UML, a realization relationship is rendered as a dashed line connecting the interface and its child, with an open arrowhead pointing to the interface.

Figure B.4 re-illustrates the generalizations and realizations used in the class diagram of a timed Lorenz system that employs SURROGATE and STRATEGY patterns. Comparing to the original diagram in Figure 7.1, one may notice the omission of other relationships between classes except generalizations and realizations.

In this system, classes Euler_Integrate and Runge_Kutta_Integrate are the actual strategies used in time integration for Lorenz and Time_Lorenz, respectively. Both Euler_Integrate and Runge_Kutta_Integrate are concrete implementations of abstract class ***Strategy***, thus both have a realization relationship with ***Strategy*** as shown in the figure. For the class hierarchy tree started with ***Surrogate*** as the root, the relationships between the parent/child pair can be said as follows: ***Integrand*** realizes ***Surrogate***, Lorenz realizes ***Integrand*** with concrete implementation, and Timed_lorenz generalizes Lorenz.

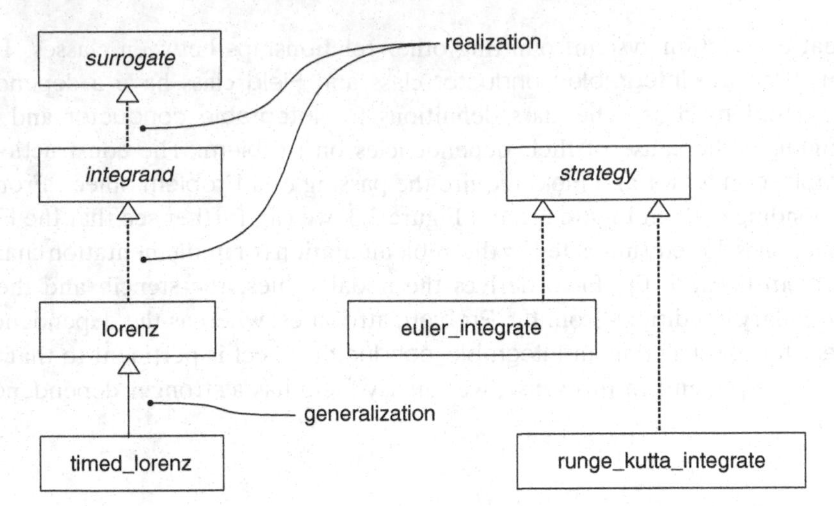

Figure B.4. Class generalizations and realizations in SURROGATE and STRATEGY patterns for a timed Lorenz system.

B.2.2.3 Association

An association represents a link or connection between two classes. Graphically, associations add visual cues to the structural relationships between classes in a class diagram. In UML, an association is drawn as a solid line connecting two classes. Beyond the basic form, a few adornments supply additional information when applied to association relationships:

- **Name and direction:** An association name (usually a verb) is used to describe the nature of the association. An additional direction can be supplied, as a big solid triangle, to indicate the action flow. Figure B.5 simplifies the class diagram of Figure 9.1 with only association relationships depicted. As an implementation of ***FieldFactory***, the concrete class Periodic6thFactory constructs a Periodic6hOrder Field object through its implementation of create() method. Exploiting the inheritance relationship between Periodic6thFactory and *FieldFactory*, however, the method create() returns a reference to type *Field*.
- **Role:** Specific roles can be given to the two participants in an association relationship. The role name is normally written as a plain text adjacent to the end of an association line connected to the class.
- **Multiplicity:** Multiplicity is a way to specify rules or restrictions on the number of instances of a class in an association. It usually answers the question of "how many" class instances are to participate in one association. The number can be specified as exact (1), within a range (1..5), or many (1..*).
  Figure B.6 highlights the association between HermeticField and HermeticFieldPointer used in Figure 10.7. The association relationship is a "pointer to target" relationship, HermeticField being a target and HermeticFieldPointer being a pointer. In the class diagram, the role for HermeticField is specified as a target; we omitted the specification for HermeticFieldPointer since its name is self-explanatory. The diagram further specifies that for each (1) HermeticField object, many (1..*) HermeticFieldPointer objects can be created and associated with the same HermeticField object.

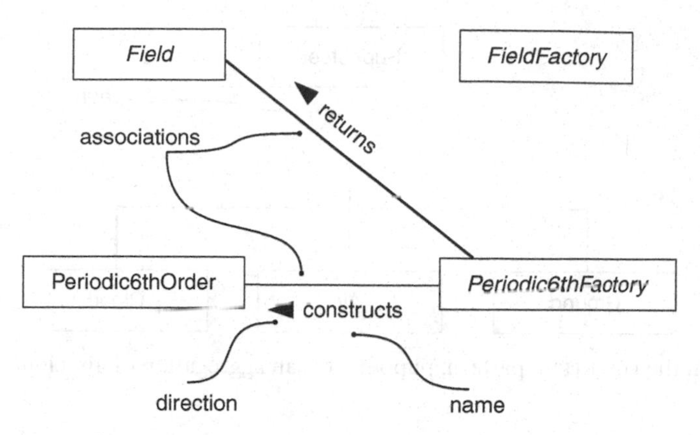

Figure B.5. Association relationships in ABSTRACT FACTORY patterns applied to a Burgers solver.

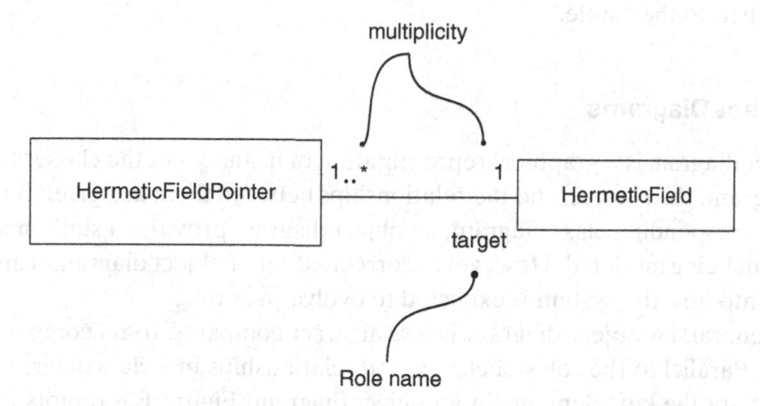

Figure B.6. HermeticField and HermeticFieldPointer class association and adornments.

- **Aggregation:** Aggregation is a special kind of association that can be best described as a "whole-part," "has-a," or "contains" relationship between two classes. An aggregation relationship can involve more than two classes, with one class (the whole) being a collection or a container of the rest of the classes (the parts). In UML, an aggregation is depicted as an open diamond attached to the class that behaves as the whole. The PUPPETEER pattern in Chapter 8 employs an aggregation relationship between the puppeteer and its three puppets: air, cloud, and ground. Figure B.7 reproduces the original class diagram as shown in Figure 8.2, except for the added adornment labeling the relationship as aggregation.
- **Composition**: Composition is a variant of aggregation with added semantics. Compared to an aggregation, a composition relationship identifies a stronger ownership between the "whole" and its "parts" for the association. In plain language, a composition best describes a relationship in which one class (the whole/composite) is composed of many others (the parts/components). Very often the lifetime of the parts coincides with that of the whole in a composition relationship. In comparison, an aggregation has a much looser ownership that allows the parts to live independently from the whole. In addition, the whole is

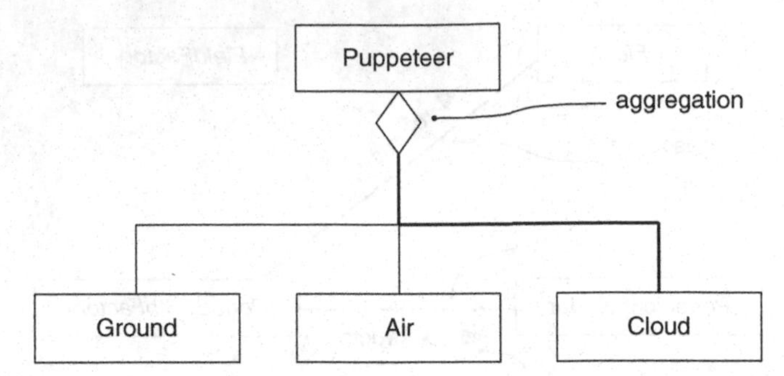

Figure B.7. In the PUPPETEER pattern, puppeteer is an aggregation of air, cloud, and ground.

also responsible for the disposition (e.g., creation and destruction) of its parts in a composition[2]. In UML, a composition is represented using a filled diamond attached to the whole.

B.3 Object Diagrams

An object diagram is a graphical representation of instances of the classes found in a class diagram, their states, and the relationships between them at a given time. Much like its corresponding class diagram, an object diagram provides a static snapshot of the system being modeled. However, a correlated set of object diagrams can provide insights into how the system is expected to evolve over time.

In general, an object diagram is less abstract compared to its companion class diagram. Parallel to the roles of classes and relationships in a class diagram, objects and links are the key elements in an object diagram. Figure B.8 replots the object diagram of Figure 2.12, showing that the fin object of type Conductor is composed of an object diff of type Differentiator. The diagram elements are as follows:

- **Objects:** An object is an instance of a class. Whereas a class represents an abstraction of similar "things," an object gives a concrete instance of one. Similarly to a class, an object is also depicted using a box in UML. However, an object can be readily distinguished from a class by the fact that the text contained in the name compartment is underlined. Also the text is usually composed of two components: an object name followed by a class name, separated by a colon (:). Sometimes an object can be anonymous, that is, the object name before the colon is omitted. This usually characterizes a temporary object returned from a function call. In rare situations, one can also render objects without associated class abstractions. These objects are called orphans, and readers can refer to Booch et al. (1999) for more details.

  In addition to the name compartment, an object can have a compartment showing attributes and their values. Collectively they represent the particular

[2] It can be confusing to decide whether an aggregation or a composition relationship should be used in a design. The difference between them is subtle, and the design errors caused by making a wrong choice are normally very minor.

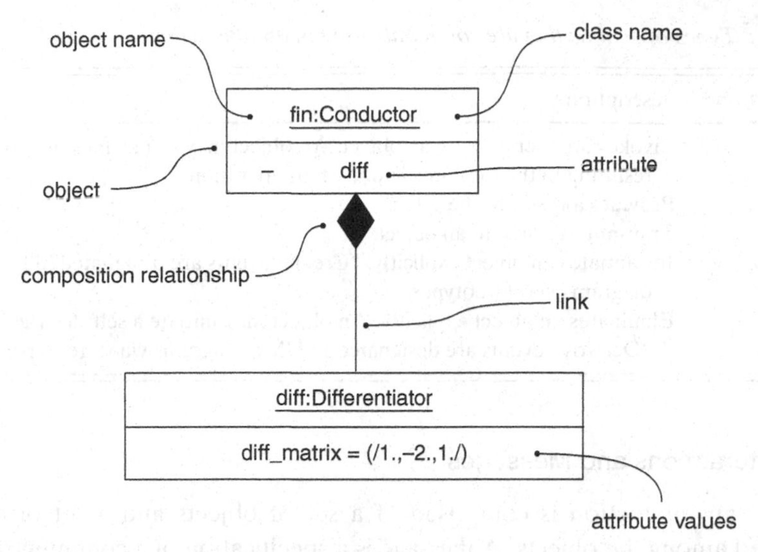

Figure B.8. An object diagram that shows the object fin of class Conductor is composed of an object diff of class Differentiator.

state that the object is in. Figure B.8 contains two objects: the heat conductor fin and the differentiator diff. Their respective types are Conductor and Differentiator. Further we note that fin contains a private component, named diff, whereas fin is in its initial state with its attribute diff_matrix being assigned to a specific set of values.

- **Links**: Analogous to an object being an instance of a class, a link is an instance of an association. As with associations, links show relationships between objects in an object diagram. In UML, a link is rendered using a line connecting two objects, just like an association in class diagrams. In addition, adornments used on associations, that is, names, multiplicities, roles, and aggregations/compositions, can also be applied to links. In Figure B.8, one can observe the fact that fin is a composition of diff.

Both the object diagrams and the class diagrams described in B.2 belong to the structural modeling category. Both types of diagrams are commonly used to capture the static view of a system. In the next section, we introduce sequence diagrams that focus primarily on modeling the dynamic aspects of the system.

B.4 Sequence Diagrams

A sequence diagram is a kind of interaction diagram that depicts the interactions between different objects and the ordering of these interactions. It is commonly used to model the dynamic aspects of a system – to show the details of an operation involving a set of objects. Objects, messages (with data exchanges), and sequences of actions are the key elements of an interaction diagram. A sequence diagram emphasizes the time ordering among processes during an operation.

Table B.1. *Types of actions that are commonly used in an interaction*

Type of action	description
call	Invokes an operation on an object. An object may send a message to itself, resulting in the local invocation of an operation.
return	Provides a result to the caller.
send	Transmits a signal to an object.
create	Instantiates an object explicitly. "Create"events are designated in UML diagrams via stereotypes.
destroy	Eliminates an object explicitly. An object may initiate a self-destruction. "Destroy" events are designated in UML diagrams via stereotypes.

B.4.1 Interactions and Messages

In UML, an interaction is comprised of a set of objects and a set of messages exchanged among the objects. A message is a specification of a communication and the information exchange between two objects. Occasionally an object can also send a message to itself, for example an invocation of a recursive function call to itself. In concrete terms, a message in an interaction can usually be mapped into an action that involves only two objects. The receipt of a message is normally considered an instance of an event during which an action is taken and a state change results as a consequence. Also the data passed between the two objects are an inherent part of the message. Even in a complex operation that involves multiple objects, it is always possible to decompose the whole process into sequences of actions, each only involving two objects. These ultimate actions are drawn as messages in UML diagrams.

In UML, a message is depicted using a straight line between the two objects with an arrow indicating the direction of invocation. In the case a self-call, that is, an object sends a message to itself, a connecting line (either curved or with straight segments) is used to show that the object is both the sender and the receiver. There are a few kinds of particular actions that are commonly used, as described in Table B.1.

Taking again the heat conduction example used in Chapter 3, in particular the forward Euler computation of `fin + fin%t()*dt` as shown in Figure 3.5(a), we illustrate the use of UML interactions via a small section from that figure.

Figure B.9 details the call sequences during the computation of the second-order spatial derivative on the T_field component of fin. As can be traced clearly by the sequence numbers encoded in the messages, the object fin initiates the computation by sending a message xx() to T_field, followed by the creation and initiation of a temporary object by T_field using the action «create ». Next a laplacian() call to stencil is made by the temporary. The last action shown in the diagram is stencil returning the laplacian back to the temporary field object. For brevity, we omit the obvious steps needed to return the computed value of xx() all the way back to fin in the diagram.

B.4.2 Sequence Diagrams

Sequence diagrams are amongst the most common interaction diagrams used in UML. Communication diagrams are another popular kind of interaction diagram.

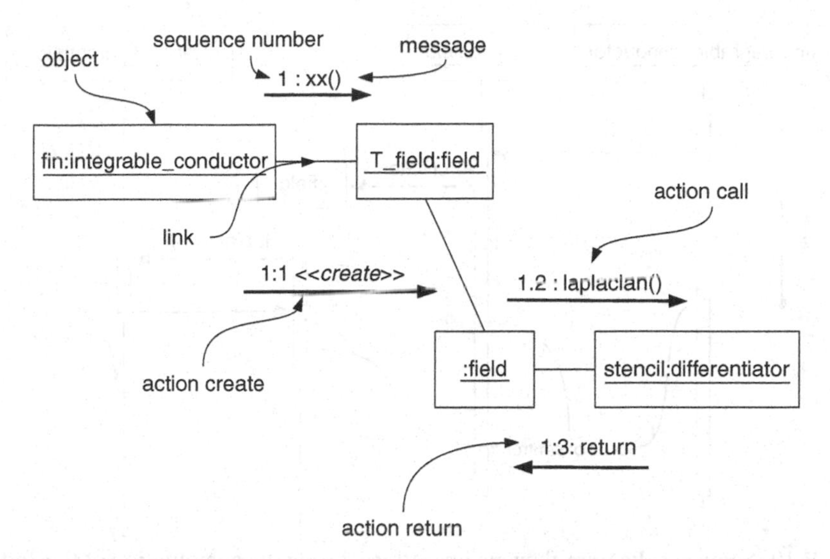

Figure B.9. Interactions among objects during the computation of the 2nd-order spatial derivative for forward Euler method.

Figure B.9 is a communication diagram, and it contains all the elements normally found in a communication diagram. Semantically a communication diagram and a sequence diagram are equivalent in that one can convert from one type of diagram to the other without losing any information.

We choose to use a sequence diagram in our examples based on two factors. First sequence diagrams are more explicit in revealing the time ordering among different actions during an operation. In a complex operation, particularly when many function calls are made between the same couple of objects at different times, the sequences of calls can become cluttered in a communication diagram, but can always be shown clearly in a sequence diagram. The second factor is that the sequence diagram has analogues in other fields. For example, the Message Sequence Chart (MSC) used in the telecommunication industry is nearly identical to a sequence diagram.

We redraw the interactions of Figure B.9 using a sequence diagram in Figure B.10. By comparing these two figures, one can observe that the vertical axis in a sequence diagram denotes the time. The higher a message is drawn, the earlier the event occurs.

Next, every object in the sequence diagram has an object lifeline: a dashed vertical line showing the time period during which the object exists. Most objects will exist during the whole operation. So they are aligned at the top of the diagram, with their lifelines extended from the top to the bottom of the diagram. However, objects can also be brought into life by dynamic creation «create».

Last, every action is drawn with a *focus of control*, a tall thin rectangle, to indicate the duration of the action. Usually the top of a rectangle is aligned with a message that starts the action. The bottom of the rectangle indicates the completion of an action, and may be aligned with another message, for example, a return message. A focus of control may also provide hints as to the duration of the action: A short focus of control usually implies a rapid completion of an action. Compared to the

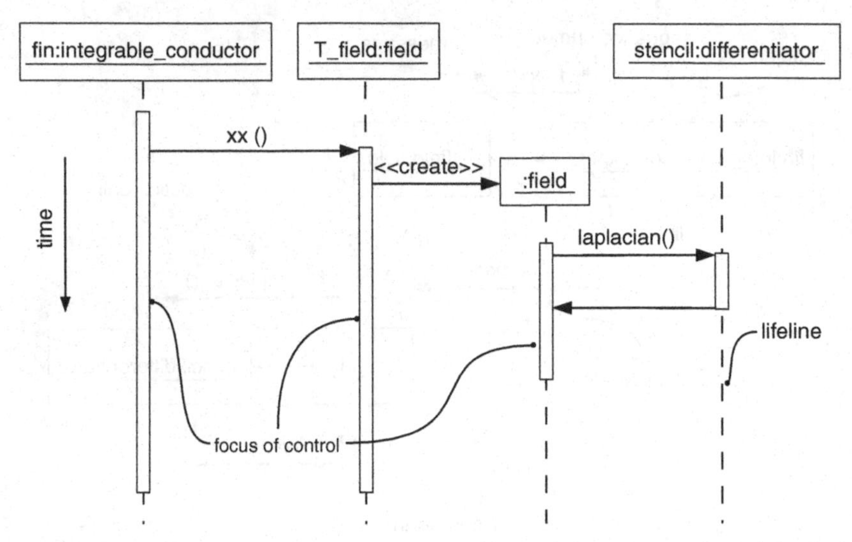

Figure B.10. Sequence diagram showing operations during the computation of the 2nd-order spatial derivative for forward Euler method.

communication diagram, we can see that a sequence diagram places greater emphasis on the time ordering of messages.

B.5 The Object Constraint Language

The Object Constraint Language (OCL), first developed in 1995 during a business modeling project within IBM, is a language based on mathematical set theory and predicate logic but is designed to be understood by people who are neither mathematicians nor computer scientists (Warmer and Kleppe 2003). OCL alleviates problems with incompleteness, informality, and imprecision that arise in models written in UML. In combination with UML, OCL expressions can make models more precise and detailed. OCL appears at first to be a constraint language, but in UML 2, one can use OCL to write not only the constraints but also any information about the elements in the UML diagrams, such as the initial values, derivation rule, or definition of an instance. This makes OCL a query language as well. The current section focuses on three main OCL expressions, or rules: those that add initial values, invariants, and pre-/postconditions.

B.5.1 The Context of OCL Expression

An OCL expression augments a UML diagram with information about the model. The link between the UML diagrams and the OCL expression is the context of OCL expression, which specifies the element in the UML diagrams for which the OCL expression is defined. The contents of the context are usually the class, interface, datatype, component, or operation that are defined in the UML diagrams. When the OCL expression is incorporated in the UML diagram, such as in the Figure 10.6, the context definition is shown by a dotted line that links the element in a UML diagram and the OCL expression. When the OCL expression is given in a separate text file, the context is denoted by the keyword "context" followed by the name of the element in UML diagram. OCL expressions depend on the context of the expression and can have many different functions; for example, the expression can be an initial value when the context is an attribute.

B.5.2 Initial Value Rule

Initial value rules add information regarding the initial values of attributes and association ends in the UML diagram. The context of an initial value rule is the UML diagram element that holds the relevant attributes or association ends. For example, Figure 10.5 shows the HermeticField class has a "temporary" attribute for which the default initial value is always "true". In OCL, an initial value of "temporary" can be defined as follows:

```
context : HermeticField :: temporary
init:  true
```

B.5.3 Specifying Object Invariants

An OCL expression can be added to a UML diagram to specify an invariant that is a constraint for an object and must be true for the object throughout its complete life. The context of the invariants can be a class, an interface, or a type in the UML diagram. Any attribute of the context can be used in invariant expressions. For example, because the `GuardedField` object defined in Figure 10.6 must be guarded against premature finalization, the "guarded" state can be formalized in the following OCL expression:

```
context : GuardedField
inv: depth > 1
```

B.5.4 Adding Pre- and Postconditions to Operations

Preconditions and postconditions are boolean expressions that constrain operations. A precondition of an operation must be true in order for an operation to execute. A postcondition must be true after an operation terminates; otherwise, the operation has failed. The pre- and postconditions only describe the applicability and the effect of the operation without specifying the implementation.

The context of pre- and postconditions is the class name that holds the operation, followed by a double colon with the operation name, parameters of the operation, and the returned type. The operation name, parameters of the operation, and the returned type are usually defined in the UML diagram. The OCL expression of pre- and postconditions follows the key words "pre" and "post" each return a boolean value:

```
context: Field :: d_dx(direction : Integer) : Field
pre: self.fourier.allocated() xor self.physical.allocated
post: self.temporary implies
(not self.fourier.allocated()) and (not self.physical.allocated())
```

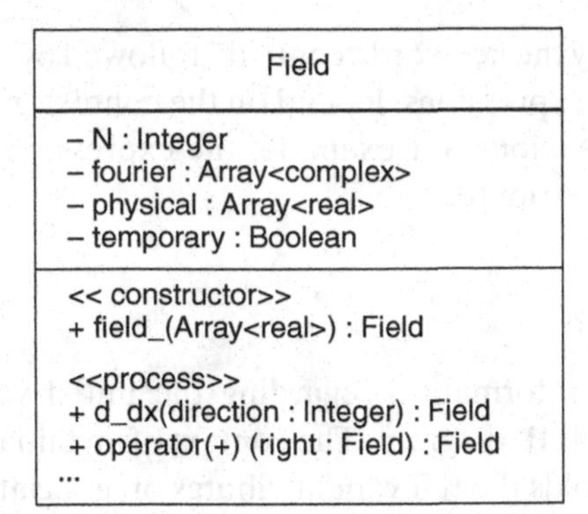

Figure B.11. Field class diagram of the turbulence solver of Rouson et al. (2006).

where `self` refers explicitly to the contextual instance and can be omitted when the contextual instance is obvious. In this example shown in Figure B.11, `self` refers to the instance of `Field` class for which the operation `d_dx` is called. In the precondition, `xor` is the exclusive or operation, which evaluates to true only when exactly one operand is true. The precondition specifies that the contextual instance must have allocated values in either Fourier space or physical space but not both. The `implies` operation in the postcondition indicates that the OCL expression is true when both boolean operands before and after `implies` are true. If the boolean operand before `implies` is false, the OCL expression is always true. The postcondition states that neither the Fourier nor physical space values of the contextual instance can be allocated if the contextual instance is a temporary object.

Bibliography

Akin, E. (2003). *Object-Oriented Programming via Fortran 90/95*. Cambridge University Press, Cambridge.

Alexander, C. (1979). *The Timeless Way of Building*. Oxford University Press, New York.

Alexander, C., S. Ishikawa, and M. Silverstein (1977). *A Pattern Language: Towns, Buildings, Construction*, Volume 2 of *Center for Environmental Structure Series*. Oxford University Press, New York.

Alexander, C., M. Silverstein, S. Angel, S. Ishikawa, and D. Abrams (1975). *The Oregon Experiment*. Oxford University Press, New York.

Allan, B., R. Armstrong, D. E. Bernholdt, F. Bertrand, K. Chiu, T. L. Dahlgren, K. Damevski, W. R. Ewasif, T. G. W. Epperly, M. Govindaraju, D. S. K. J. A. Kohl, M. Krishnan, G. Kumfert, J. W. Larson, S. Lefantzi, M. J. Lewis, A. D. Malony, L. C. McInnes, J. Nieplocha, B. Norris, S. G. Parker, J. Ray, S. Shende, and T. L. W. andn S. Zhou (2006). A component architecture for high-performance scientific computing. *Intl. J. High Perform. Comput. Appl. 20*(2), 163–202.

Allan, B., B. Norris, W. R. Elwasif, R. C. Armstrong, K. S. Breuer, and R. M. Everson (2008). Managing scientific software complexity with Bocca and CCA. *Scientific Programming 16*(4), 315–327.

Aris, R. (1962). *Vectors, Tensors, and the Basic Equations of Fluid Mechanics*. Dover Publications. Mineola, NY.

Artoli, A., A. Hoekstra, and P. Sloot (2006). Mesoscopic simulation of systolic flow in the human abdominal aorta. *Journal of Biomechanics 39*, 837–884.

Atlas, D., K. R. Hardy, and K. Naito (1966). *Optimizing the radar detection of clean air turnbulence, Journal of Applied Meterology 5*(4), 450–460.

Balay, S., K. Buschelman, V. Eijkhout, W. Gropp, D. Kaushik, M. Knepley, L. C. McInnes, B. Smith, and H. Zhang (2007). Petsc users manual. Technical Report ANL-95/11 Rev. 2.3.3, Argonne National Laboratory.

Barker, V. A., L. S. Blackford, J. Dongarra, J. D. Croz, S. Hammarling, M. Marinova, J. Waniewsky, and P. Yalamov (2001). *LAPACK95 Users Guide*. Society for Industrial and Applied Mathematics.

Barrett, D. (1998). *Polylingual systems: An approach to seamless interoperability*. Ph. D. thesis, University of Massachusetts, Amherst, MA.

Barton, J. and L. R. Nackman (1994). *Scientifc and Engineering C++: An Introduction with Advanced Techniques and Examples*. Addison-Wesley, New York.

Beck, K. and W. Cunningham (1987). Using pattern languages for object-oriented programs. In *OOPSLA-87 Workshop on the Specification and Design for Object-Oriented Programming*.

Benton, E. R. and G. W. Platzman (1972). A table of solutions of the one-dimensional burgers equation. *Quarterly of Applied Mathematics 30*, 195–212.

Berger, M. J. and P. Colella (1989). Local adaptive mesh refinement for shock hydrodynamics. *J. Comput. Phys 82*, 64–84.

Bernard, P. S., J. M. Thomas, and R. A. Handler (1993). Vortex dynamics and the production of Reynolds stress. *Journal of Fluid Mechanics 253*, 385–419.

Bhatnagar, P., E. Gross, and M. Krook (1954). Model for collision processes in gases. *Phys. Rev. 94*, 511–525.

Blackford, L. S., J. Choi, A. Cleary, E. D'Azevedo, J. Demmel, I. Dhillon, J. Dongarra, S. Hammarling, G. Henry, A. Petitet, K. Stanley, D. Walker, and R. C. Whaley (1997). *ScaLAPACK Users' Guide*. Society for Industrial and Applied Mathematics.

Blackford, L. S., J. Demmel, J. Dongarra, I. Duff, S. Hammarling, G. Henry, M. Heroux, L. Kaufman, A. Lumsdaine, A. Petitet, R. Pozo, K. Remington, and R. C. Whaley (2002). An updated set of basic linear algebra subprograms (blas). *ACM Trans. Math. Soft. 28*(2), 135–151.

Blilie, C. (2002). Patterns in scientific software. *Computers in Science and Engineering 4*(4), 48–53.

Booch, G., J. Rumbaugh, and I. Jacobson (1999). *The Unified Modeling Language User Guide*. Addison-Wesley, New York.

Boyd, J., J. Buick, J. Cosgrove, and P. Stansell (2004). Application of the lattice boltzmann method to arterial flow simulation: Investigation of boundary conditions for complex arterial geometries. *Australasian Physical and Engineering Sciences in Medicine 27*, 207–212.

Bientinesi, P., Gunnels, J. A., M. E. M. E. S. Q.-O., and van de Geijn, R. A. (2005). The science of deriving dense linear algebra algorithms. *ACM Transactions on Mathematical Software 31*, 1–26.

Burgers, J. M. (1948). A mathematical model illustrating the theory of turbulence. *Adv. Appl. Mech.* (1), 25–27.

Canuto, C., M. Y. Hussaini, A. Quarteroni, and T. A. Zang (2006). *Spectral Methods: Fundamentals in Single Domains*. Springer-Verlag, Berlin/Heidelberg.

Canuto, C. M. Y. Hussaini, A. Quarteroni, and T. Zang (2007). Spectral Methods. Evolution to Complex Geometries and Applications to Fluid Dynamics. Springer-Verlag, Berlin/Heidelberg.

Chaitin, G. J. (1996). How to run algorithmic complexity theory on a computer: Studying the limits of mathematical reasoning. *Complexity 2*(1), 15–21.

Chapman, B., G. Jost, R. van der Pas, and D. J. Kuck (2007). *Using OpenMP: Portable Shared Memory Parallel Programming*. The MIT Press, Cambridge, MA.

Chivers, I. D. and J. Sleightholme (2010). Compiler support for the Fortran 2003 and 2008 standards. *ACM Fortran Forum 29*(1), 29–32.

Clarke, L. A. and D. S. Rosenblum (2006). A historical perspective on runtime checking in software development. *ACM SIGSOFT Software Engineering Notes 31*(3), 25–37.

Cole, J. D. (1951). On a quasilinear parabolic equation occuring in aerodynamics. *Q. Appl. Math. 9*, 225–236.

Collins, J. B. (2004). Standardizing an ontology of physics for modeling and simulation. In *Proceedings of the 2004 Fall Simulation Interoperability Workshop*, Orlando, FL. Paper 04F–SIW–096.

Cosgrove, J., J. Buick, S. Tonge, C. Munro, C. Greated, and D. Campbell (2003). Application of the lattice boltzmann method to transition in oscillatory channel flow. *Journal of Physics 36*, 2609–2630.

Dalhgren, T. L. (2007) Performance-Driven Interface Contract Enforcement for Scientific Components. Ph.D. Dissertation, University of California, Davis (also Technical Report UCRL-TH-235341, Lawrence Livermore National Laboratory).

Dahlgren, T. L., T. Epperly, G. Kumfert, and J. Leek (2009). Babel users' guide. Technical Report 230026, Lawrence Livermore National Laboratory, Livermore, CA.

Davidson, P. (2001). *An Introduction to Magnetohydrodynamics*. Cambridge University Press, New York.

de Bruyn Kops, S. M. and J. J. Riley (1998). Direct numerical simulation of laboratory experiments in isotropic turbulence. *Phys. Fluids 10*(9), 25–27.

Decyk, V. K. and H. J. Gardner (2006). *A factory pattern in Fortran 95*, Volume 4487, pp. 583–590. Springer, Berlin/Heidelberg.

Decyk, V. K. and H. J. Gardner (2007). Object-oriented design patterns in Fortran 90/95. *Computer Physics Communications 178*(8), 611–620.

Decyk, V. K., C. D. Norton, and B. K. Szymanski (1997a). Expressing object-oriented concepts in Fortran 90. *ACM Fortran Forum 16*(1), 13–18.

Decyk, V. K., C. D. Norton, and B. K. Szymanski (1997b). How to express C++ concepts in Fortran 90. *Scientific Programming 6*(4), 363–390.

Decyk, V. K., C. D. Norton, and B. K. Szymanski (1998). How to support inheritance and run-time polymorphism in Fortran 90. *Computer Physics Communications 115*, 9–17.

Deitel, P. J. (2008). *C++: How to Program*. Prentice Hall, Upper Saddle River, NJ.

Dijkstra, E. W. (1968). Go-to statement considered harmful. *Communications of the ACM 11*(3), 147–148.

Doviak, R. and D. Zrnic (1984). *Doppler radar and weather observations*. Academic Press, New York.

Eichenberger, A. E., P. Wu, and K. O'Brien (2004). Vectorization for simd architectures with alignment constraints. *ACM SIGPLAN Notices 39*(6), 82–93.

Fenton, N. E. and N. Ohlsson (2000). Quantitative analysis of faults and failures in a complex software system. *IEEE Transactions on Software Engineering 26*(8), 797–814.

Frisch, U., B. Hasslacher, and Y. Pomeau (1986). Lattice gas cellular automata for the navier-stokes equations. *Phys. Rev. Lett. 56*, 1505–1508.

Gamma, E., R. Helm, R. Johnson, and J. Vlissides (1995). *Design Patterns: Elements of Reusable Object-Oriented Software*. Addison-Wesley, New York.

Gardner, H. and G. Manduchi (2007). *Design Patterns for e-Science*. Springer-Verlag, Berlin/Heidelberg.

Grant, P. W., M. Haveraaen, and M. F. Webster (2000). Coordinate free programming of computational fluid dynamics problems. *Scientific Programming 8*, 211–230.

Gray, M. G., R. M. Roberts, and T. M. Evans (1999). Shadow-object interface between Fortran 95 and C++. *Computing in Science and Engineering 1*(2), 63–70.

Hatton, L. (1997). The T Experiments: Errors in scientific software. *IEEE Computational Science and Engineering 4*(2), 27–38.

He, X. and L.-S. Luo (1997). Theory of the lattice boltzmann: from the boltzmann equation to the lattice boltzmann equation. *Phys. Rev. E 56*, 6811–6817.

Henshaw, W. D. (2002). Overture: An object-oriented framework for overlapping grid applications. Technical Report 147889, Lawrence Livermore National Laboratory (also 32nd AIAA Conference on Applied Aerodynamics, St. Louis, Missouri, June 24–27, 2002).

Heroux, M. A., R. A. Bartlett, V. E. Howle, R. J. Hoekstra, J. J. Hu, T. G. Kolda, R. B. Lehoucq, K. R. Long, R. P. Pawlowski, E. T. Phipps, A. G. Salinger, H. K. Thornquist, R. S. Tuminaro, J. M. Willenbring, A. Williams, and K. S. Stanley (2005). An overview of the trilinos project. *ACM Transactions on Mathematical Software 31*(3), 397–423.

Hopf, E. (1950). The partial differential equation $u_t + uu_x = \mu u_{xx}$. *Commun. Pure Appl. Math. 3*, 201–230.

Huang, K. (1963). *Statistical Mechanics*, Wiley, Hoboken.

J3 Fortran Standards Technical Committee (1998). Technical report. Technical Report ISO/IEC 15581:1998(E), International Organization for Standards/International Electrotechnical Committee, Geneva, Switzerland.

Karniadakis G. E. and Sherwin S. J. (2005). *Spectral/hp Element Methods for Computational Fluid Dynamics (Numerical Mathematics and Scientific Computation)*. Oxford University Press, New York.

Kirk, S. R. and S. Jenkins (2004). Information theory-based software metrics and obfuscation. *J. Sys. Soft. 72*(2), 179–186.

Knuth, D. (1974). Structured programming with go to statements. *Computing Surveys 6*(4), 261–301.

Koplik, J. and H. Levine (1993). Vortex reconnection in superfluid helium. *Physical Review Letters 71*, 1375–1378.

Krawezik, G., G. Alléon, and F. Cappello (2002). *SPMD OpenMP versus MPI on a IBM SMP for 3 Kernels of the NAS Benchmarks*, Volume 2327/2006 of *Lecture Notes in Computer Science*, 515–518.

Lewis, J. P. (2001). Limits to software estimation. *ACM SIGSOFT Soft. Eng. Notes 26*(4), 54–59.

Lieber, B., A. Stancampiano, and A. Wakhloo (1997). Alteration of hemodynamics in aneurism models by stenting: Inuence on stent porosity. *Annals of Biomedical Engineering 25*(3), 460–469.

Linde, G. J. and M. T. Ngo (2008). WARLOC: A high-power coherent 94 GHz radar. In *Aerospace and Electronic Systems, IEEE Transactions*, Milan, Italy, pp. 1102–1117.

Liu, Z., B. Chapman, T. Weng, and O. Hernandez (2003). *Improving the Performance of OpenMP by Array Privatization*, Volume 2716/2003 of *Lecture Notes in Computer Science*, 244–259.

Long, K. (2004). Sundance 2.0 tutorial. Technical Report 2004-4793, Sandia National Laboratories.

Lorenz, E. N. (1963). Deterministic Nonperiodic Flow. *Journal of the Atmospheric Sciences*, 130–141.

Lu, J., S. Wang, and S. Norville (2002). Method and apparatus for magnetically stirring a thixotropic metal slurry. *U.S. Patent No. 6,402,367 B1*.

Maario, B., S. Viriato, and C. Francisco (2007). Dynamics in spectral solutions of burgers equation. *Journal of computational and applied mathematics 205*(1), 296–304.

Machiels, L. and M. O. Deville (1997). Fortran 90: An entry into object-oriented programming for the solution of partial differential equations. *ACM Transactions on Mathematical Software 23*, 32–42.

Malawski, M., K. Dawid, and V. Sunderam (2005). Mocca – towards a distributed cca framework for metacomputing. In *Proceedings of the 19th IEEE International Parallel and Distributed Processing Symposium (IPDPS'05) – Workshop 4 – Volume 05*. IEEE Computer Society.

Manz, B., L. Gladden, and P. Warren (1999). Flow and dispersion in porous media: Lattice-boltzmann and nmr studies. *AIchE Journal of Fluid Mechanics and Transport Phenomena 45*(9), 1845–1854.

Markus, A. (2003). Avoiding memory leaks with derived types. *ACM Fortran Forum 22*(2), 1–6.

Markus, A. (2006). Design patterns and Fortran 90/95. *ACM Fortran Forum 26*(1), 13–29.

Markus, A. (2008). Design patterns in Fortran 2003. *ACM Fortran Forum 27*(3), 2–15.

Martin, K. and B. Hoffman (2008). *Mastering CMake 4th edition*. Kitware, Inc.

Martin, R. C. (2002). Agile software development, principles, patterns and practices. Prentice Hall, Upper Saddle River, NJ.

Mattson, T. G., B. A. Sanders, and B. L. Massingill (2005). *Patterns in Parallel Programming*. Addison-Wesley, New York.

Metcalf, M., J. K. Reid, and M. Cohen (2004). *fortran 95/2003 explained*. Oxford University Press, New York.

Mittal, R. (2005). Immersed boundary methods. *Annual Reviews of Fluid Mechanics 37*, 239–261.

Moin, P. (2001). *Fundamentals of Engineering Numerical Analysis*. Cambridge University Press, New York.

Morris, K. (2008). *A direct numerical simulation of superfluid turbulence*. Ph. D. thesis, The Graduate Center of The City University of New York.

Morris, K., R. A. Handler, and D. W. I. Rouson (2010, [in review]). Intermittency in the turbulent ekman layer. *Journal of Turbulence*.

Morris, K., J. Koplik, and D. W. I. Rouson (2008). Vortex locking in direct numerical simulations of quantum turbulence. *Physical Review Letters 101*, 015301.

Mulcahy, J. (1991). Pumping liquid metals. *U.S. Patent No. 5011528*.

Norton, C. D. and V. K. Decyk (2003). Modernizing Fortran 77 legacy codes. *NASA Tech Briefs 27*(9), 72.

Numrich, R. W. (2005). Parallel numerical algorithms based on tensor notation and Co-Array Frotran syntax. *Parallel Computing, 31*(6), 588–607.

Numrich, R. W. and J. K. Reid (1998). Co-array fortran for parallel programming. *ACM Fortran Forum 17*(2), 1–31.

Oliveira, J. N. (1997). http://www3.di.uminho.pt/ jno/html/camwfm.html.

Oliveira, S. and D. E. Stewart (2006). *Writing Scientific Software: A Guide to Good Style*. Cambridge University Press, New York.

Pace, D. K. (2004). Modeling and simulation verification challenges. Technical Report 2, Johns Hopkins APL Technical Digest.

Press, W. H., S. A. Teukolsky, W. T. Vetterling, and B. P. Flannery (1996). Numerical recipes in fortran 90: The art of parallel scientific computing, volume 2. Cambridge University Press, New York.

Press, W. H., W. T. Vetterling, B. P. Flannery, and S. A. Teukolsky (2002). Numerical Recipes in C++: The Art of Scientific Computing, 2nd edition. Cambridge University Press, New York.

Pressman, R. (2001). *Software Engineering: A Practitioner's Approach, 5th Ed.* McGraw-Hill, New York.

R. van Engelen, L. W. and G. Cats (1997). Tomorrow's weather forecast: Automatic code generation for atmospheric modeling. *IEEE Computational Science and Engineering*.

Rasmussen, C. E., M. J. Sottile, S. S. Shende, and A. D. Maloney (2006). Bridging the language gap in scientific computing: the chasm approach. *Concurrency and Computation: Practice and Experience 18*(2), 151–162.

Reid, J. K. (2005). Co-arrays in the next fortran standard. *ACM Fortran Forum 24*(2), 4–17.

Rouson, D., K. Morris, and X. Xu (2005). Dynamic memory de-allocation in fortran 95/2003 derived type calculus. *Scientific Programming 13*(3), 189–203.

Rouson, D., X. Xu, and K. Morris (2006). Formal constraints on memory management for composite overloaded operations. *Scientific Programming 14*(1), 27–40.

Rouson, D. W. I. (2008). Towards analysis-driven scientific software architecture: The case for abstract data type calculus. *Scientific Programming 16*(4), 329–339.

Rouson, D. W. M., H. Adalsteinsson, and J. Xia (2010). Design patterns for multiphysics modeling in Fortran 2003 and C++. *ACM Transactions on Mathematical Software 37*(1), Article 3.

Rouson, D. W. I., S. C. Kassinos, I. Moulitsas, I. Sarris, and X. Xu (2008a). Dispersed-phase structural anisotropy in homogeneous magnetohydrodynamic turbulence at low magnetic reynolds number. *Physics of Fluids 20*(025101).

Rouson, D. W. I., R. Rosenberg, X. Xu, I. Moulitsas, and S. C. Kassinos (2008b). A grid-free abstraction of the navier-stokes equation in fortran 95/2003. *ACM Transactions on Mathematical Software 34*(1), Article 2.

Rouson, D. W. I. and Y. Xiong (2004). Design metrics in quantum turbulence simuluations: How physics influences software architecure. *Scientific Programming 12*(3), 185–196.

Rubin, F. (1987). 'GOTO considered harmful' considered harmful. *Communications of the ACM 30*(3), 195–196.

Satir, G. and D. Brown (1995). *C++: The Core Language*. O'Reilly & Associates, Inc., Sebastopol, CA.

Schach, S. (2002). *Object-Oriented and Classical Software Engineering*. Addison-Wesley, New York.

Schwarz, K. W. (1985). Three-dimensional vortex dynamics in superfluid ^{4}He: line-line and line-boundary interactions. *Physical Review B 31*, 5782–5804.

Schwarz, K. W. (1988). Three dimensional vortex dynamics in superfluid ^{4}He: homogeneous superfluid turbulence. *Physical Review B 38*, 2398–2417.

Shalloway, A. and J. R. Trott (2002). *Design Patterns Explained*. Addison-Wesley, New York.

Shannon, C. (1948). A mathematical theory of communication. *The Bell System Technical Journal*.

Shepperd, M. and D. C. Ince (1994). A critique of three metrics. *Journal of Systems and Software 26*, 197–210.

Spalart, P. R., R. D. Moser, and M. M. Rogers (1991). Spectral methods for the navier-stokes equations with one infinite and two periodic directions. *J. Comp. Phys. 96*(297), 297–324.

Stevens, W. P., G. J. Myers, and L. L. Constantine (1974). Structured design. *IBM Systems Journal 13*(2), 115.

Stevenson, D. (1997). How goes cse? Thoughts on the IEEE cs workshop at Purdue.

Stewart, G. (2003). Memory leaks in derived types revisited. *ACM Fortran Forum 22*(3), 25–27.

Stokes, G. (1851). On the effect of the internal friction of fluids on the motion of pendulums. *Trans. Camb. Phil. Soc. 9*(Part II), pp. 8–106.

Strang, G. (2003). *Introduction to Linear Algebra.* Wellesley Cambridge Press, Wellesley, MA.

Stucki, L. G. and G. L. Foshee (1971). New assertion concepts for self-metric software validation. In *Proceedings of the International Conference on Reliable Software*, pp. 59–71. ACM SIGPLAN and SIGMETRICS.

Tatarskii, V. (1961). *Wave propagation in a turbulent medium.* McGraw-Hill, New York.

Thomas, H. (1949). Elliptic problems in linear difference equations over a network. Technical report, Watson Sci. Comput. Lab. Rept., Columbia University, New York.

Vandevoorde, D. and N. M. Josuttis (2003). *C++ Templates, the complete guide.* Addison Wesley, New York.

Vinen, W. F. and J. J. Niemela (2002). Quantum turbulence. *Journal of Low Temperature Physics 128*(5–6), 167–231.

Wallcraft, A. J. (2000). Spmd openmp versus mpi for ocean models. *Concurrency: Practice and Experience 12*, 1155–1164.

Wallcraft, A. J. (2002). A comparison of co-array fortran and openmp fortran for spmd programming. *The Journal of Supercomputing 22*(3), 231–250.

Warmer, J. and A. Kleppe (2003). *The Object Constraint Language: Getting Your Models Ready for MDA, 2nd Ed.* Addison-Wesley, New York.

Won, J., S. Suh, H. Ko, K. Lee, W. Shim, B. Chang, D. Choi, S. Park, and D. Lee (2006). Problems encountered during and after stentgraft treatment of aortic dissection. *Journal of Vascular and Interventional Radiology 17*, 271–281.

Xu, X. and J. Lee (2008). Application of the lattice boltzmann method to flow in aneurysm with ring-shaped stent obstacles. *Int. J. Numer. Meth. Fluids 59*(6), 691–710.

Zhang, K., K. Damevski, V. Venkatachalapathy, and S. G. Parker (2004). Scirun2: A cca framework for high performance computing. In *Ninth International Workshop on High-Level Parallel Programming Models and Supportive Environments (HIPS'04)*, pp. 72–79.

Index

Printed in the United States
By Bookmasters

Printed in the United States
By Bookmasters